David Murfin is a physicist and technical information retrieval specialist with a lifelong interest in warship design. His previous forays into print are some short scientific papers, a major revision of Dodd's classic *Dictionary of Ceramics* for the Institute of Materials, *A Directory of British Cruiser Designs 1860–1960* (2011) and 20 years editing the monthly *World Ceramic Abstracts*.

Conrad Waters is a barrister by training but a banker by profession. He is the author of numerous articles on modern naval matters and editor of the annual *World Naval Review* (Seaforth Publishing). He also edited Seaforth's *Navies in the 21st Century,* shortlisted for the 2017 Mountbatten Award. He is currently writing a history of the Royal Navy's 'Town' class cruisers of the Second World War.

Mike Williams has contributed a number of articles to warship, including a major feature on the loss of the battlecruiser *Queen Mary* (1996). More recently he has turned his attention to the Imperial Japanese Navy.

Tom Wismann is a Marine Engineer by education and former Lieutenant-Commander in the Danish Maritime Home guard. He has written and published a series of books on ships of the Royal Danish Navy and modern fortifications. He has for many years been a member of the board of the Danish Naval Historical Society and editor of the society's journal.

Dr Alan D Zimm is an Operations Research Analyst at the Applied Physics Laboratory, John Hopkins University, Maryland. A former naval officer, he served as a nuclear power–qualified surface warfare officer in carriers, cruisers, destroyers and hydrofoils. The author of *Attack on Pearl Harbor: Strategy, Combat, Myths, Deceptions,* he has received many writing awards, including the US Naval Institute's Naval History Author of the Year for 2016.

WARSHIP 2018

WARSHIP 2018

Editor: **John Jordan**

Assistant Editor: **Stephen Dent**

OSPREY
PUBLISHING

Title pages: A fine view of HMS *Achilles* taken at Melbourne on 18 February 1938. Note the single 4in Mark V HA guns abreast the funnel and the Supermarine Walrus amphibian atop the catapult; the latter was embarked in place of the original Fairey Seafox when the ship was transferred to New Zealand in 1936. *Achilles* was one of the three British cruisers which engaged the German *Admiral Graf Spee* in the Battle of the River Plate in December 1939 – see the articles by Alan D Zimm and William J Jurens.
(Allan C Green Collection, State Library of Victoria, H91-108-2384)

OSPREY PUBLISHING
Bloomsbury Publishing Plc
PO Box 883, Oxford, OX1 9PL, UK
1385 Broadway, 5th Floor, New York, NY 10018, USA
E-mail: info@ospreypublishing.com
www.ospreypublishing.com

First published in Great Britain in 2018

A catalogue record for this book is available from the British Library.

ISBN: HB 978-1-4728-2999-3; eBook 978-1-4728-3000-5;
ePDF 978-1-4728-3001-2; XML 978-1-4728-2998-6

18 19 20 21 22 10 9 8 7 6 5 4 3 2 1

Typeset by Stephen Dent
Printed in China by C&C Offset Printing Co Ltd.

CONTENTS

EDITORIAL

Perceptive readers will have noted that this year's annual appears for the first time under the Osprey imprint. Since the sale to Bloomsbury four years ago, Conway titles have been commissioned and managed through the Bloomsbury Special Interest department. Following the purchase by Bloomsbury of Osprey Publishing, which had its own existing premises in Oxford and its own editorial staff, it was decided to consolidate the two lists, and the military and naval books formerly published under the Conway imprint will henceforth be published by Osprey.

The change of name on the cover will have little impact on the annual itself, which retains the same editorial team that it has had for the past fourteen years, and which will continue to be published in the same format. However, one of the immediate benefits of the change is that Osprey have agreed an extra sixteen pages this year – something we have been trying to achieve for some time. This has enabled us to clear a 'logjam' of feature articles which had been building up over the past two years. Last year we had to hold Stephen McLaughlin's article on Russian Monitor designs (see below), and even with the extra pages this year we have had to hold three other feature articles which will now be published in *Warship 2019*. It is a sign of the continuing success of the annual that we are still attracting so many submissions, often from established authors, and the increase in page extent recently agreed means that there will be less of a delay in publication, which benefits both our contributors and our readers.

December 2019 will mark the 80th anniversary of the Battle of the River Plate. The passing of time provides an opportunity for new perspectives, and we are publishing two major articles on different aspects of the battle this year. Dr Alan D Zimm, author of a recent major book on Pearl Harbor, has written a critique of the tactics adopted respectively by Commodore Harwood and the German Captain Langsdorff. Neither officer comes out of it particularly well, despite the accolades which Harwood received from the politicians and press of the day. The companion article by William J Jurens, an acknowledged authority on gunnery and ballistics, analyses the hits made on *Admiral Graf Spee* by the British cruisers and their consequences. The lessons which emerge focus on the practical difficulties faced by a surface raider operating in mid-Ocean, far from its base support, when even apparently trivial damage such as the flooding of a flour store or the wrecking of a galley might make it difficult to sustain the crew during a lengthy transit.

Both of the above authors are new to *Warship*, as is Tom Wismann, whose article on the cruiser *Niels Iuel* ('A funny little Danish warship') leads this years annual. Well-written, well-researched and beautifully illustrated, this is a classic *Warship* feature. Other major features by returning authors include Hans Lengerer's article on the IJN 'command cruiser' *Oyodo*, the article by Stephen McLaughlin on Russian monitor designs of the early 20th century mentioned previously, a survey of modern replenishment vessels by Conrad Waters, and a comprehensive account of Royal Navy trade protection cruiser design 1905–1920 by David Murfin. In the third of a series of articles on the early French armoured cruisers, Luc Feron focuses on the revolutionary *Jeanne d'Arc*, the first major French vessel to be designed by the distinguished naval architect Louis-Emile Bertin, and a ship which inspired a new generation of fast armoured cruisers both in France and abroad. And the feature section concludes with the customary drawing/photo contribution by A D Baker III, which this year focuses on USS *Huntington*, one of ten large, powerful armoured cruisers authorised following the Spanish-American War of 1898.

These 'technical' features are counter-balanced by two articles with a more historical focus. Enrico Cernuschi, who has a background in cryptography, questions some of the inflated claims which have been made for the impact on the naval war in the Mediterranean of Ultra intercepts, and reveals the little-known Italian successes in this field, which were achieved by a handful of individuals on a 'shoe-string' budget. And Mike Williams returns to *Warship* with an account of the little-known late career of the IJN destroyer *Amatsukaze*. After losing her bow to a submarine torpedo, this ship was patched up and used as an escort for mercantile convoys carrying oil, precious metals and repatriated personnel from South-East Asia to Japan during the dark days of 1945. Constantly harassed by USAF aircraft and US Navy submarines, the convoys and their escorts sustained devastating losses as the Americans effectively made the South China Sea a 'no-go' area for Japanese maritime operations.

The stand-out item in this year's Warship Notes is a contribution by Hans Lengerer on the 15.5cm 3rd year gun which equipped the cruisers of the *Mogami* class as built, and was subsequently employed for the main battery for *Oyodo* and the secondary battery of *Yamato* and her sister *Musashi*. Other items include a note by Stephen McLaughlin which attempts to explain the mysterious loss of the German *U-56* in November 1916, and a piece by Kenneth Fraser on the naming of US warships.

The Gallery this year features a series of rare photographs of two former German U-boats, *UB-125* (UB III type) and *U-55* ('Mobilisation' type), being dismantled at Sasebo Navy Yard, Japan, in 1921. Many of the lessons learned from these submarines were incorporated in the new generation of IJN submarines designed and built during the 1920s.

Next year's *Warship* will include major articles on Japan's Six-Six Fleet by Hans Lengerer, a second article by Kathrin Milanovich detailing the powder accidents and explosions which plagued the IJN during the early years of the 20th century, an account of the fate of the ex-German destroyers after the First World War by Aidan Dodson, and Philippe Caresse's technical history of the *Brennus*, arguably the French Navy's first modern battleship but a vessel plagued with stability problems.

Finally, it is with great sadness that we announce the death of Keith McBride after a prolonged illness. Keith's articles on the Royal Navy, with a primary focus on the Victorian era, were a staple of *Warship* both as a quarterly and as an annual. An obituary detailing his contributions to *Warship* follows; his in-depth knowledge and his ability to engage the reader will be greatly missed.

John Jordan
April 2018

Keith Donald McBride

31 July 1931 – 7 February 2017

Just as the 2017 edition of *Warship* was going to press we heard that Keith McBride, a regular contributor to the annual for over twenty years, had died. Since it was too late to print an obituary in that edition, one is instead included here.

Keith was born in Hillingdon in west London and lived all his life in nearby Hanwell and Brentford. His father had been a military bandsman, serving in the Marines, the 9th Queen's Lancers and the Household Cavalry; after leaving the British Army he worked for the Post Office as a sorter. Keith's interest in naval ships was inspired by listening to his father's stories of seeing action during the Battle of Jutland.

Keith attended Drayton Manor Grammar School and later University College, London, where he read history, graduating in 1953. After National Service he passed the exams for the Home Civil Service and worked as an Executive Officer in the Ministry of Defence. Although his great interest was in naval ships, he actually worked for the Air Ministry.

A great reader and member of several public libraries, Keith had a wide range of interests which included politics, astronomy, archaeology, sport, and photography. However naval history was his biggest interest and from university days he researched and wrote articles on the subject.

Although he worked for the Ministry of Defence, Keith was in fact a Quaker, attending his local Meeting from 1979. This unusual pairing reflected his views, which could be summed up as being for defence, though against war. A shrewd observer of the political scene, both domestic and international, Keith opposed the Iraq War of 2003.

After retiring early on health grounds when he was 54, Keith occupied himself doing research at the Public Record Office (now the National Archives), the National Maritime Museum and elsewhere in London, the fruits of which resulted in numerous articles in *Warship* (both as a quarterly and as an annual), *Warship International* and *The Mariner's Mirror*. He also acted as a volunteer helper in an adult education class in English for speakers of other languages.

Between the April 1987 quarterly and the 2012 annual editions, Keith had sixteen articles published in *Warship*, all concerned with British warships from the late nineteenth century through to the 1930s, and characterised by diligent and careful use of sources, and an equal ease with 'technical' and 'operational' history. The majority of Keith's contributions were published before the current editorial team took over *Warship*, however a previous Editor describes him as having been in many ways the 'model' contributor, with his submissions always on time and on spec, without any of the lengthy subsequent correspondence that inevitably accompany some articles in any publication with multiple authors. The current Assistant Editor met him just once, at a book launch, and recalls a thoroughly affable character who very much enjoyed the company of fellow enthusiasts.

Keith never married. His father died when he was 25 and he remained living with his widowed mother and unmarried younger sister until their deaths.

Thanks to Bessie White, Mary Blackwell and Rob Gardiner for information included in this obituary.

Keith McBride's Articles for *Warship*

Warship vol 44 (April 1987) 'The *Diadem* Class Cruisers'
Warship vol 46 (April 1988) 'The First County Class Cruisers of the Royal Navy, Part I: The *Monmouths*'
Warship vol 47 (July 1988) 'The First County Class Cruisers of the Royal Navy, Part II: The *Devonshires*'
Warship vol 48 (Oct 1988) 'The Wreck of HMS *Bedford*'
Warship 1989 'The First Flowers'
Warship 1990 'The Weird Sisters' [*Courageous*, *Glorious* & *Furious*]
Warship 1991 'British 'M' Class Destroyers of 1913–14'
Warship 1992 'After the *Dreadnought*'
Warship 1993 'Super-Dreadnoughts: the *Orion* Battleship Family'
Warship 1994 '"The Hatbox": HMS *Argus*'
Warship 1995 'On the Brink of Armageddon: Capital Ship Development on the Eve of the First World War'
Warship 1997–1998 '"Eight six-inch Guns in Pairs": The *Leander* and *Sydney* Class Cruisers'
Warship 2000–2001 '*Nile* and *Trafalgar*: The Last British Ironclads'
Warship 2001–2002 '"The Wobbly Eight": *King Edward VII* Class Battleships, 1897–1922'
Warship 2005 '*Lord Nelson* and *Agamemnon*'
Warship 2012 'The Cruiser Family *Talbot*'

NIELS IUEL:

'A funny little Danish warship'

Niels Iuel was designed as a heavily-gunned and heavily-armoured low-freeboard monitor-type coast defence battleship. Completed as a hybrid cruiser intended to 'show the flag', she ended her days as a training ship in the German *Kriegsmarine* during the tumultuous days of May 1945. **Tom Wismann** investigates the conception and career of this unusual ship.

'A funny little Danish warship': that was how a British newspaper described *Niels Iuel* when the ship visited a British port, and with some justification – depending on your interpretation of the word 'funny'.

The Royal Danish Navy (RDN – until 1814 the combined navy of the Danish-Norwegian Kingdom) has existed since 1510 and is one of the oldest national navies in the world. During the period 1510–1807 the RDN was a major power in the Baltic, North Sea and in the North Atlantic. This ended with the catastrophic British bombardment of Copenhagen which lasted for three days and nights in early September 1807. Denmark was compelled to capitulate to this overwhelming British show of force, and the peace terms were harsh: the Danes had to hand over their entire fleet to the British. 'Albion' departed Copenhagen on 21 October, exactly two years after the Battle of Trafalgar, with loot comprising 16 ships of the line, 15 frigates and corvettes, and 14 minor warships, plus equipment, tools and goods for the fleet loaded into 92 merchant ships. The British had also wreaked havoc at Orlogsværftet (the Royal Naval Dockyard, Copenhagen – also often called Holmen) where ships under construction were destroyed on the slipways.

An aerial view of Holmen, where Orlogsværftet (the Royal Naval Dockyard) was situated, taken in 1916–17. To the right of centre the hull of *Niels Iuel* can be seen under construction on the slipway. The deck plates have not yet been fitted. Danish torpedo boats are moored to the piers on the left, and on the right is the hulked former cruiser-corvette *Fyen*, which was used as a barracks until the mid-1960s. (Danish Armed Forces Photo Gallery)

With the loss of Norway in 1814, Denmark lost influence and the economy was ruined. Over the following years the Navy only recovered slowly, and would never regain anything close to its former might. The RDN, however, continued to keep up with technological developments, and was still a regional naval power, which during the Schleswig wars of 1848–50 and 1864 maintained control of home waters.

Over a period of 45 years from 1863 to 1908, eleven heavily-gunned and more or less heavily-armoured coast defence battleships were built for the RDN. The first, *Rolf Krake*, was built by R Napier & Sons at Glasgow on the Clyde, but the other ten were built at the Orlogsværftet. The first eight ships were all of different design with displacements ranging from 1,340 tons to 5,400 tons. Their armament ranged from 60lb smoothbore guns to 35.5cm (14in) rifled guns. These ships formed the backbone of the RDN of the period and would, in wartime, together with the coastal fortifications, provide the heavy gun back-up to defend extensive defensive minefields. From 1890 these minefields were protected by torpedo boats, and from 1909 also by submarines.

Niels Iuel: The Last of the Danish Coast Defence Battleships

After building the three well-armed and -armoured monitor-type ships of the *Herluf Trolle* class (1899–1908, displacement 3,600 tons, two 24cm and four 15cm guns) a commission set to work to decide future Navy force structures. The rise of the Imperial German Navy and the arms race with the Royal Navy influenced Danish defence politics, but there were also the customary political discussions regarding how much should be spent on the armed forces. The defence agreement of 1909 stipulated that a new armoured ship should be built for the Navy. As funding was tight the displacement was to be limited to 3,800 tonnes,[1] not much greater than the previous *Herluf Trolle* class.

Several different designs were drawn up conforming to this specification. They ranged from a lightly-armoured cruiser armed with 15cm guns to a full-blown monitor armed with two single 30.5cm (12in) guns. In 1913 the Navy Ministry, basing its decision on the work done by the commission, opted for a design based on the *Herluf Trolle* class.

Name

Niels Juel was a famous Danish admiral who, on 30 June 1677 with 25 ships of the line armed with a total of 1,267 guns, defeated a Swedish fleet of 36 ships of the line armed with 1,800 guns in the Battle of Køge Bight.

When the ship was launched, and also when first commissioned, the name on both sides of the stern was spelled 'Niels Juel'. However, a short time after commissioning the spelling of the surname was changed to 'Iuel' following a request from the old admiral's ancestors. (When the admiral was alive he had spelled his name variously as 'Iuel', 'Juel' or 'Juell', there being no strict rules for spelling at that time.)

Designation

As first conceived the ship was designated *Panserskib* ('armoured ship'). When completed the type changed to *Artilleriskib* ('gunship') or *Orlogsskib* ('warship').

Original design

The ship was designed for deployment in the inner Danish waters. A very low freeboard of only one metre was therefore adopted to present as small a target as possible. Only the bow was raised slightly, with a forecastle height of four metres to improve seakeeping qualities. Amidships was a superstructure for an armoured conning tower, bridge, funnel and the secondary armament. The main armament was located in single turrets fore and aft.

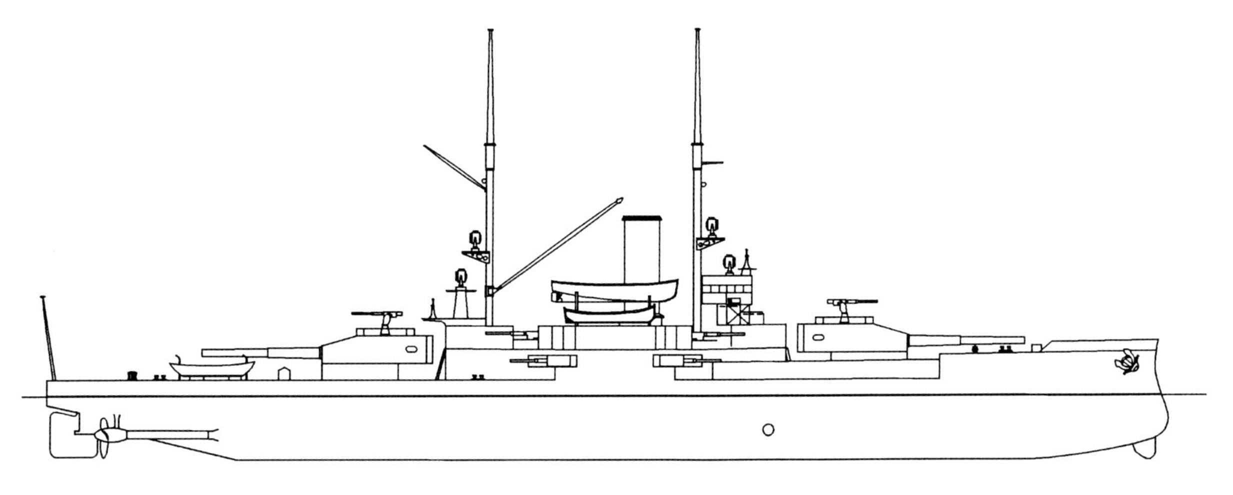

Niels Iuel was designed as a coast defence battleship of the monitor type. The low freeboard of only one metre made the ship a small target, and the two single 30.5cm guns gave the ship a heavy punch for her size. Had this type of ship, which had side armour 195mm (7.6in) thick, come up against a large light cruiser of the British *Glorious* class (side armour: 76mm/3in) the outcome would have been interesting. (Tom Wismann)

The date is 3 July 1918, and *Niels Iuel* is ready for launch. The pavilion on the right is for the Danish royal family. At this time *Niels Iuel* still has the appearance of the 'low-slung' monitor type she was designed as; the ship was launched without the side armour fitted. After almost four years on the slipway it took more than 75 minutes to get the ship into the water; in the end all went without a hitch. (Danish Armed Forces Photo Gallery)

The hull was built of steel and divided by nine transverse bulkheads into ten watertight compartments. In the ten compartments there were all-in-all 115 watertight spaces, of which 52 were in the double bottom.

Construction

Niels Iuel was laid down on 21 September 1914, but even before this date her construction was influenced by events related to the outbreak of the First World War on 28 July.

The period from the assassination of Archduke Franz Ferdinand of Austria on 28 June to the outbreak of war was a time of intense international tension. On 5 August the Danish government declared the country's neutrality, which according to international law the Danish armed forces had to enforce. Denmark had therefore already begun to mobilise 65,000 men on 29 July, formed into a force called *Sikringsstyrken*; with this reinforcement the manpower of the RDN increased to 4,000.

Orlogsværftet started at once to equip all ships which were not already fully stocked with war supplies, and Denmark mobilised every ship and boat in the fleet that was able to sail. The main part of the fleet was ready for combat on 1/2 August and was formed into two squadrons: the first squadron was stationed in The Sound, close to the capital of Copenhagen, and the second squadron in The Great Belt area. It was quite an achievement for the Navy to be combat-ready at such short notice.

From this moment to the end of hostilities on 11 November 1918, Orlogsværftet was hard at work keeping the fleet at sea and undertaking the necessary dockings, overhauls, repairs and maintenance. New-build ships, mainly new torpedo boats, submarines and minor warships, plus a myriad of other tasks to keep the fleet on a near-war footing took a heavy toll on the available manpower.

Hard-pressed as Orlogsværftet was, work on *Niels Iuel* only progressed slowly, with manpower and materials in short supply; it took almost four years before *Niels Iuel* was ready for launch on 3 July 1918.

Launch

The launching of a new warship has traditionally been a day of festivities at Orlogsværftet. At 1400 the Sixtus battery fired a 27-gun salute to mark the arrival of the royal family. A crowd of notable personalities, both military and civil, plus workers from Orlogsværftet had gathered to witness the great moment, and a multitude of Danish flags were fluttering in the wind. After a short speech by the Dean of Holmen Church (the Naval Church since 1619) about the famous admiral after whom the new ship was to be named, everything was ready for the launch. On a sign from His Majesty King Christian X the stops were pulled – but nothing happened! After almost four years on the slipway the hull was stuck firm. After more than an hour of hard labour, and a considerable amount of embarrassment on the part

When the politicians decided that a heavily-armed monitor would be a provocation for foreign powers, it was decided to complete the ship to a new design as a training cruiser intended to show the flag. An additional deck was constructed on top of the monitor hull to provide accommodation for a larger crew, and the increase in freeboard resulted in improved sea-keeping qualities. This photo was probably taken in 1921. (Danish Armed Forces Photo Gallery)

of the authorities, Sixtus could at 1515 begin the salute to mark the launch. Slowly, almost hesitatingly, the hull gained speed and ran down the slipway, and there were resounding 'Hurrahs' from all those attending.

When the hull was launched the armour had yet to be fitted and the superstructures were yet to be built (see the photo on the facing page).

A change of plans

After the launch, work continued to progress slowly, and when hostilities ended on 11 November 1918 it ground to a complete halt.

The politicians who before August 1914 had denied all possibilities of a European war at once wanted to save money by cancelling the ship. The First World War would be the last war – 'the war to end all wars'; universal disarmament was the order of the day. The hard work done by the Navy in the four years of neutrality was immediately forgotten, and the expertise gained was to no avail. The old saying 'In times of war, but not before, god and the sailors we adore ashore, but when war is over and peace remitted, god is forgotten and the sailor is quitted' was as true as ever.

As the conversion of the ship to a ferry or a cargo ship for peaceful purposes was deemed impractical, consideration was given to scrapping the hull. In the end wiser counsels prevailed, as a lot of effort and money had already been invested in the hull and machinery. The idea of changing the original design from a heavily-gunned coast defence battleship to a training cruiser armed with lighter guns and capable of showing the flag was put forward. For many politicians the big 30.5cm guns were seen as offensive weapons which foreign powers might consider provocative. Instead it was decided to complete *Niels Iuel* to a new design. The original hull and machinery were used, and an additional deck was built on top of this. This provided accommodation for the larger crew implied by the training cruiser role, and the increased freeboard also gave the ship improved seakeeping qualities, allowing her to operate in open waters. The original main armament of two 30.5cm guns was changed to ten 15cm guns. The new drawings were approved in 1920, and construction of the ship was resumed to the new design.

The *Niels Iuel* commissioned for the first time in May 1923, and embarked on a work-up cruise on 28 May. The tour took the ship to Bergen (Norway), The Faroe Islands (part of the Danish Kingdom), Leith (UK) and Gothenburg (Sweden).

Machinery

The machinery was located in five spaces extending over a length of 21.8m – 29% of the ship's length between perpendiculars. From fore to aft were: the auxiliary machinery room, the forward boiler room, the after

Characteristics, as completed 1923

Displacement	3,800 tons standard
	4,100 tonnes normal
Dimensions	length 87.0m pp, 90.0m oa
	beam 16.3m, draught 5.0m
Propulsion	Four boilers for super-heated steam built by Orlogsværftet to Yarrow design:
	two oil-fired (four burners)
	two coal-fired (one grate)
	Two VTE reciprocating steam engines on two shafts; 6,026ihp = 16.1 knots max
Bunkerage	oil 223 tonnes; coal 244 tonnes
Endurance	6,000nm at 9 knots
Armament	ten 15cm/45
	two 57mm/30
	two 45cm TT (4 torpedoes)
Complement	310/369

boiler room, the port and starboard engine rooms – the two engine rooms were located abreast in two separate compartments divided by a longitudinal bulkhead. History would later show that dividing these large spaces by longitudinal bulkheads was bad for damaged stability, but the practice was not uncommon at that time.

The auxiliary machinery room housed two 60hp single action four-stroke diesels from the manufacturer Holeby Dieselmotor Fabrik A/S (Danish engine manufacture), each driving a 40kW dynamo. The dynamos were constructed by the Danish company A/S Titan, Copenhagen; they were six-pole compound models supplying 112V x 125A at 360rpm (DC).

The forward boiler room housed two oil-fired water-tube boilers with superheaters of Yarrow design built at Orlogsværftet. Each boiler was fitted with four oil burners of the pressure jet type working at pressures between 7–14kg/cm^2. The supply of steam was controlled by varying the number of burners, the oil pressure, nozzle sizes and the amount of air for combustion. The fuel was kept at constant temperature for optimal running. The two boilers in the after boiler room were also of Yarrow design but were coal-fired. Each boiler had a single hearth and the grate size was 8.5m^2. The boilers produced superheated steam at a temperature of 275°C (527°F), which for a short period could be raised to 350°C (662°F) without damage to the superheaters. They were protected against galvanic corrosion by a Cumberland anti-corrosion system.

Each of the engine rooms housed a vertical triple-expansion (VTE) steam engine designed and built by Orlogsværftet. The nominal power rating for each of the engines was 3,000ihp. The diameter of the high-pressure (HP) cylinder was 0.61 metres, the intermediate cylinder (IP) had a diameter of 0.95 metres and the low-pressure (LP) cylinder had a diameter of 1.52 metres. Each of the engines propelled a shaft with a three-bladed fixed-pitch bronze propeller made by Stone (London, UK) with a diameter of 3.35 metres; the propellers rotated

Two triple expansion steam engines built at Orlogsværftet delivered 6,026ihp and gave *Niels Iuel* a maximum speed of 16.1 knots. The engine seen here is in the workshop at Orlogsværftet prior to installation. (Danish Armed Forces Photo Gallery)

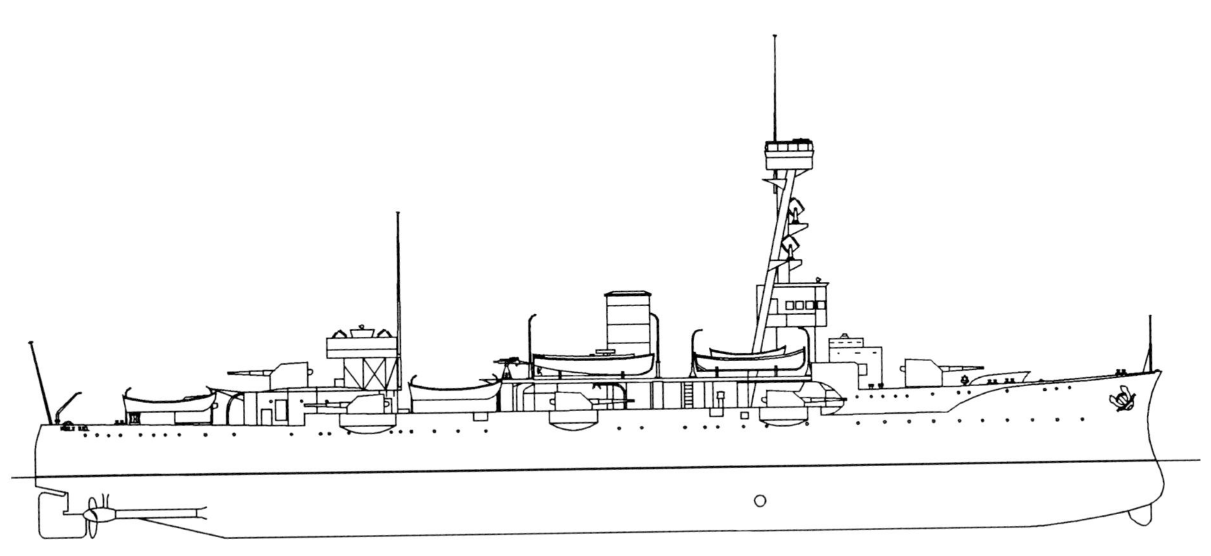

Niels Iuel as completed in 1923 with ten 15cm and two 57mm HA guns. The ship has retained the two 45cm underwater torpedo tubes which were a feature of the original design. (Tom Wismann)

outwards. The designed speed of 14.5 knots was achieved with 3,962ihp during the endurance trial, and 16.1 knots with 6,026ihp was attained during the full power trial.

Two of the three 70kW turbo-dynamos were located in the starboard engine room and the third in the port engine room. All were designed and built by Brown Boveri & Cie, Switzerland: the dynamo was a four-pole compound type producing 112V x 134A at 5,000rpm. There were ca 540 electrical outlets (mainly for lighting)

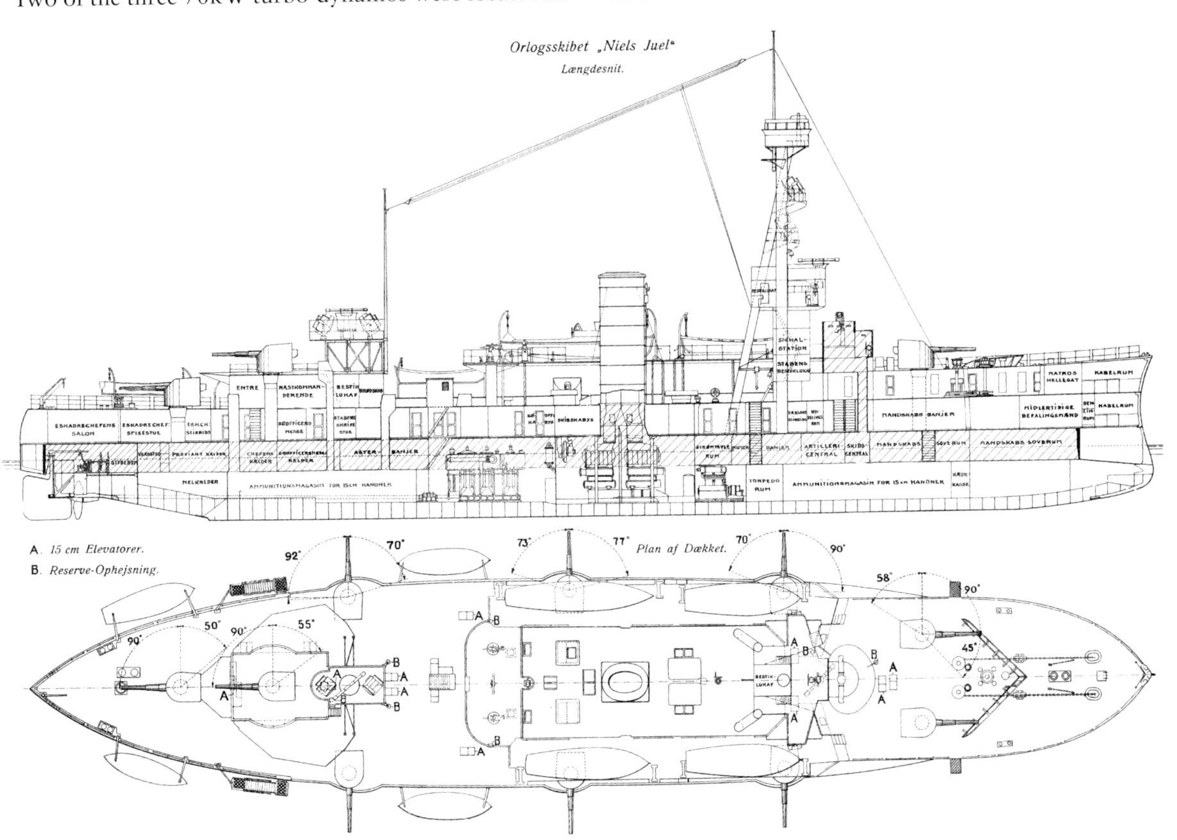

Inboard profile and deck plan of *Niels Juel* as completed in 1923. The armoured belt (see hatching at waterline) is relatively narrow – in the original monitor design it extended to the upper deck. Note the broad arcs on the single guns. Even when the ship was completed the HA armament of two 57mm guns (after end of the shelter deck) was outdated, and would have been ineffectual against the aircraft of the day. (National Archives of Denmark)

distributed throughout the ship, and 40 electrical motors with a total power of 37kW. Emergency lighting was by a lead battery consisting of 17 elements; it supplied 34V and had a capacity of 525Ah for three hours' lighting.

The refrigerating plant was sized to provide cooling for the magazines and the cold store rooms. Cooling was generated by three carbon dioxide refrigerating plants, of which two were located in the steering engine compartment and the third in a separate space to starboard in the auxiliary engine room

The ship had a single balanced rudder with a surface area of 9.5m². The steering gear was powered by a two-cylinder steam engine, located on the rear bulkhead in the port engine room and operated via a chain drive.

Bunkers and endurance

There was bunkerage for 223 tonnes of fuel oil and 244 tonnes of coal. The ship had a range of 6,000nm at 9 knots.

Protection

The armour was for the most part of the Krupp Cemented type (KC) and was manufactured by the American company Bethlehem Steel. When a Danish commission visited the factory in 1919 to witness the proving trials, the armour was penetrated by the shells fired at it and therefore not accepted. The Navy's Head Armourer also discovered that Bethlehem Steel had used shells with softer noses than specified in the contract. As the contracted weight of the armour was a crucial element in the design of the hull and was key to draught and stability, the plates were eventually accepted, but at a reduced price!

The side armour extended from five metres from the bow to two metres from the stern. It had a height of 2.10, with one metre above and 1.10m below the waterline. The thickness amidships was 195mm tapering to 155mm at the ends. The armour belt was secured to a 50mm wooden backing and a 13mm inner steel skin, and was closed fore and aft by armoured transverse bulkheads with thicknesses of 175mm and 165mm respectively. The armoured deck rested on the upper edge of the belt, and was of special shipbuilding steel with a thickness of 55mm. Forward of the transverse bulkhead the deck armour was continued flush with the lower edge of the side armour to strengthen the ram bow.

At the forward end of the superstructure there was an armoured conning tower with navigation and fire control positions on two levels. The upper – fire control – position had slit windows overlooking the navigation position, which enveloped the fire control position to the front and sides. An armoured tube for the electrical cabling and voice-pipes led from the conning tower down to the ship control centre and the transmitting station below the armoured deck. The sides of the tower had a thickness of 170mm, the roof and deck a thickness of 40mm, and the communications tube 100mm.

The boiler uptakes were protected by 75mm of armour, and the ammunition hoists by 16mm plating.

Niels Iuel early in her career, probably in Iceland during the royal cruise in 1926 or 1930. The square deckhouse abaft the funnel is a smoking lounge for the royal family specially built for these occasions. (R Steen Steensen, Danish Armed Forces Photo Gallery)

Armament

The Navy wanted the main armament to be a further development from the *Herluf Trolle*'s two single 24cm (9.4in) guns, with four 24cm guns in two twin hydraulically operated turrets. As the RDN had no experience with twin hydraulic turrets, the Director of Shipbuilding and Machinery and the Director of Ordnance went on a study tour of several foreign gun manufacturers. Some of the latter suggested that the RDN opt for 30.5cm (12in) guns instead of 24cm, as the weight and price of two 30.5cm single turrets was nearly the same as two twin 24cm turrets. The Ministry of the Navy welcomed this suggestion, although it was thought that four 24cm might have a better chance of securing an early hit on the enemy because of their greater rate of fire.

As a result of these investigations, bids were sought at the beginning of 1914 for the delivery of two single turrets armed with 30.5cm guns from British, French and German gun manufacturers. Much to the surprise of those concerned the Swedish Bofors company submitted an unsolicited bid at a price lower than the other contenders, even though Bofors had never before produced a naval turret for a 30.5cm gun. The RDN and the Ministry was unconvinced by the Swedish proposal, which was considered a risk, and the order was placed with the German Krupp company, which had submitted the second lowest bid, in July 1914, less than a month before the First World War broke out. The contract for the two turrets, which included the guns and 150 semi-armour piercing (SAP) shells and 100 exercise shells plus two 57mm sub-calibre guns for practice firings to be mounted atop each of the turrets, amounted to 1,566,000 Marks. In the event the turrets and guns were never delivered due to the outbreak of war.

As designed, the secondary battery was to comprise eight single 10.5cm (4.1in) guns, but this was amended in 1917 to 12cm (4.7in) as a result of the experience obtained by the belligerent nations.

Only two weeks after the end of hostilities on 11 November 1918, Krupp contacted the Danish authorities to enquire if Denmark still wanted the turrets and guns delivered. That was not the case; it was now completely unacceptable to Danish politicians to purchase 30.5cm guns, as such guns were seen as offensive weapons. The response was that Denmark might be interested instead in ten to twelve 15cm guns, provided that the first instalment for the 30.5cm turrets (which had already been paid in 1914) could be part of the deal.

In July 1919 the Director of Ordnance visited the Bethlehem Steel Corporation in the United States to witness proof firings against the side armour for the *Niels Iuel*. He had hoped that during the trip he might make contacts with a view to obtaining 6in (15.2cm) guns at a bargain price from allied surplus stocks. Unfortunately this turned out not to be possible.

Bofors, Armstrong and Vickers were therefore contacted and asked if they would be able to supply modern 6in or 15cm guns. Both British companies could deliver the guns, but could not include follow-the-pointer systems as requested. A French company offered 15cm guns but with older-type mountings, and Bofors offered the same 15cm model as mounted in the Danish coast defence battleships of the *Herluf Trolle* class. None of this was satisfactory.

Krupp, which was not allowed to deliver complete guns according to the Armistice regulations, was again contacted and asked if they could deliver partly-finished guns which could be completed in Denmark. Krupp was not allowed to deliver the guns, but out of the blue the Danes received an offer from Bofors. (The offer may have been the result of secret talks between Krupp and Bofors.) In February 1920 the British Armstrong company offered 6in (15.2cm) guns but of an older model. Even though the Armstrong guns could be delivered at a price of 100.000 Danish kroner (kr) each compared to 146.000 Swedish kr in the Bofors bid, it was the latter company which received the order for twelve guns, of which two were to be spares. Bofors used drawings and materials supplied by Krupp, and the deal took account of the prepaid amount.

In the summer of 1922 the guns were test-fired at the Bofors proving range in Sweden, and delivered to the RDN in March 1923. The guns were installed on board in the following months.

Antiaircraft guns

According to both the 1914 and the 1920 designs, *Niels Iuel* was to have been fitted with two 75mm anti-aircraft guns. However, in the event guns of this kind, with their associated time-fused ammunition, proved much too expensive. Instead the ship was completed with two 57mm anti-aircraft guns. These guns remained on board until the modernisation of 1935–36, only to be reinstated in 1939 as saluting guns.

As part of the 1935–36 modernisation ten 20mm RK M/31 in five twin mountings were installed and test-fired on 22 August 1936. In April 1937 the anti-aircraft armament was augmented with 14 8mm machine guns L/75 M/37 in twin mounts; the MGs were belt fed from a 100-round box magazine. Both the 20mm and 8mm were produced by the company Dansk Rekylriffel Syndikat.

In the winter of 1940–41 two single 40mm L/60 M/37 Bofors guns were installed. They had originally been fitted on submarines of the 'H' class, which each had to give up one of their two guns. The guns were test-fired on *Niels Iuel* in January 1941.

In the winter of 1941–42 the ten 20mm R.K. M/31 in twin mounts and the 14 8mm machine guns were replaced by ten single 20mm L/60 M/41.

Torpedoes

The ship was originally designed with two fixed athwartships submerged torpedo tubes for 45cm Type H torpedoes (see data table for characteristics), and these torpedo tubes were retained in the 1920 design. The torpedo room, which could stow four torpedoes, was situated on the lower deck between the forward maga-

Armament: Guns and Torpedoes

15cm (149.1mm) P.K. L/45

Gun Data	
Manufacturer	Bofors (Sweden)
Length in calibres	45
Weight of barrel	6,115kg
Breech mechanism	horizontal sliding block
Ammunition type	separate
Projectiles	*Brisantgranat* (46kg)
Propellant	13.3kg
Muzzle velocity	835m/s
Maximum range	17,800m

Mounting	
Designation	?
Protection	shield: front 50mm / sides 10-20mm
Weight of mounting	7,273kg
Elevation	-10° / +30°
Loading angle	any angle
Firing cycle	5–7rpm

Notes:

Brisantgranat	High Explosive (HE)

40mm L/60 R.K. M/37

Gun Data	
Manufacturer	Bofors, Sweden
Length in calibres	60
Weight of gun	?kg
Breech mechanism	?
Ammunition type	cartridge (?kg)
Projectiles	HE (0,90kg)
Propellant	?kg
Muzzle velocity	881m/s
Maximum range	7,160m

Mounting	
Designation	?
Protection	none
Weight of mounting	?
Elevation	-5° / +90°
Loading angle	any angle
Firing cycle	120rpm

Notes:

R.K.	Rekyl Kanon

57mm A.B.K. L/30

Gun Data	
Manufacturer	Søartilleriet
Length in calibres	30
Weight of barrel	135kg
Breech mechanism	?
Ammunition type	cartridge
Projectiles	*Brisantgranat / Antiballongranat*
Propellant	0.39kg / 0.39kg
Muzzle velocity	530m/sec, 500m/sec
Maximum range	7,500m

Mounting	
Designation	?
Protection	none
Weight of mounting	368kg
Elevation	-7° / +70°
Loading angle	any angle
Firing cycle	ca 16rpm

Notes:

A.B.K.	*Anti Ballon Kanon* (Anti Balloon Gun)
Brisantgranat	High Explosive shell
Antiballongranat	Anti Balloon shell
Søartilleriet	Naval Ordnance Dept

20mm R.K. M/31 (twin)

Gun Data	
Manufacturer	Dansk Rekylriffel Syndikat
Length in calibres	60
Weight of barrel	55kg
Breech mechanism	?
Ammunition type	cartridge (0.30kg)
Projectiles	HE (0.138kg), bullet (0.160kg)
Propellant	0.033kg
Muzzle velocity	795m/sec, 765m/sec
Range	?m

Mounting	
Manufacture	Søartilleriet
Designation	?
Protection	none
Weight of mounting	440kg
Elevation	-5° / +90°
Loading angle	any angle
Firing cycle	ca 250rpm

Notes:

R.K.	Rekyl Kanon
Søartilleriet	Naval Ordnance Dept

[Dansk Rekylriffel Syndikat was a Danish private Company]

20mm R.K. L/60 M/41

Gun Data	
Manufacturer	Dansk Rekylriffel Syndikat
Length in calibres	60
Weight of gun	55kg
Breech mechanism	?
Ammunition type	cartridge (0.30kg)
Projectiles	HE (0.128kg)
Propellant	0.040kg
Muzzle velocity	850m/s
Range	?m
Mounting	
Designation	?
Protection	none
Weight of mounting	226kg
Elevation	-10° / +90°
Loading angle	any angle
Firing cycle	ca 500rpm
Notes:	
R.K.	Rekyl Kanon

8mm R.G. L/75 M/37 (twin)

Gun Data	
Manufacturer	Dansk Rekylriffel Syndikat
Length in calibres	75
Weight of gun	9.5kg
Breech mechanism	?
Ammunition type	cartridge (?kg)
Projectiles	bullet (0.0127kg)
Propellant	0.0039kg
Muzzle velocity	760m/s
Range	?m
Mounting	
Designation	?
Protection	none
Weight of mounting	65kg
Elevation	-15° / +90°
Loading angle	any angle
Firing cycle	1,000rpm
Notes:	
R.G.	Rekyl Gevær

Torpedo Type H

Diameter	45cm
Length	5.68m
Weight	768kg
Warhead	121.5kg
Pressure in air vessel	180kg/cm2
Range	8,000m at 27knots

The torpedo room was situated below the waterline abaft the forward magazine. A special series of eight torpedoes designated type H was specially manufactured for *Niels Iuel* by Søminevæsnet from 1919–22. The extended time of manufacture was the result of tight budgets for the armed forces. The torpedo room had space for four torpedoes: two in the tubes and two stowed against the bulkheads. Above the spare torpedo on the right and in the left centre of the photo are cylinders for the compressed air used in the torpedoes and for firing them. (Kay Jørgensen)

zine and the auxiliary engine room. Eight of these torpedoes were manufactured at Orlogsværftet between 1919 and 1922.

Modernisation 1935–36

As there was no possibility of obtaining sufficient funding for a new major warship in the early 1930s, the Navy started planning for a modernisation of *Niels Iuel*. The modernisation was carried out at Orlogsværftet from 1935 to 1936. Upgrading the fire control system was the first priority. This, together with a general overhaul, gave *Niels Iuel* not only improved fighting capabilities but also a more modern appearance.

The old tripod foremast with its fixed gunnery direction platform was replaced by a heavy pole mast with a central gunnery director tower on two levels. The lower level, which rotated, had a 6-metre base coincidence rangefinder, and on the upper level was the central sight, constructed by the Dutch Hazemeyer Company, at that time one of the world's leading companies in this field. The trainable central sight was on a fixed pedestal mount, and all ten guns could be directed and centrally fired either manually or by gyro.[2]

Data from the gunnery director were transmitted electrically to a plotting table in the transmitting station. The latter was situated below the armoured deck and was fitted with an analogue computer which handled all data inputs, including own-ship and target speed, headings, wind speed and direction, barometric pressure, type of shell and muzzle velocity. Calculation of deflection and elevation for the guns was performed here and transmitted to the guns via a follow-the-pointer system.

The three original 3-metre rangefinders were replaced by 6-metre models purchased from the German company Zeiss.[3] There was a coincidence RF in the gunnery director tower, a stereoscopic RF in a tub atop the wheelhouse on the pole mast, and a coincidence RF atop the after deckhouse. The obsolete 57mm anti-aircraft guns were replaced by ten 20mm anti-aircraft guns (see above), and the searchlights moved to new positions.

At the stern on either side of the hull two sets each of two pressure vessels were fitted. These contained chemicals for generation of a smoke-screen. Each set contained one bottle containing siliciumtetrachloride at a pressure of 40kg/cm^2 (589psi) and another containing ammonia at 10kg/cm^2 (147psi). The contents were combined to produce a smoke-screen with a duration of six minutes.

Finally, damage control provision was upgraded.

Service History

1923

As soon as the ship was ready for service she embarked on a shakedown cruise lasting from 28 May to 6 August 1923. Visits were made to the following ports: Bergen (Norway), Thorshavn on the Faroe Islands (a part of the Danish Kingdom), Leith (UK), Gothenburg (Sweden) and Odense (Denmark). Crown Prince Frederik embarked in the ship for this cruise.

The next cruise took *Niels Iuel* far from home waters, course being set for South America. The cruise began in Copenhagen on 21 October and the ship returned home on 23 February 1924. The following ports and places were visited on the way out: Dartmouth (UK), Cadiz (Spain), Madeira (Portugal), the Cape Verde Islands, Rio de Janeiro (Brazil) and Buenos Aires (Argentina). The ship then returned to Denmark via Montevideo (Uruguay), Pernambuco (Brazil) and the Azores (Portugal). After leaving the latter islands the ship ran into a heavy gale which lasted for several days, and at times the ship had to reduce speed. The gyrocompass broke down, and there were numerous other mechanical failures, of which the most serious was when the rudder chains broke. The situation was critical and for a time the ship had to be steered by the two propellers alone, while emergency steering was rigged under the most dangerous circumstances. Several of the ship's boats were carried away by the waves and others were smashed at their davits. After the storm *Niels Iuel* called in at Plymouth to lick her wounds and to repair the worst of the damage.

1924

In the summer *Niels Iuel* became flagship of the gunnery training squadron and in the autumn flagship of the general training squadron. The year ended with participation in a minor exercise in December.

1925

Niels Iuel was flagship of the training squadron from 25 June to 25 July, visiting Helsingfors (Finland), Reval (today Tallin, Estonia), Riga (Latvia) – the first Danish Warship to visit this country – Danzig and Lübeck (Germany).

1926

Niels Iuel embarked on a cruise to the Faroe Islands and Iceland (part of a Union with Denmark until 1944), with the royal family on board. Several cabins and reception rooms had to be specially prepared to accommodate the royal family.

1927

Niels Iuel was flagship of the training squadron.

1928

Niels Iuel was equipped as a royal yacht for a royal visit

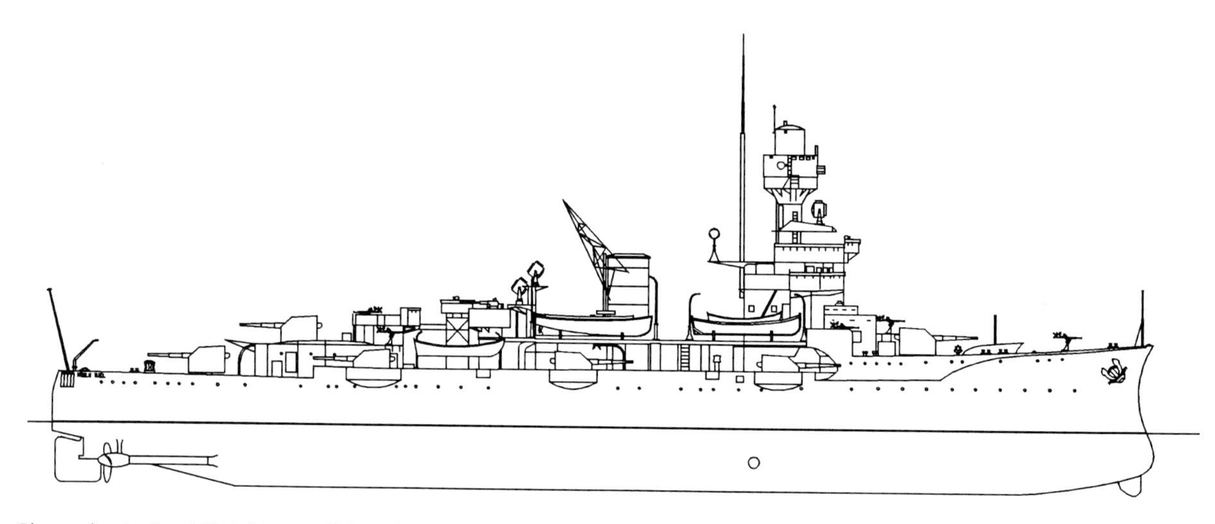

The modernisation 1935–36 gave *Niels Iuel* a much more modern look and also a first-rate gunnery control system. The base length of the rangefinders was increased from 3 metres to 6 metres. The anti-aircraft armament was augmented with the addition of ten 20mm guns in five twin mounts and fourteen 8mm machine guns, also in twin mounts. The drawing shows the AA armament as it was after the winter of 1941–42 when it comprised two 40mm L/60 M/37 and ten single 20mm L/75 M/41. (Tom Wismann)

The original tripod foremast is removed during the ship's reconstruction at Orlogsværftet. (Danish Armed Forces Photo Gallery)

to Finland, on which she was escorted by the cruiser *Heimdal*. At the same time she was a training ship for officer and petty officer cadets. Later in the year she resumed her role as flagship of the training squadron.

1929

Niels Iuel served as a training ship for officer and petty officer cadets on a cruise to the Mediterranean, calling at Marseille (France), Barcelona (World exhibition, Spain), Naples and Amalfi (Italy), Tripoli (Libya) and Lisbon (Portugal). After returning home she again served as flagship of the training squadron.

1930

From 22 May *Niels Iuel* was equipped as a royal yacht for a royal visit to the Faroe Islands and Iceland, on which she was escorted by the fishery protection vessel *Fylla*. During the course of this cruise the ship also served as a training ship for officers and petty officer cadets.

1931

Niels Iuel embarked on a training cruise to the Mediterranean for officer cadets. She visited Constantinople (today Istanbul, Turkey) and Odessa

Niels Iuel recommissioned the day after her modernisation was completed, on 9 July 1936. The crew is mustered on the upper deck, and steam has been raised for the first time in four years. The gun tubs are still missing their twin 20 mm mounts, but the after 6-metre rangefinder is in place. (Danish Armed Forces Photo Gallery)

Niels Iuel leaves Naval Station Holmen for her first trial run on 10 July 1936. She would be involved in a protracted period of trials and work-up after four years alongside. (Danish Armed Forces Photo Gallery)

(Soviet Union) – she was the first Danish warship ever to enter the Black Sea – returning via Piraeus (Greece), Genoa (Italy), Algiers and Bordeaux (France). She decommissioned on 3 September.

1932–36

She was out of commission from 1932–34, then underwent a major reconstruction at Orlogsværftet (see above) during 1935–36. She then spent the rest of 1936 working up.

1937

On 10 May *Niels Iuel* left for the Coronation Review at Spithead (Portsmouth, UK) to celebrate the coronation of King George VI. She returned home on 25 May in company with the Finnish coast defence battleship *Väinämöinen*.

She then took part in an exercise which ended with a visit to Hälsingborg (Sweden) on 17 September. The remainder of the squadron, which was commanded by Rear Admiral Trap, comprised the torpedo boats *Glenten*, *Høgen* and *Ørnen*, the minelayer *Lossen*, the mine vessel *Kvintus*, the submarine tender *Henrik Gerner* and five submarines.

1938

Niels Iuel accompanied the torpedo boat flotilla on a visit to Åbo/Turku (Finland), arriving on 5 August 1938 and departing on the 8th. From 2 September she was with the training squadron on a visit to Sønderborg (Denmark).

1939

A planned cruise to the USA to represent Denmark at the World Exhibition in New York in mid-May, while the ship was acting as an training ship for officer cadets, had to be cancelled due to international tensions. The deteriorating situation led to the mobilisation of the elderly coast defence battleship *Peder Skram*, and the two ships operated together from May to July. In late August *Niels Iuel* was at Holmen preparing for a cruise to Oslo (Norway). On 25 August the ship was instead ordered to fuse all shells. On 1 August the *Sikringsstyrken* was established as in August 1914. However, in August 1939 the RDN was only a shadow of the force mobilised in August 1914. Conscripts from the years 1937–38 were called up to complete the crew of *Niels Iuel*, and on 4 September she took station in the Aarhus area, together with the major part of the operational fleet.

1940

In early January *Niels Iuel* deployed to Copenhagen due to the ice situation in Danish waters. As the port was ice-bound, most of the crew were sent on leave. On 8 April they were again called up, and at 0400 on 9 April German Forces occupied Denmark.

The Occupation of Denmark

During the 1920s and 1930s the government, backed by votes from the majority of the population, had disarmed the country to such an extent that a defence of Danish territorial integrity was no longer feasible. Danish forces, which were considerably smaller and less capable than in August 1914, made only a token stand against the German invasion forces which crossed the border in the southern part of Jutland. Eleven Danish soldiers lost their lives; no doubt the armed forces would have continued to fight had they been ordered to, even though most of the tools had been taken from them over the past 20 years.

Niels Iuel was not ready for war on 9 April. On the morning of that day she was at Holmen, some 1,000 metres from the German troop transport *Hansestadt Danzig* at the Langelinie pier on the other side of the Harbour, which was disembarking a German battalion which was to occupy key points in Copenhagen. Nothing was done to oppose the German forces; the Government had put its trust in a non-aggression pact made with Germany in 1939.

From 9 April 1940 to 29 August 1943 Denmark cooperated with the Germans in what might be called a 'peaceful occupation'. Denmark was allowed to keep part of its armed forces, and all-in-all life went on as it had in peacetime. Denmark became a German 'show-room' for how peaceful an occupation could be.

However, by the summer of 1943 relationships between the Danish population and the German forces had been deteriorating for the best part of a year, and a small Danish resistance movement was becoming more and more active. The *Führer* in Berlin wanted the German forces to put away the velvet gloves and disarm the Danish armed forces. Thus on 29 August 1943 the policy of peaceful cooperation effectively ended.

The Skirmish on 29 August 1943

On 29 August *Niels Iuel* was at Holbæk at the bottom of Holbækfiord, conducting a training cruise. At 0410 the ship received a radio signal to be on high alert. The CO, Commander Carl Westermann, ordered 'action stations'

Outside Copenhagen harbour on 10 July 1936: the daily routine of trials and testing could now begin. It would take some time for the gunnery department to familiarise itself with all the new equipment. Note the four pressure vessels on each side of the stern: they contained chemicals for the generation of a smoke screen. (Danish Armed Forces Photo Gallery)

Gunnery trials with combat rounds in 1939. The gun crews are closed up and the main battery is trained on the same target as the director control tower. (Danish Armed Forces Photo Gallery)

and steam was raised in all four boilers. At 0412 the ship received a coded signal: 'Make your way to Swedish waters!'

Niels Iuel initially had difficulty getting off the jetty due to the wind conditions and low water, but at 0550 the ship got underway. Just before she left, a police officer had arrived with information that the warships at Holmen had scuttled themselves as German forces moved in to prevent them from falling into enemy hands, and that German troops had occupied the government and parliament buildings. Westermann planned to pass Hundested (see map) and then follow the swept channel to the east; it would then look as if the ship was headed for Copenhagen. Once she had passed the Kattegat South lightship, *Niels Iuel* would turn east and make a dash for Swedish waters.

Heading north in Holbækfiord, *Niels Iuel* first observed German reconnaissance planes at 0610, and at 0630 the ship left Holbækfiord and entered Isefjord, proceeding at twelve knots. At 0644 the German Admiral Dänemark, Vice Admiral Wurmbach, was informed that aircraft had observed *Niels Iuel* and were following her movements. At 0730 everything on board *Niels Iuel* was ready for combat. The 15cm shells had been fused, and ready-use ammunition was in place for the main battery and the anti-aircraft guns. At about the same time the German forces delivered a note to the Danish coastal police[4] at Hundested at the mouth of Isefjord – *Niels Iuel* had to pass this point in order to exit the fiord. The note stated that the entrance to the fiord had been mined by the *Kriegsmarine* during the night. It was also stated that all traffic to and from Isefjord was prohibited until further notice.

In the meantime, Wurmbach summoned the C-in-C of the Danish Navy, Vice Admiral Vedel, to his office in Copenhagen, and strongly recommended that he radio *Niels Iuel* ordering the ship to stay in Isefjord. Admiral Wurmbach had already prepared the following signal:

> Mines have been laid in the mouth of Isefjord and strong German Naval forces are present. *Niels Iuel* is to anchor and await further orders, signed 29 August 1943, 0730. Vedel.

The German admiral was well aware that there were no minefields at the entrance to Isefjord, as the vessels ordered to lay the minefield had been delayed at the mine depot in Kiel. A bluff of war! Vedel agreed to send the signal, being convinced that *Niels Iuel*'s CO would act according to his own judgment and would refuse to obey the order knowing that the radio transmitter could well be in German hands. In the event it was decided to send the Danish Cdr Pontoppidan to Hundested in a German seaplane, from where he could be transferred to *Niels Iuel* by a small Danish navy cutter.

As *Niels Iuel* reached Lynæs Sand close to the mouth

Unusually, Denmark was allowed to keep part of its armed forces even though the country was occupied by Germany. This policy was abruptly ended on 29 August 1943, when the German *Wehrmacht* disarmed the Danish forces on an order from Hitler. *Niels Iuel* tried to reach Swedish waters, less than two hours steaming from the mouth of Isefiord. After bombing and strafing from German aircraft the CO decided to run the ship aground and destroy as much equipment as possible to deny the enemy the use of the ship. (Svend Kieler)

of Isefjord, three German naval vessels, one large and two smaller, were observed outside the fiord at a distance of 18,000 metres. When the distance was down to 15,000 metres the ships were identified as a German auxiliary cruiser, probably a minelayer, and two motor torpedo boats. In fact, it was the German torpedo boat *T17* and two MTBs. German Stukas and bombers were constantly circling overhead.

The coastal police were arranging to get the received message concerning the minefields outside Isefjord to *Niels Iuel* using the small patrol cutter *P31*. The message was delivered to *P31* at 0735, which at the same time spotted *Niels Iuel* coming up from the south. *P31* would approach *Niels Iuel* when the ship passed Hundested. At 0815 the coastal police again contacted *P31*, which still was at Hundested, with a message from Vice Admiral Vedel that the cutter should await the arrival of Cdr Pontoppidan, who was to be taken to *Niels Iuel* as soon as he arrived.

At 0835, as *Niels Iuel* passed Hundested, the coastal police signalled by light: 'German minefield to the north

Action stations at one of the ten 20mm AA guns prior to the attack on 29 August 1943. Five Danish sailors were wounded, one fatally. The 20mm gun was a Danish design produced in Denmark by the company Dansk Rekylriffel Syndikat. (Danish Armed Forces Photo Gallery)

On the morning of 29 August between 0855 and 0935 outside Hundested, German planes attacked the ship five times with bombs and machine guns. In all four bombs were dropped, and one of the last two detonated at a distance of only 5-10 metres from the ship's side. The shock waves damaged the old ship, so the CO decided to run the ship aground and in this way render it of no use to the enemy. (Danish Armed Forces Photo Gallery)

of Isefjord'. The signal was received on board, but Westermann decided to proceed, as he was of the opinion that the minefield had not been laid so close to land that it blocked his intended route; he thought that *Niels Iuel* had a good chance against the German sea and air forces currently in the area. *P31*, which had left Hundested at 0830, was now so close to *Niels Iuel* that it could inform the ship about the probable minefield and the arrival of Cdr Pontoppidan using hand signals and a megaphone.

As *Niels Iuel* had earlier been spotted by a German coastal lookout post and reported to Admiral Wurmbach as coming north at high speed and with all guns manned, the admiral had asked the *Luftwaffe* to have planes readied in the area to stop the ship if necessary. At 0855 one of the planes attacked with machine guns and dropped two bombs, which exploded at a distance of approximately 30 metres. Westermann had the impression that this was only meant as a warning and not as a full-blooded attack, so he did not give the order to open fire with the anti-aircraft guns. Course was then changed to the south, but while the ship was in the middle of her turn the plane attacked again with machine gun fire, which resulted in some of the gun crews spontaneously returning fire. Westermann ordered the gun crews to stop firing and the crews was ordered to take cover so as not to take casualties from the aircraft's machine guns.

After the second attack the gunnery officer in the director control tower asked for permission to open fire on the German ships outside the mouth of Isefjord. Westermann refused, as he was not of the opinion that a state of war existed between Germany and Denmark.

Between 0910 and 0920 one of the planes attacked twice with machine gun fire, resulting in hits on the ship. At 0935 came the fifth and final attack: one plane attacked with machine guns and dropped two bombs which exploded in the water at a distance of 5–10 metres from the ship's port side amidships. This attack resulted in serious damage to the elderly vessel and five of the crew were wounded. This time Westermann saw it as a serious attack on his ship and ordered the anti-aircraft guns manned.

When *Niels Iuel* was built little attention had been paid to the shock/mining effect from aircraft bombs exploding close by in the water. The electrical lighting went out, first intermittently, then completely. As the lighting could not be restored, the crew below decks had to work by emergency lamps and hand-held torches. The loss of electrical power also resulted in the failure of the primary and secondary gunnery control systems; the ammunition hoists and the gyro compass failed, so the ship had to be conned from under the open sky. The shock of the explosions caused the forward rangefinder to break in half, and in the magazines 15cm fused shells fell out of their racks. The engines of the two armed torpedoes started running hot inside their tubes, creating a potentially serious situation. Most of the floor plates in the engine spaces were catapulted into the air and could not be re-seated, as the outer plating of the ship and a number of bulkheads had been deformed.

When this damage was reported to the CO he decided that even if he could pass the minefield without hitting a mine by holding close to the Zealand coast and managed to evade German naval forces, it was unlikely that he would be able to reach Swedish waters. It was clear that German aircraft would continue to make determined attacks on *Niels Iuel* which could lead to considerable loss of life.

Following the last attack *Niels Iuel* slowly headed south and when the ship was off Hundested Cdr Pontoppidan arrived and was put on board by the cutter *K7*. The Commander handed over the message purporting to be from Vice Admiral Vedel, and Westermann enquired where the signal had been written. When Cdr Pontoppidan replied that it had been written at Hotel Phønix in Copenhagen, which served, for the duration, as the headquarters of the German Naval Command, he put two and two together. Westermann concluded was that it would now be impossible for the ship to reach Sweden. He therefore decided to run the ship aground and destroy as much of her equipment as possible so as to render her unusable for the German *Kriegsmarine*.

Cdr Pontoppidan left *Niels Iuel* at 1040 and the ship proceeded at maximum speed on a southwesterly course until she grounded at 1048 on the slightly raised bottom off the Annebjerggård farmhouse a few kilometres south of the town of Nykøbing Sjælland (see map). Four minutes later she dropped anchor. After this the German planes that had been circling the ship disappeared.

Niels Juel is hard aground in Isefiord. Note the national flag painted on the sides of the bow; there were similar identification markings on the ship's decks. The line running through the flag and along the hull is a degaussing cable to protect the ship against magnetic mines. (Danish Armed Forces Photo Gallery)

The wounded were taken by the ship's motor boat to the hospital at Nykøbing Sjælland. Unfortunately one of the wounded, chief gunner HE Adreasen, died of his wounds in hospital on 2 September. When later in the day the boat returned to *Niels Iuel* it brought a note from the German CO at Nykøbing Sjælland stating that any of the crew who went ashore would be shot! If the CO wanted to negotiate with the German CO he was allowed to enter Nykøbing by the ship's motor boat, flying a white flag.

At 1040 the CO mustered the ship's company on the upper deck, where he told the crew that he was proud of their behaviour during the skirmish. The crew answered by sounding a 'Hurrah' and singing the national anthem. They then set about sabotaging as much equipment as possible. The breeches of the 15cm guns were taken apart and thrown overboard, together with some of the shells and much other equipment. At 1110 a demolition charge in the shaft tunnel was set to explode, but unfortunately the detonator malfunctioned and no spare was available. Instead the magazines were flooded and the sea-cocks in the engine and boiler rooms were opened and smashed. After half an hour the water had risen so high in the machinery spaces that the water covered engines, dynamos and switchboards.

At 1535 the Danish Lt-Cdr Overbye, accompanied by a German staff officer, reached *Niels Iuel* in a German seaplane bringing orders from Vice Admiral Vedel, and the state of the ship was reported to both Admiral Vedel and the German Admiral Dänemark.

The following morning at 0520 two German MTBs came alongside and put a boarding party on board. The commander of the boarding party at once demanded that Westermann haul down the Danish colours and that the officers should give up their side arms. The answer was that the Danish flag would be lowered at 0700 and that no Germans were wanted on the quarterdeck at this time. This was accepted by the German officer.

At 0700 the crew mustered on the quarterdeck, where the CO conducted the ceremony and made a short speech. To the sound of the bugle the Dannebrog (the Danish ensign) was lowered, followed by the crew singing the national anthem. This was the sad end for the career of *Niels Iuel* in the RDN.

Under New Management

On 30 August 1943 the German authorities consulted the Danish salvage company Svitzer to enquire if they would do a survey of *Niels Iuel*. The work started on 31 August when the Svitzer salvage vessel *Garm* arrived on the scene. For several days divers from *Garm* did their survey and reported as follows to the Svizter main office: 'The ship is grounded on a sandy bottom, no visible damage to hull, rudder or propellers, hull filled with water to 1.5m below the armoured deck'. It was also noted that items of equipment from the ship, parts of the guns and some ammunition were strewn on the bottom around the ship.

In September 1944 the former *Niels Iuel* was commissioned into the *Kriegsmarine* under the name *Nordland*, and was used as a stationary training vessel for the rest of the war. The ship was probably stationary due to German fuel shortages during 1944–45. It is possible that one reason for the salvage of *Niels Iuel* was that the ship had two coal-fired boilers, and coal was not in such short supply as oil. (Danish Armed Forces Photo Gallery)

For some unknown reason the Svitzer company did not secure the contract to salvage the ship. *Niels Iuel* was salvaged by the Germans in October 1943 and towed directly to Kiel for repair. During the repair the armament was removed. Eight of the ten guns went to German shore batteries in Denmark: four at Frederikshavn and four at Løkken. What happened to the last two guns is not known.

In September 1944 the former *Niels Iuel* was commissioned into the *Kriegsmarine* as a training vessel with the name *Nordland*. She served from the autumn of 1944 until February 1945 as a stationary platform at Stolpmünde (now Ustka, Poland) on the Baltic coast. On 18 February steam was raised and *Nordland* sailed under her own power for Kiel to escape the approaching Red Army. After two days of steaming without being harassed by allied or Russian planes, *Nordland* arrived at Kiel. Here the training of the cadets went on – business as usual. The training continued until 3 May, four days after Hitler had committed suicide in his Berlin bunker. Early in the morning the CO ordered steam raised, and later in the day the ship, with a skeleton crew, steamed out into the Ekernförde Bay, where demolition charges were detonated. The German crew was taken off by small boats and all made it safe back to dry land.

In the chaotic years after the war, unauthorised salvage divers partially dismantled the wreck. In 1952 the remains were sold by the Danish state to the German company Firma Eisen und Metall KG, Hamburg. In the following years the ship was dismantled on the bottom and the steel lifted by crane.

Today the wreck of *Niels Iuel* is hard to find and to recognise, as the salvage company salvaged everything which protruded above the sea bed; its location is 54°28.20' N 010°04.36' E, at a depth of approximately 28 metres.

Sources:

Flaadens virksomhed under verdenskrigen 1914–1919 (Copenhagen 1920).

Maskineriet i Flaadens skibe (Copenhagen 1934).

'Nordland ex Niels Juel' (www.bubblewhatcher.de).

Skibsbygning og Maskinvæsen ved Orlogsværftet 1692 – 6 oktober 1942 (Copenhagen 1942).

Christiansen, Henrik, *Orlogsflådens skibe gennem 500 år*, Vol 3.

Fuchs, Werner, 'Die letzen Seekadetten der Kriegsmarine auf der Nordland und ihre letze Fahrt', Jahrbuch der Marine 19??.

Hendriksen, Knud, *Operation Safari: 29 August 1943* (Copenhagen 1993).

Gyldenkrone, EM, *Håndbog i Søartilleriets materiel* (Copenhagen 1923).

Kristensen, Kenneth, *Niels Juels Kanoner: fra skib til kystbefæstning* (Bangsbo Museum & Arkiv 2013).

Larsen, Kay, *Vore Orlogsskibe fra halvfemserne til nu* (Copenhagen 1932).

Madsen, Kaj Toft, *Danske Torpedoer 1868–2008* (Copenhagen 2008).

Nørby, Søren, 'To gange sænket af egen besætning', Marinehistorisk Tidsskrift No 4/2003 (Copenhagen 2003).

Steensen, R Steen, *Vore Panserskibe 1863–1943* (Copenhagen 1968).

Thiede, Sven Egil, Dansk Søartilleri 1860–2004, Vol II (Copenhagen 2004).

Endnotes:

1 Denmark adopted the metric system in 1907, and it has been assumed for the purposes of this article that after that date displacements (other than Washington standard) were expressed in metric tons (tonnes) rather than long tons.

2 In 'manual' operation the Control Officer in the director tower could fire the guns by blowing into a tube, thereby triggering an electrical circuit. In theory this minimised reaction time, as the brain/mouth delay was thought to be less than the brain/hand delay involved in pushing a button. In gyro firing the firing circuit was activated when the ship passed a pre-set point during its roll.

3 The original 3-metre rangefinders were likewise manufactured by Zeiss and delivered in 1923; they were of the stereoscopic type. Two of these rangefinders were removed in 1929 and reinstalled in *Peder Skram* and *Olfert Fischer* (*Herluf Trolle* class). In their place *Niels Iuel* received a12ft (3.66m) coincidence rangefinder purchased from Barr & Stroud (delivered 1926), which was installed in the gunnery director tower. In 1930 the 3.66-metre model was replaced by a 5-metre Zeiss coincidence model purchased for *Peder Skram* in 1928.

4 The coastal police was established during the occupation following a German request on 1 March 1941 for a special watch to be kept on certain coastal areas: it was a Danish-manned, uniformed force.

THE BATTLE OF THE RIVER PLATE: A TACTICAL ANALYSIS

In the first of a pair of articles on the Battle of the River Plate, **Alan D Zimm** focuses on the tactics adopted by the British Commodore Henry Harwood and his German adversary, Captain Hans Langsdorff, and presents an analysis using the battle charts together with the rules for prewar wargaming simulations employed by the British and US Navies.

'In this sombre dark winter ... the brilliant action of the Plate ... came like a flash of light and colour on the scene, carrying with it an encouragement to all who are fighting – to ourselves and our allies.'[1]

The Second World War did not start auspiciously for the Royal Navy: the aircraft carrier *Courageous* sunk by a U-boat, the 17,000-ton armed merchant cruiser *Rawalpindi* sunk with most of her crew by German battlecruisers, and the battleship *Royal Oak* sunk in the Home Fleet's sanctuary of Scapa Flow. Against those embarrassments the scuttling of the 'pocket battleship' *Graf Spee* following an engagement with the heavy cruiser *Exeter* and light cruisers *Ajax* and *Achilles* was welcome news.

Seizing the propaganda opportunity, the British Ministry of Information quickly produced an account[2] designed to encourage the population, give heart to Britain's sailors, and begin the process of undermining Adolf Hitler's confidence in his surface navy. Offered to the public at two pence a copy, it was a stirring tale of a British David against the German Goliath. This interpretation has largely carried over into our time.

The British used two metrics to back their claim that they were the inferior force: shell weight of a single broadside, and displacement. For broadside weight, the British quoted *Graf Spee* at 4,700lb compared with the 3,136lb of the three cruisers, a 50 per cent German advantage. The measure was for a single shot from each gun; however, a more realistic computation would have been shell weight *per minute*. *Graf Spee's* battery could fire 13,121 pounds per minute; the three cruisers 21,760lb/min. Moreover, a shell's destructive power is not directly proportional to its weight. The US Naval War College calculated that the 'destructive effect of an explosion varies inversely as the square of the distance away', so 'the radius of the destructive effect will thus vary as the square root of the figure of weight.' By this standard an 11in shell weighing more than six times a 6in shell is nevertheless only three times as damaging. Taking into account these considerations, the British had a firepower advantage of perhaps 125 per cent.

Graf Spee's displacement was listed as 'over 10,000 tons'; the tuppenny broadsheet emphasised she was larger than any of the British ships. *Exeter's* displacement was 8,550 tons, and each of the two *Leanders* ca 7,000 tons, for a total of approximately 22,550 tons. The Germans could therefore justifiably claim that the tuppenny broadsheet should have credited the British with approximately twice the German's total tonnage and firepower.

Admiral Graf Spee underway in the English Channel in April 1939. During her commerce raiding operations in the South Atlantic and the Indian Ocean she sank nine merchant ships. (US Naval History & Heritage Command, NH 80973)

Tactical Analysis

Both the US and the Royal Navies used their wargaming systems as tools to determine the best battle tactics.

Armour and Fire Effects, US Naval War College 1939

According to these rules, *Graf Spee's* 11in guns penetrate the British armour at all ranges, as do *Exeter's* 8in guns against *Graf Spee's* armour. The 6in guns of the light cruisers penetrate *Graf Spee's* deck at ranges in excess of 21,000 yards, or her belt at under 12,000 yards with a 90-degree angle of incidence ('normal'). If *Graf Spee* presents her belt armour at 15 degrees or more off normal, the shells will not penetrate. *Graf Spee* had a 'normal' Immunity Zone against 6in shell between 12,000 and 21,000 yards; off normal, from 0 to 21,000 yards.

One important measure of effectiveness was 'Fire Effect', the percentage of damage that ships could inflict at a given range. This takes into account accuracy, dispersion, ballistics, angle of fall, armour penetration, destruc-

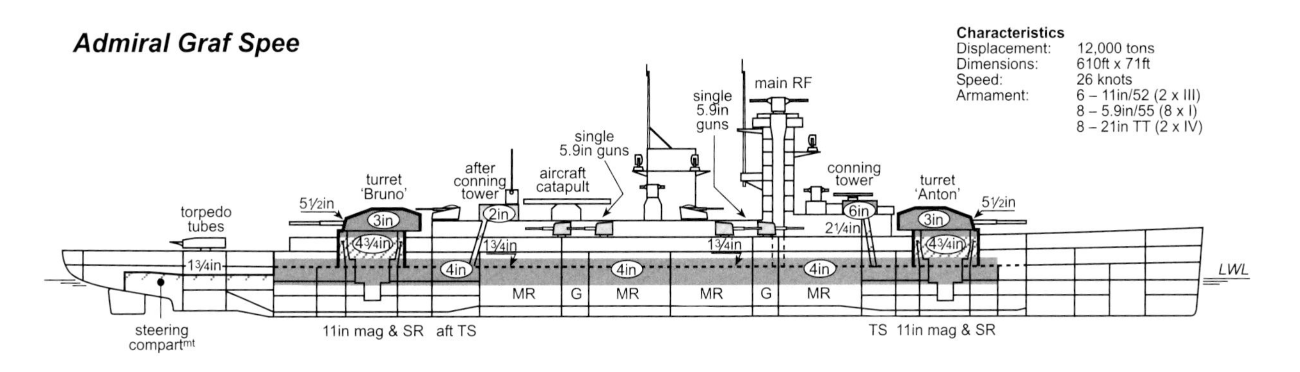

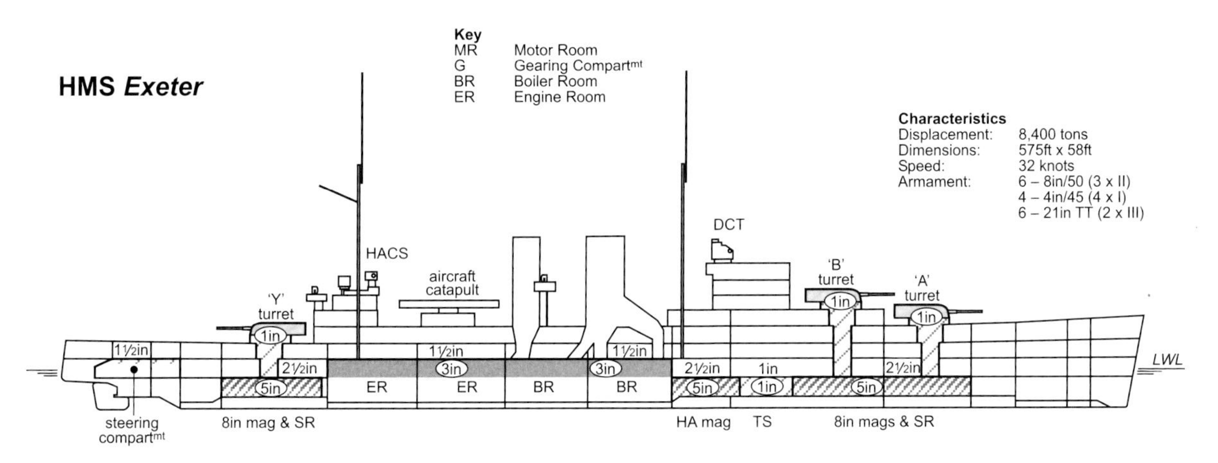

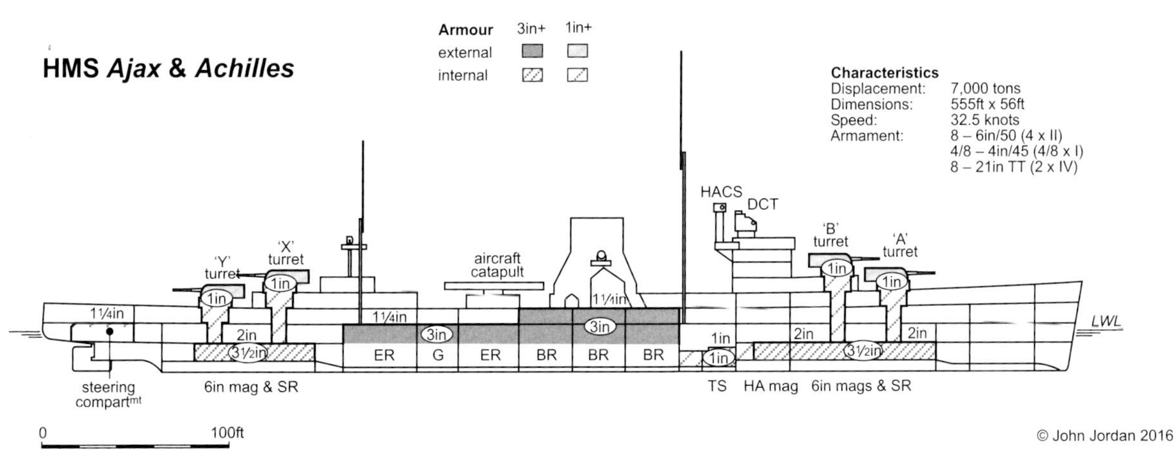

In order to facilitate a comparison with the British ships, armour thickness and gun calibre for *Graf Spee* are given in approximate imperial measurements. Calibres were 28cm and 15cm for the main and secondary guns respectively. The armour belt was 100mm, and the main armoured deck 45cm; Whitley suggests that this was reinforced to 70mm over the magazines, although firm evidence is lacking. The main turrets had 140mm plates on the face and 75-80mm on the sides; barbettes were 125mm. The face and sides of the conning tower were 150mm.

The protection drawings have been simplified in the interests of clarity. For armour plate (cemented or NC) only the thickness of the armour itself is recorded (not the DI backing), and the thicknesses of transverse bulkheads have been omitted. The sides of the armoured boxes for the magazines on *Exeter* comprised two strakes: an upper strake of 5in cemented armour and a lower strake of 2½in NC, both secured to ½in DI.

tive power of the shell, ship survivability, and a host of other factors. Smaller guns generally show well at shorter ranges due to their greater rate of fire. At longer ranges they cannot take advantage of their rate of fire because of the delays involved in spotting, and they will score fewer hits as their angle of fall becomes steeper and the 'danger space' shrinks. Larger guns perform better at longer ranges, in part due to their flatter trajectories.

Using the US Naval War College Maneuver Rules (USNWCMR), comparative Fire Effect calculations were

A close-up of the massive turret 'Bruno' taken during April/May 1939, when *Graf Spee* deployed to Spain and Morocco. The 28cm/52 SK C/28 gun fired a 300kg (661.4lb) shell. Note the prominent armour belt and the lower row of scuttles, which must have had an adverse impact on habitability when the ship operated in tropical waters, and was also a vulnerability in terms of reserve buoyancy in the damaged condition. (NHHC, NH 80897)

performed with *Graf Spee* against Division 1 (*Ajax* and *Achilles*), and Division 2 (*Exeter*). The calculations assumed spotting from the fire control tops. Only *Graf Spee* had fire control radar, which was not used. The British had aircraft for spotting; their travails will be discussed later.

Graf Spee held a significant advantage over *Exeter*. At close range *Graf Spee's* Fire Effect is 66 per cent greater, expanding to over 250% greater at 22,000 yards. One on one, *Graf Spee* should defeat *Exeter* comfortably. *Graf Spee* was superior to the two light cruisers at ranges greater than 12,000 yards by virtue of her larger guns and thicker armour, while Division 1 had the advantage under 11,000 yards. The combined British force had a total Fire Effect from 1.5 to 3 times that of *Graf Spee* at close ranges. Over 21,000 yards *Graf Spee* delivers 1.5 to 7 times the British Fire Effect. This advantage at long range is mitigated by the low hit percentage – *Graf Spee* did not carry enough ammunition to sink all three cruisers at more than 19,000 yards, even under ideal conditions.

The overall advantage was with the British. With a 9-knot speed advantage,[3] the British controlled the range at which the battle would be fought. Between 8,000 and 20,000 yards, the British could deliver a total Fire Effect up to three times that of *Graf Spee*.

German Tactics from the USNWCMR

The British speed advantage limited German options: the British could choose to fight at their selected range, or disengage and shadow. German tactics had to be based in part on luck and in part on exploiting British mistakes. If Captain Eric Langsdorff's objective was to escape, one possibility would be a long-range duel. If plunging fire could hit the cruisers' engineering spaces and cut their speed, *Graf Spee* could outrun them.

Running a USNWCMR experiment where the British attempt to close the range as fast as possible starting at the maximum range of the 11in gun, four 11in hits were scored after firing 180 rounds. If the engineering spaces constituted 50 per cent of the target area and magazines a further 20%, an 'expected value' calculation would suggest three critical hits. If optimally distributed all three ships might be disabled. However, British salvo chasing and manoeuvring would make hits a matter of luck. And if two of the cruisers were slowed, the undamaged ship could disengage, shadow, and await reinforcements. Thus, unless luck intervened, a long-range

Exeter as she appeared on completion in 1930–31. Designed for commerce protection, she was more heavily armoured than the earlier 10,000-ton British 'treaty' cruisers. She was essentially unmodified when she took part in the Battle of the River Plate. (NHHC, NH 60817)

An unusual overhead view of the light cruiser *Ajax* taken at Coco Solo in Panama in 1939, shortly before the outbreak of the Second World War. Turret 'Y' is concealed beneath the awning rigged over the quarterdeck, and her Seafox reconnaissance aircraft is atop the catapult. The original single 4in Mk V HA guns have been replaced by the Mk XVI in twin mountings Mark XIX; her sister *Achilles* would retain the single mountings until 1943. Note the proximity of the after 'austere' fire control station to 'X' turret. With the turret trained on forward bearings, the personnel manning that station would have been exposed to considerable shock and noise. (NHHC, NH 50344)

gunnery duel was not a promising tactic. Alternatively, the Germans could hope the British would close to decisive ranges.

Visibility was unlimited with a moderate wind from the southeast. The conditions were ideal for laying smoke screens. If the British divided their forces *Graf Spee* could split the battlefield with smoke, blocking the line of fire from the light cruisers. She could then concentrate on the outclassed *Exeter*, hoping to dispose of her quickly before suffering significant damage. A close-range engagement would also help protect *Graf Spee's* engines from 8in plunging fire. Eliminating *Exeter* would make the odds more even. *Graf Spee* could then fight the light cruisers between 13,000 and 17,000 yards, within her 6in Immunity Zone. *Graf Spee* held the advantageous lee position. Smoke from her gunfire would be blown away from her line of sight to the targets.

British Tactics from the USNWCMR

With their speed advantage the British can dictate the terms of the battle. Strategically, their objective is to sink *Graf Spee*, or to cripple her so that she can either be intercepted by one of the eight other forces hunting for raiders, or rendered unseaworthy and interned in a neutral port.

The USNWC rules suggest that *Ajax* and *Achilles* should close to under 12,000 yards, where they can smother the German ship with a high volume of hits. They should close rapidly, to avoid a chance hit that might disable their propulsion machinery. Engaging at long range was likely to be unsuccessful, as their shells could not penetrate their opponent's deck armour unless fired from over 21,000 yards, at which range hits would be scarce.

Exeter should engage at ranges where her guns could penetrate *Graf Spee*'s deck armour and place a plunging projectile in the German ship's engineering spaces. Alternatively, she could engage at closer ranges where she could penetrate *Graf Spee*'s belt armour, but the shell trajectories would make a disabling engine room hit less likely. She would also be unlikely to survive *Graf Spee*'s fire for long. At longer ranges, it would be a roll of the dice whether *Exeter* or *Graf Spee* would first hit the other's propulsion machinery. With *Exeter*'s more rapid rate of fire, the odds of landing such a hit would be in her favour.

The USNWCMR would suggest that the British ships should operate independently to complicate the German fire control, but on the same side of *Graf Spee*, allowing only half her secondary battery to engage.

British Tactics from CB 3011 War Game Rules

The British official naval war game rules were CB 3011 of 1929. These rules did not have the full sophistication and resolution of the USNWCMR, as they were designed for quick play by a smaller number of participants. They were used to develop tactics against pocket battleships at the staff college.

CB 3011 and the USNWCMR agree that *Exeter* was outclassed by *Graf Spee* at all ranges. However, they show a surprising difference in the match-up against *Ajax* and *Achilles*. At 12,000 yards and less, the USNWCMR has the Fire Effect of the light cruisers 1.5 to 3.5 times greater than CB 3011; at 13,000 yards and above CB 3011 is 3 to 35 times greater than the USNWCMR. CB 3011 expects many more hits at long range and fewer at close range than the USNWCMR, by a very wide margin.

CB 3011 also awards an additional 10 per cent bonus above 12,000 yards and a 25% bonus over 15,000 yards for 'Concentration Fire'. This was a technique whereby two or three ships used the lead ship's fire control solution (corrected for distance from the leader) and fired simultaneously. Theoretically, this would reduce the confusion when multiple ships fired on the same target and tried to differentiate their fall of shot.

Another gunfire technique, not included in CB 31011, was 'Flank Marking'. At long range, a firing ship could only tell if its salvo was long, short, or straddle, by whether the bases of the visible shell splashes were cut off by the hull of the target. Distances over or short could not be determined. In Flank Marking, a ship observing from a different angle would estimate the distance long or short of the salvo's Mean Point of Impact (MPI) and transmit it, allowing the firing ship to adjust accordingly.

CB 3011 included armour penetration, with little difference from the USNWCMR tables. CB 3011 also shows a different cruiser gun, the 6in/45 Mark XII, which could penetrate *Graf Spee's* deck beyond 17,000 yards rather than 21,000 yards. There was the possibility of confusion.

In the USNWCMR shells that could penetrate armour were given destructive effects greater than non-penetrating shells. In CB 3011 there was no difference in destructive value. Instead, penetrating hits were given a 1 in 30 chance of exploding a ship's magazine, and a greater (umpire-specified) chance to reduce the target's speed by penetrating machinery spaces. For a 'large cruiser' (such as *Graf Spee*) requiring 24 6in hits to sink, a magazine explosion from penetrative hits could end the battle before reaching the figure of 24, possibly even on the first hit.

Guided by these expectations, it would make sense to drive *Exeter* in to close range. Her rate of fire was higher than the German's 11in, so she could expect to hit faster. The German ship would more likely blow up first.

If the light cruisers engaged at 18,000 yards, CB 3011 predicts that *Graf Spee* would be destroyed in 30 minutes (21 minutes if fire from *Exeter* is taken into account). In the rules manoeuvring did not affect the hit rate; that might have been the conclusion of the British naval community in general, and explains some of the conduct of the battle.

The surprising conclusion is that the two different sets of wargaming rules would recommend the British commander use almost diametrically opposite tactics. The British commander at the River Plate, Commodore Henry Harwood, spent two years at the Staff College developing tactics against pocket battleships. He was considered an expert, and he undoubtedly used CB 3011 to develop and test his tactics.

The Battle of the River Plate

Track Charts

The British report included a track chart, drawn to scale, which was probably taken from *Achilles'* Admiralty Fire Control Table (AFCT) plot. *Graf Spee's* track is based on the estimates made in the AFCT, not German testimony. The German report, on the other hand, was written some months after the battle, without reference to the original logs and gunnery records, which were destroyed with all other classified documents prior to the scuttling. The Allies obtained the report after the war; it included a track chart of the battle.

There are inconsistencies between the two charts, most significantly between 0620 and 0640, and there is a mismatch of six to eight minutes at some critical points, possibly due to memory lapses or recording errors. It is likely that the changes of course for *Graf Spee* shown on the German chart are accurate (in direction, if not in time). These course changes induced the AFCT in *Achilles* to hunt for a new target course and speed solution, overshooting and undershooting in course and speed until sufficient rangefinder bearing and range data were accumulated upon which to generate a new solution. This explains some of the wild course swings the British attributed to *Graf Spee*, for example, between 0620 and 0646. Unfortunately, this track has been accepted uncritically by historians, some of whom have actually tried to assign tactical meaning to the swings. The following analysis uses a best estimate for *Graf Spee's* track drawn from both charts.

Battle Chronology

This chronology should not be considered precise, as sources conflict. Times, courses, speeds, target angles, bearings and ranges are approximated from the track charts and reports, and may be off by some minutes. Courses are in degrees true, speed in knots, ranges in yards. (T+number) indicates the time after firing began at 0617. 'A' , 'B' , and 'Anton' are forward turrets, 'X' , 'Y', and 'Bruno' are after turrets. 'A' arcs denote broadside bearings where all main battery turrets can fire. The remarks (in italics) are the author's; they provide explanation and analysis.

0552 ***Graf Spee***: Warship masts spotted. Conditions: clear, visibility unlimited. Wind from the southeast, moderate. Seas placid. *Graf Spee* sounds battle stations.

0600 ***Graf Spee***: Cleared for action. Targets are identified as the heavy cruiser *Exeter* and two destroyers.

Langsdorff presumes they are escorting a convoy. He rejects advice to withdraw in compliance with his orders restricting him from engaging warships. He decides to attack immediately 'in order to close to effective range before the enemy can work up to maximum speed, since it appeared to be out of the question that three shadowers can be shaken off.'[5]

0605 ***Graf Spee***: Challenges via signal light.

This was probably a ruse, to gain time to close the range while the British puzzled over the signal.

0610 ***Graf Spee***: The smaller warships are identified as light cruisers of the *Leander* class. Full speed is ordered.

With her fouled bottom and worn engines, maximum speed available was 24 knots. The British Admiralty Fire Control Table shows a maximum speed made good during the battle as about 22 knots.

0612 ***Graf Spee***: Ordered course/speed 115/24. As her diesels accelerate, smoke comes from the funnel.

0614 ***Ajax***: Smoke is sighted. A signal is initiated ordering *Exeter* to close and identify.

0615 ***Exeter***: Signals that smoke is sighted.

0616 ***Exeter***: Signals, 'I think it is a pocket battleship', then 'Enemy in sight bearing 322.' Action Stations is sounded.
Ajax* and *Achilles: Sound Action Stations.
Graf Spee: Passes through resonance speed of 21 knots. Heavy vibration in turret Anton causes a screw to back out of a motor controller.

0617 (T+0) ***Graf Spee***: Turret Bruno opens fire on *Exeter*, range 21,500, target speed 17. Turret Anton reports it cannot traverse and 'the barrels will not move any more.'

This problem had occurred before. The remedy was to replace a screw and put the middle barrel in manual elevation control.

Graf Spee's *first four salvos used base-fused delay shells for better spotting; instantaneous nose-fused shells were then fired until expended. The first four salvos took five minutes.*

After the battle, an intelligence agent in a bar in Montevideo overheard a German petty officer say the engine and hull vibration made the guns 'difficult to handle.' It is doubtful that vibration caused the loading cycle to increase from 24 seconds to more than 60 seconds. The gunnery officer was methodically establishing the range, with a time of flight close to one minute.

0618 (T+1) ***Exeter***: 11in hit No 1 strikes the ammunition embarkation hatch abaft 'B' turret and exits the starboard side without detonating.

The first hit on Exeter *had an angle of descent of 13.5 degrees fired from 19,000 yards arriving from Red 100 (260 degrees relative, just abaft the port beam), indicating the hit was received before the ship turned to close* Graf Spee.

Track Charts

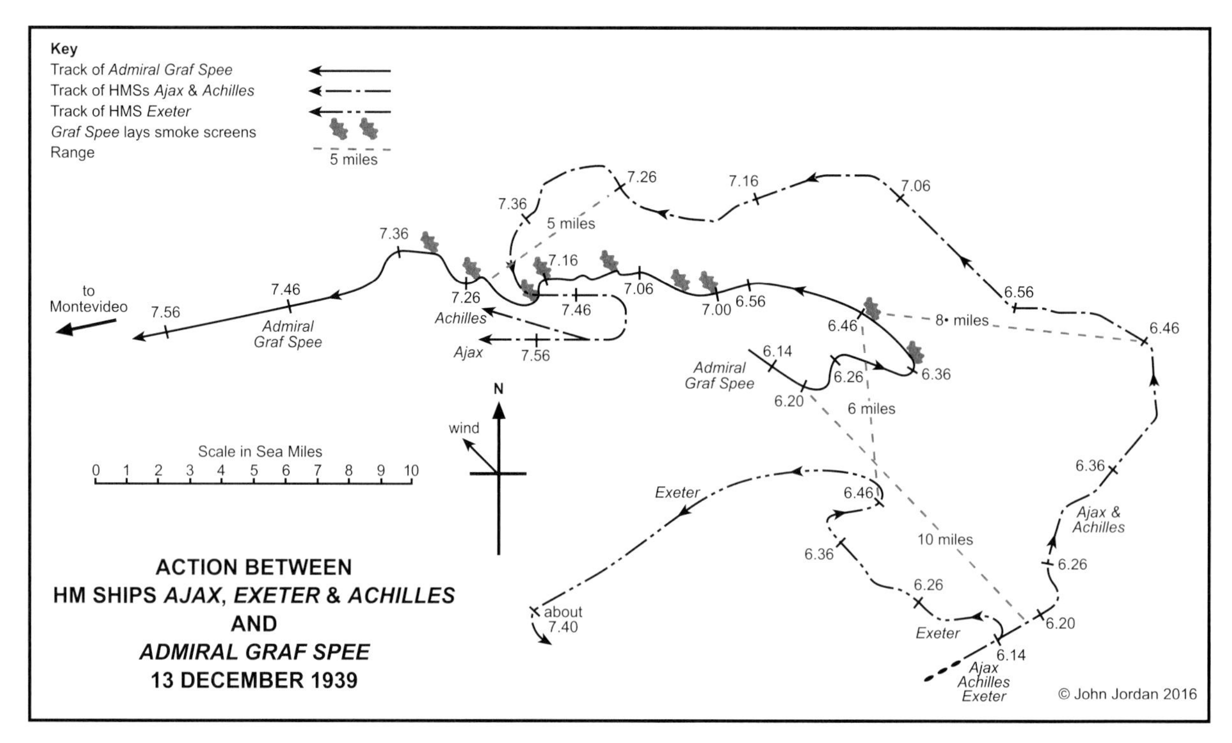

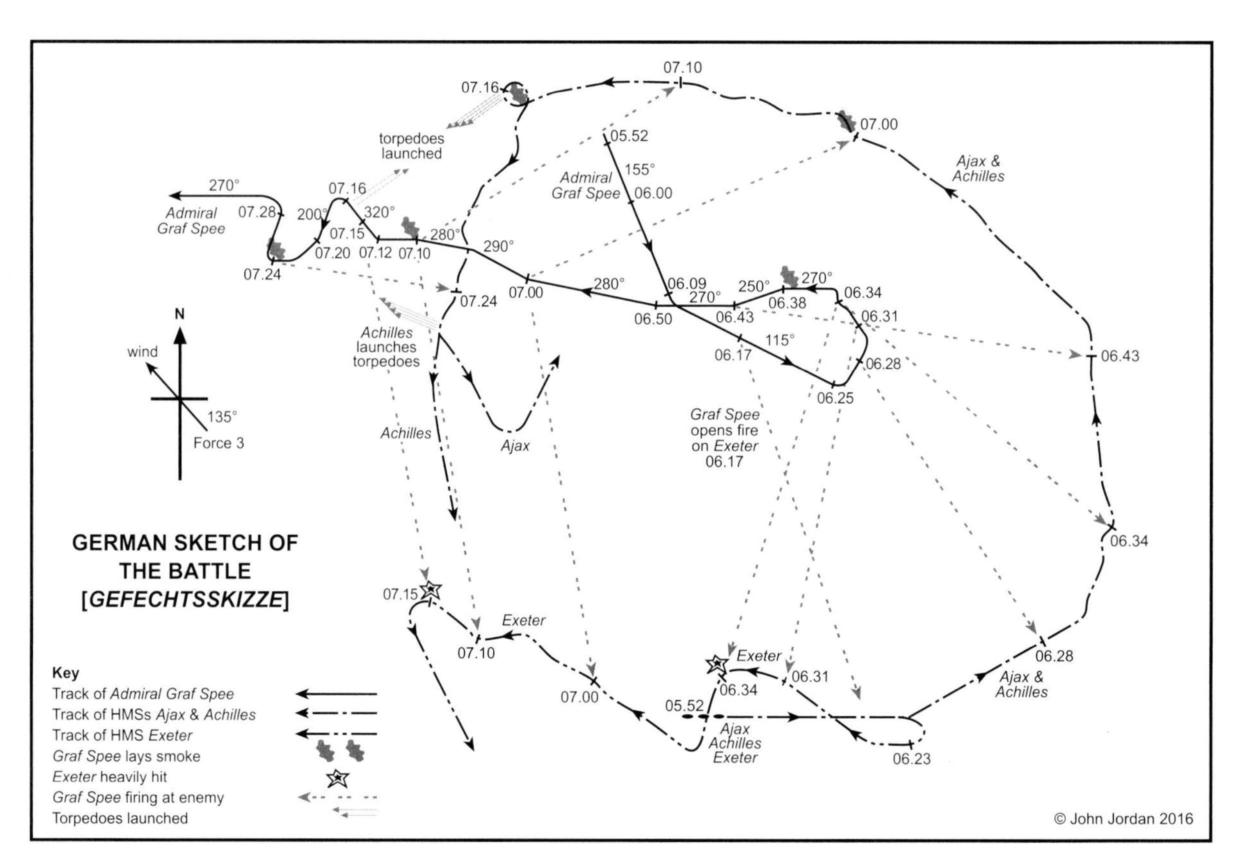

The British track chart report included in the British report of the action, drawn to scale, was probably taken from *Achilles*' Admiralty Fire Control Table (AFCT) plot. The German track chart was drawn from memory after the event, without reference to the original logs and gunnery record; it is not to scale and the original was hand drawn. The anomalies between the two charts are detailed in the text.

Exeter: Departs formation, orders course/speed 320/20, accelerating. She is closing bow on to the target.
Ajax* and *Achilles: Maintain 060/14.

0620 (T+3) *Graf Spee*: Turret Anton returns to service. Six-gun broadsides are fired at *Exeter*. The Torpedo Officer warns Langsdorff that, with cruisers off both bows, maintaining course will make them vulnerable to torpedo attack. Langsdorff orders a slow turn to port to course 030.

The range was far too great for torpedoes, a fact that Langsdorff, a torpedo specialist, would well know. The turn was to forestall a potential problem, not a response to an immediate threat.

The turn placed Graf Spee *on a course parallel with the light cruisers, with* Exeter *off her starboard quarter, her armour presenting an oblique angle to the heavy cruiser. The light cruisers had a port bearing drift, range 20,500, slightly closing and gaining bearing, and the heavy cruiser a starboard bearing drift, range 18,000, closing quickly.*

Exeter: Opens fire with her forward turrets. Permission for the after turret to fire is denied to avoid damaging the aircraft on the catapult preparing to launch. Range 18,700. Ranging salvos are fired at 30-second intervals. Full speed is ordered. She begins a shallow zig-zag to confuse *Graf Spee*'s gunnery.

Captain Bell was immediately driving down the range. This indicates that Harwood intended to destroy Graf Spee *with the tactics suggested by CB 3011, with* Exeter *in close and the light cruisers engaging outside 15,000 yards.*

Ajax* and *Achilles: Harwood hoists a preparatory signal for Division 1 to come to course 340, speed 25.

The track chart shows Division 1 at 18 knots (accelerating from 14 knots). The British ships had earlier put boilers on standby to reduce fuel expenditure. It would take time to bring these boilers up to pressure and cut them into the steam mains.

Achilles: Action Stations closed up (remarkably, in four minutes). Captain Parry orders the conning officer ('Pilot' in British terminology) to open to three or four cables (600 to 800 yards) from the flagship, keep a loose formation, and weave without using much rudder when the pocket battleship fires at them.

The interval between ships allowed employment of Concentration Fire and room for Achilles *to manoeuvre if* Ajax *did something unexpected. It would also allow* Graf Spee *to shift targets from one ship to the other quickly, as the two ships' tracks in the fire control plot would not differ significantly. The facility of German fire control to shift targets quickly against ships in formation was demonstrated at the Denmark Strait where, after* Hood*'s magazine exploded,* Bismarck *shifted targets and quickly hit* Prince of Wales.

Exeter: An 11in near-miss blasts splinters through the Walrus aircraft, kills crewmen at the starboard torpedo tubes, and cuts communications to gunnery, steering, and After Control. A hydraulics line to the steering telemotor is severed.

0621 (T+4) *Exeter*: Straddles *Graf Spee* with her third salvo. She initiates a port turn to new course 270.

The turn opens the 'A' arcs, and disrupts her fire control solution just after a straddle.

Achilles: Opens fire. Time of flight for the 6in shells is 61 seconds.

Graf Spee *would move more than three ship lengths before the shells arrive, indicating the criticality of an accurate fire control solution at such long range.*

0622 (T+5) *Graf Spee*: Starboard 5.9in battery opens fire against *Ajax* and *Achilles*.

Graf Spee*'s secondary battery was totally ineffectual for the entire battle. Afterwards, a faulty inclinometer was blamed. The 5.9in salvos were ragged, scattered, and irregular. The USNWCMR assess a 50 per cent penalty to the fire of an open-mount secondary battery if the main battery is firing. The 5.9in guns were mounted behind open-back shields, so their crews were subjected to the shock and deafened by the 11in muzzle blasts. Also, the USNWCMR assess a 20 per cent bonus for 'not being under equivalent fire.' The 5.9in battery put* Ajax *and* Achilles *'under equivalent fire', occasionally interrupting their line of sight with shell splashes from near-misses and wetting the ships' fire control instruments; this was its only contribution to the battle.*

Ajax* and *Achilles: Execute course 340. Range 19,000.

This course will close the range, but also put 'X' and 'Y' turrets against the stops.

0622.5 (T+5.5) *Exeter*: Permission granted to fire the after turret.

The damaged aircraft is later jettisoned.

0623 (T+6) *Exeter*: Splinters from an 11in shell riddle the funnel and searchlights.

Ajax: The flagship signals 'open fire' and 'speed 25 knots.' *Ajax* opens fire. The British subsequently reported that she generated a 'high rate of fire, combined with great accuracy.' The light cruisers are firing one broadside every 15 seconds.

Considering the swings shown in the fire control plot, the claim of 'great accuracy' is exaggerated. Graf Spee*'s account does not mention early hits. By USNWC estimates, if full power is available (all boilers hot and available to be cut into the steam main), a cruiser could accelerate from 14 to 25 knots in eight minutes; CB 3011 specifies six minutes. The 'weave' induces rudder drag and slows acceleration.*

0624 (T+7) *Graf Spee*: 8in shell hit from *Exeter's* fifth or sixth salvo destroys a 4.1in mount, cuts down the gun crew, and destroys the boiler supplying steam to the ship's fresh water distilling plant.

The destruction of the distilling plant, and later the fuel oil purifier for main engine diesel fuel were major considerations in Langsdorff's post-battle assessment that Graf Spee *was incapable of returning to Germany via the North Atlantic.*

Exeter: 11in Hit No 2 strikes 'B' turret after its eighth salvo. The 1in shield is destroyed and the turret gutted. Splinters sweep the bridge; everyone but the commanding officer and two others are killed or seriously wounded. Communications to the wheel house are wrecked. The ship's head drifts to starboard. The after turret goes against the stops and ceases fire.
Graf Spee: the hit on *Exeter* is seen as 'a column of fire rising almost as high as her mast.'

0625 (T+8) *Exeter*: 11in hit No 3 fired from 16,000 yards, relative bearing green 030 (030 relative), hits the starboard sheet anchor. The Torpedo Officer notices the ship's drift to starboard and orders After Steering to bring the ship back on course.
Ajax* and *Achilles: Initiate Concentration Fire. The first salvos are long.

A flaw in Concentration Fire was that, in 16-shell salvos with tight groupings, 'over' and 'straddle' splashes might not be seen, blocked by 'short' shell splashes. There is thus a tendency to see salvos as 'short,' and call for an 'up' correction when straddling, resulting in a tendency to fire long.

0628 (T+11) ***Graf Spee***: Shifts turret Anton to Division 1.
Exeter: Helm control is regained. A slow turn to port allows 'X' turret to re-engage. 11in hit No 4 from 13,000, Green 090 (090 relative), strikes the forecastle deck forward of the breakwater. The ship is badly holed and begins flooding.
Ajax* and *Achilles: Believe they are straddling *Graf Spee* with 16-gun salvos. Range 18,000 and closing slowly. Rate of fire is four salvos a minute.

At this rate, the light cruisers will expend their ammunition in fifty minutes.

This is a critical decision point in the battle. Exeter *is rapidly closing on* Graf Spee, *while the light cruisers are 5,000 yards farther out. The two British divisions are diverging. Harwood is relying on Flank Marking and Concentration Fire to make his light cruisers' fire effective at long range. He may also fear that they could not survive close in. Here was Langsdorff's golden opportunity: he could reverse course and close on* Exeter, *which* Graf Spee *outclasses in a 1:1 fight. The chance to isolate and destroy* Exeter *was Langsdorff's best chance to reduce the odds against him.*

A best estimate is that Graf Spee *split her fire from 0630 to 0632, shifted all guns back to* Exeter *at 0632, and again split fire from 0635 to 0640. Shifting and splitting fire hampers accuracy. The guns must re-establish the range each time their target is shifted, and the after director and rangefinder are less accurate than the main director and main rangefinder on the forward Fire Control Tower. In the USNWCMR, a three-gun salvo is assigned a penalty associated with the difficulties in spotting a small salvo and the chance that one wild shot could significantly change the spotted MPI.*

0630 (T+13) ***Graf Spee***: Turret Anton shifts to *Ajax*.
Exeter: Initiates slow turn to starboard, keeping the arc for 'X' turret open and closing the range rapidly. She takes additional near-miss splinter damage. She is flooding forward, down by the bow, with a seven-degree list.
Ajax* and *Achilles: *Graf Spee* straddles *Ajax* three times. Harwood initiates a turn to starboard to throw off *Graf Spee*'s gunnery. The cruisers are salvo chasing. The new course opens the range.

Salvo chasing takes advantage of a weakness in analogue fire control computers. When a salvo is long, a 'down' spot is applied. If the target ship turns to open the range while the enemy is correcting to a shorter range, the succeeding salvo will land short. An 'up' spot would then be applied while the ship now turns to decrease the range. In each case the ship turns in the direction of the salvo splashes, thus 'chasing' the salvos. Neither USNWCMR nor CB 3011 account for salvo chasing – a significant deficiency.

0631 (T+14) *Exeter*: Fires her starboard torpedo tubes in local control.

Range 13,000, target angle 120 degrees, target speed 20. The British 21in Mark VII torpedo had a maximum range of 7,000 yards with a speed of 29 knots, so in effect Exeter *fired at a receding target with a torpedo with insufficient range to intercept. The Torpedo Officer must have feared that his ship was in a critical condition; he did not want to be sunk with all torpedoes on board, a common attitude in the culture of torpedo specialists.*

Ajax* and *Achilles: Their gunfire is 'not scoring any hits.'

0632 (T+15) ***Graf Spee***: Turns to port, new course to the northwest. Turret Anton cannot bear on Division 1 and is shifted back to *Exeter*. *Graf Spee* makes smoke.

Smoke screens would interfere with the light cruisers' line of sight for the next 20 minutes, reducing the effectiveness of their fire to near zero. Here was another opportunity for Langsdorff to turn on the isolated Exeter, *with the British divisions separated by 12,000 yards and steering divergent courses.*

Harwood continues to the north on a course where the wind drift of the smoke matches his movement. He does not act to clear his line of fire. From 0626 to 0646 the British and German track charts are irreconcilable. Time for this turn is based on gunnery shifts, and is approximate.

0633 (T+16) ***Ajax* and *Achilles***: Turn to port, regaining their course from before the evasive manoeuvre. Range is rapidly opening.
Exeter: The range is steady as *Exeter* and *Graf Spee* are on nearly parallel courses.

0635 T+18) ***Graf Spee***: Turret Bruno and the secondary battery shift fire to Division 1. Her fire is intermittent due to the smoke.

The wrecked forward turrets and splinter-scarred bridge of *Exeter*, photographed at Port Stanley in the Falkland Islands, where she put in for temporary repairs after the battle. Once these were complete she returned to Devonport, arriving in February 1940, for a full refit. This lasted until March the following year, when heavy air raids on the dockyard forced the ship to depart with the last of the work still being completed. *Exeter* returned to service fully repaired and with a much enhanced AA armament, Type 279 and Type 284 radar, a new aircraft catapult and various other improvements. (Cdr RD Ross, Conway Picture Library)

0636 (T+19) ***Graf Spee***: Turns to the west (time approximate). This again disrupts the British fire control. The turn may not immediately have been recognised in the *Achilles* AFCT due to smoke. She initiates a zig-zag or salvo chasing.

Langsdorff abandons his original intent to close to effective range. He runs.

Exeter: Tracking at 16 knots, range 11,100.

Ajax* and *Achilles: 28 knots, range 15,100. They continue to struggle to get on target.

According to O'Hara, Graf Spee's *radical course change 'led Harwood to suppose his gunfire was hurting his enemy'. If true, it was wishful thinking.*

0637 (T+20) ***Ajax***: Launches a Seafox aircraft, which takes station at an altitude of 3,000 feet on the disengaged bow of the flagship. The mission is gunfire spotting.

Before launch, the observer signalled to the flag deck that he would use the reconnaissance radio frequency, not the spotting frequency. The message did not make it to the wireless office. There was no communication with the aircraft for 12 minutes. Harwood's staff was not doing its job.

Graf Spee: An 8in shell penetrates her belt armour and explodes in a workshop, three feet forward of an engine room. Another 8in shell strikes the Fire Control Tower, passing through without detonating but cutting cables from the main rangefinder, placing it out of service.

The loss of the main rangefinder is a serious event. The after rangefinder is less precise, lower and cannot see well over the smoke, and (judging from the results) is manned by less skilled personnel.

0638 (T+21) ***Exeter***: Turns to starboard to unmask her port torpedo tubes.

Ajax* and *Achilles: *Graf Spee* 'put out smoke and hid behind it.'

The light cruisers are losing bearing and range on Graf Spee. *Harwood maintains his course to the northwest, steaming away from both* Graf Spee *and* Exeter. *For the next 15 minutes he maintains a range of over 17,000 yards, perhaps in the belief that his 6in guns can penetrate* Graf Spee's *deck armour at that range. He possibly felt that* Exeter *did not require support, as she was still firing four guns.*

By this time, CB 3011 predicted that Graf Spee *would*

have been sunk. The engagement is lasting longer than Harwood would have expected. Harwood maintains a position where the wind blows the smoke screen across his line of fire. Division 1 needed to be to the southeast or south of the target to eliminate this disadvantage. Harwood takes no action to address the problem. Smoke will hamper his fire until the last minutes of the battle. He has no evidence that his fire is hurting Graf Spee. *He appears unable to recognise the unfavourable tactical situation, or to take steps to remedy it.*

0639 (T+22) ***Graf Spee***: The light cruisers are astern and engaged by turret Bruno intermittently due to smoke.

0640 (T+23) ***Graf Spee***: The 5.9in battery straddles the light cruisers; no damage is inflicted. Both main battery turrets are directed against *Ajax* and *Achilles*, engaging to port.
Achilles: An 11in shell explodes short and blasts fragments over the ship. The captain and chief yeoman are wounded. In the Director Control Tower (DCT), five of the ten men are killed or wounded and the door is jammed; medical aid and replacements cannot get in. Her fire pauses, as all those in the director are stunned. The guns shift to After Control, but the two operators at this austere station just forward of 'X' turret have been subjected to the continuous blast from the guns and, as Grove states, 'were completely deafened and nearly stupid from concussion' and 'were sick' from the fumes. After only a few minutes control is switched back to the DCT.

Grove states that the ships continued Concentration Fire, discharging 32 combined broadsides in the following six minutes. Pope states that Concentration Fire ceased when Achilles' DCT was hit. Other sources place this event at 0646.

0641 (T+24) ***Exeter***: 11in hit No 5 from 10,700, Red 130 (230 relative). 'A' turret is destroyed.

With Exeter *down to one main battery turret and the British light cruisers steaming away from her, there was no reason why Langsdorff should not close on* Exeter *and finish her.* Exeter's *radio is out of action; loss of communications would terminate Flank Marking, if it was ever initiated. There is no reason for Harwood to keep his forces separated.*
Exeter: Fires her three port-side torpedoes. Range 10,000, target angle 210, target speed 20.

Torpedoes capable of 7,000 yards are again launched at a manoeuvring, receding target at a range of 10,000 yards with no chance of an intercept. The motivation may have been to get the torpedoes off the ship so that their warheads and pressurised fuel and air flasks would not be a hazard.

0642 (T+25) ***Exeter***: 11in hit No 6 from 10,700 yards, Red 45 (315 degrees relative). The shell passes through the bridge and explodes against the starboard 4in AA gun. Ammunition in the ready-use locker detonates.
Ajax* and *Achilles: Harwood signals for a turn to port, speed 30 knots.

The USNWCMR imposes a 30 per cent penalty on a ship's fire if it is within 3 knots of its maximum speed due to funnel gases. With a moderate following wind from the southeast and their course to the northeast, funnel gases from Ajax *would have been a problem for* Achilles. *Harwood remains oblivious to gunnery conditions. Harwood should see that* Exeter *is down to two main battery guns. CB 04027 of 1939, Royal Navy Fighting Instructions, advises commanders that 'In the past the unfailing support given by one British ship to another in battle has contributed largely to success in action. To-day* [sic] *the same mutual support must be given and expected between all classes of ships.' Harwood appears unmindful of this. The turn he orders provides no relief for the heavy cruiser.*
Achilles: Ten salvos are off in deflection until it is realised that the Range to Elevation and Deflection Operator in the DCT is dead, his body resting in a natural position against his instrument. He is replaced, and firing improves.

0643 (T+26) ***Exeter***: 11in hit No 7, from Red 45 (315 degrees relative), explodes on the armoured deck above the 4in magazine. An intense fire endangers the forward magazine, which is flooded, further slowing the ship. Splinters cut power cables to the transmitting station controlling gun elevation orders, disabling the AFCT. The remaining 8in turret goes to local control.

0646 (T+29) ***Ajax* and *Achilles***: Range 17,000. Harwood turns Division 1 to port to 300 degrees, approximately parallel to *Graf Spee's* course and on the pocket battleship's starboard quarter. The light cruisers will slowly close the range and gain bearing – the track chart shows them making 26 knots. The gunnery radio link is lost. *Achilles* ceases Concentration Fire (if not earlier). Her first independent fire control solution is off in deflection and well short.

Ajax's gunnery officer, assuming they are still using Concentration Fire, is confused by the scattered fall of shot. This, combined with Graf Spee's *smoke screen, begins another period of more than 25 minutes when the fire from the light cruisers is ineffectual.*
Graf Spee: Continues to make smoke and drop smoke floats. British spotters believe *Graf Spee* is zig-zagging. All main battery turrets are firing on *Exeter*.

Division 1 continues to engage ineffectually from long range. Exeter *is no longer battleworthy. Her only further possible contribution is to absorb* Graf Spee's *attention. The light cruisers have been firing for more than half an hour, with no discernible effect on the pocket battleship. More than half their ammunition is expended. Following CB 3011 expectations,* Graf Spee *should have been disabled or sunk ten minutes ago. Harwood can see* Exeter *on fire and shooting badly from only her after turret, and* Graf Spee *apparently undamaged. By now he should be questioning his assumptions and his tactics.*

Langsdorff has the light cruisers to his east at 17,000 yards and obviously reluctant to close, and the seriously damaged heavy cruiser south at 12,000 yards on a parallel course and slowing. The heavy cruiser is firing one main battery turret and is on fire. Here is another opportunity to close and finish Exeter. *Instead, Langsdorff continues west. Perhaps the deciding factor was the wind: if he turned to close* Exeter *he would lose the smoke blocking the fire from the light cruisers. Perhaps he felt he could finish* Exeter *from 12,000 yards. Perhaps he had lost all initiative, was out of ideas, and was just running and hoping.*

0649 (T+32) ***Graf Spee***: Shifts all main battery guns onto the light cruisers.

Exeter *is again reprieved*.

Ajax: Communications are established with the Seafox. The observer confuses *Achilles'* off-target salvos with those of *Ajax*, causing *Ajax* to fire long for 14 minutes.

0650 (T+33) ***Ajax* and *Achilles***: Range 18,000. Harwood turns west, bow on to *Graf Spee*, to close the range. Only his forward turrets will bear.

After more than 30 minutes of battle Harwood's light cruisers have not taken any direct hits.

0656 (T+39) ***Ajax* and *Achilles***: Range 15,000. Division 1 turns starboard to a northwesterly course to open 'A' arcs. The ships' pit logs are indicating 31 knots. The wind is almost directly from astern.

Harwood closes by only 3,000 yards before again turning away, and the range will again begin to open. He is firing from the middle of Graf Spee*'s Immunity Zone.*

0700 (T+43) ***Graf Spee***: Dumps additional smoke floats and makes smoke from her smoke generators. The main battery is shifted to *Exeter* with a reported range of over 20,000 yards, due to confusion between the remaining two main battery rangefinders. Her fire is over by thousands of yards.

Exeter: A malfunctioning shell hoist puts one of the remaining 8in guns out of action; the other's rate of fire is reduced due to a faulty rammer. Under local control, her single gun fires short.

Harwood must now see that Exeter *is close to being disabled, yet he continues to steam away.*

0703 (T+46) ***Ajax***: Realises the aircraft spots are wrong, and disregards them.

0709 (T+52) ***Ajax* and *Achilles***: Range 17,500 yards. With the target obscured intermittently, his cruisers no longer engaged by the enemy's main battery and his fire off target, Harwood turns 60 degrees to port to close the range.

From this point on the light cruisers will steadily close the range. Harwood has apparently realised after engaging for over 50 minutes that his light cruisers are not hurt and, in turn, have not hurt Graf Spee. *Expectations based on CB 3011 have been confounded: to do significant damage he must close the range. He has lost Situation Awareness of his ammunition supply.*

0710 (T+53) ***Graf Spee***: Turret Anton problems resurface. The centre and right-hand guns fire only occasionally, the left-hand gun intermittently.

The problems were not from battle damage. However, it is possible that four- and five-gun salvos from Graf Spee *would have given Harwood an indication that his fire might be having some effect.*

0712 (T+55) ***Ajax* and *Achilles***: Turn to the southwest, bow on to the target. The range drops quickly. Division 1 zig-zags, slowing the rate of closure.

0713 (T+56) ***Ajax* and *Achilles***: Harwood signals 'Proceed at utmost speed.' The British track chart shows the ships at 28 knots. Their target is intermittently obscured.

This signal relieves Achilles *of the requirement to maintain formation. She is allowed to pull ahead if she has sufficient speed. Harwood is (finally) serious about closing the range.*

0715 (T+58) ***Graf Spee***: Shifts main battery fire to the light cruisers.

Ajax* and *Achilles: *Graf Spee* is straddled.

0716 (T+59) ***Graf Spee***: Attempts to fire torpedoes at a range of 13,000 yards, target angle 010. The first torpedo is fired, but a hard turn to port puts the torpedo mount against the stops, and no others can be fired.

This shot had a remote chance to intercept, with a net closing speed of over 60 knots and a run time of six and a half minutes. Tactically, the torpedoes might force the cruisers to turn away, which Graf Spee *would desire as the British are now beginning to hit. Turning in the middle of firing a torpedo salvo is inexcusable. Langsdorff was commanding his ship from the foretop, without all the communications and staff advice available if he had fought the ship from the bridge.*

0717 (T+60) ***Ajax***: Ship is straddled by three 11in salvos. The zig-zag is continued.

Graf Spee: Captain Langsdorff is wounded by a shell fragment. The wound is not serious. Over the next 20 minutes *Graf Spee* takes the majority of her 6in hits.

0720 (T+63) ***Ajax* and *Achilles***: Turn to starboard to open 'A' arcs, range under 11,000, where a good rate of hitting could be expected. Their 6in guns are 'firing furiously'. *Ajax's* fire control problems recur, but *Achilles* is hitting. The Seafox repeatedly reports 'Good shot'.

Graf Spee: A 6in shell explosion knocks Langsdorff unconscious. The First Officer assumes command until Langsdorff comes round. His officers would later state that after this wound Langsdorff's decision-making became, as Grove states, 'erratic and less considered'.

0724 (T+64) ***Ajax*****:** Harwood orders *Ajax* to prepare to fire torpedoes.

Range 10,000, target angle 110, target speed 18. With the long-range slow setting the torpedoes would require a 30-degree lead angle and 34,000 yards to intercept – another impossible shot. Harwood, a torpedo specialist, should have been able to run the calculation in his head.

0725 (T+68) ***Ajax*****:** 11in AP shell from 9,000 yards hits aft. It passes through 'X' turret's working chamber and explodes in the flag officer's quarters. 'Y' turret is scoured by fragments and jammed in train. Both after turrets are out of action.

This is the first direct hit suffered by the light cruisers.

0727 (T+70) ***Ajax*****:** Cuts across *Achilles'* line of fire to fire torpedoes.

With Achilles *trailing to port, it is hard to see how* Ajax *could have cut across her bow unless, under the 'utmost speed' signal,* Achilles *had moved off to starboard, possibly to clear* Ajax's *funnel smoke. The torpedoes porpoised after launch. Several authors postulate that this was seen by the Germans, prompting* Graf Spee's *next course change.*

Graf Spee**:** Turns to port to the southwest. This places the two light cruisers off the pocket battleship's stern and opens the range. Smoke floats are released. The prevailing wind blows the smoke between the combatants.

0728 (T+71) ***Ajax*** **and** ***Achilles*****:** Turn to port to the southwest, matching the pocket battleship's previous turn.

Graf Spee**:** Turns to starboard to place the light cruisers on her beam. The range drops rapidly.

Graf Spee's *0727 and 0728 turns are too extreme on the British track chart, perhaps another AFCT artifact.*

0729 (T+70) ***Graf Spee*****:** Turns to the northwest.

0731 (T+74) ***Ajax*****:** An ammunition hoist in 'B' turret fails. *Ajax* is now firing from only three forward guns. The Seafox reports sighting a torpedo that will pass ahead of the cruisers.

Historians speculate that this was not a torpedo, but porpoises. However, torpedoes leave distinctive tracks for thousands of yards, clearly visible from the air in calm seas. It seems more likely that this was either an actual torpedo fired but not recorded in Graf Spee's *report, or a mistaken sighting by an aircrew that was not having a good day.*

0732 (T+75) ***Graf Spee*****:** Range 9,000, and she is being hit repeatedly. She turns away to port, placing the cruisers on her starboard quarter. More smoke is released; with the wind drift, the smoke will pass between the ships.

Ajax **and** ***Achilles*****:** Turn to port to evade the reported torpedo. Their guns are now engaged to starboard. Speed made good drops to 24 knots due to the turns and continued zig-zags.

Grove reports that Harwood stated that he turned away from the reported torpedo in order 'not to take any chances', and to engage the enemy off his starboard side. He now was on course to cross Graf Spee's *stern, to 'cross the T.' Harwood's pre-battle instructions stated he intended to cross the enemy's stern. It has the added advantage of getting upwind of the smoke. Arguably zig-zagging should have been discontinued, as it would not be effective at such short range: the time of flight was too short for a ship to duck a salvo.*

0733 (T+76) ***Achilles*****:** Straddles the *Graf Spee* repeatedly.

0738 (T+81) ***Ajax*** **and** ***Achilles*****:** Range 8,000. Harwood is informed that *Ajax* is down to 20 per cent of her ammunition. Surprised at this, Harwood decides to break off the action. An 11in shell severs *Ajax's* mainmast, the last German hit of the battle.

*Harwood later justified the decision to withdraw, stating: '*Graf Spee's *shooting was still very accurate and she did not appear to have suffered much damage. I therefore decided to break off the day action and try to close in again after dark.' The claim is questionable:* Graf Spee's *fire was hardly 'still very accurate,' having scored only one hull hit on the light cruisers. The key issue was that the British fire had failed to slow* Graf Spee *or materially affect her main battery.*

The supposed intention to resume the action at night was not attempted. Harwood was content to shadow Graf Spee *astern until she passed into Uruguayan national waters. The comment that he intended to 'close in again after dark' was possibly an afterthought to embellish the propaganda account, designed to imply British determination and fighting spirit.*

Harwood's surprise at the report of low ammunition reflects a serious lapse in Situation Awareness. His light cruisers began firing four salvos a minute at 0628 increasing to 5.3 salvos a minutes at 0640, equivalent to less than 50 minutes of available ammunition. Smoke reduced his rate of fire for twenty minutes or so, during which he apparently became oblivious to his ammunition state. He ordered the charge 52 minutes after firing began, sending his ships into a critically dangerous situation without knowing if they had the means to finish the fight. He thereby left himself open to an accusation of poor resource management and decision-making. It is possible that he was suffering from a form of Target Fixation, locked on to the Graf Spee *to the exclusion of all other data. He was clearly ill-served by his staff, who should have been monitoring the ship and advising him.*

0740 (T+83): ***Ajax*** **and** ***Achilles*****:** Make smoke and break off the action.

Graf Spee**:** Disengages. Langsdorff remarks: 'We must run into port, the ship is not now seaworthy for the North Atlantic.'

Graf Spee moored in Montevideo following the battle. Splinter damage is visible. Note the burned-out Arado floatplane atop the catapult. Note also how much lower the after main rangefinder is compared with the forward main rangefinder on the control tower. One consequence of this was reduced performance through smoke, and this may have been responsible for overestimation of the range to *Exeter* near the end of the battle. (NHHC, NH 59656)

Analysis

One technique in tactical analysis is to recreate a battle using wargaming simulations. The ship movements and gun assignments are replicated. Comparing the simulation results with the actual battle provides valuable insights.

The Battle using USNWCMR

The USNWCMR simulation duplicates the duration of the battle and the damage to the British ships remarkably well. In the simulation *Ajax* had half her main battery destroyed and the *Exeter* lost two thirds of her guns to direct battle damage and was in local control for her remaining guns; both results are a match. *Achilles* was untouched in the simulation; in the real battle she took some dangerous splinter damage but no direct hits. This lends credence to the wargame: the Fire Effects in the USNWCMR did a good job in modelling the battle.

Graf Spee took more damage in the USNWCMR, down to 23 per cent Life. A better match to the actual battle results would have been 70 per cent Life remaining. There are several possible explanations for this discrepancy.

Suspicion might fall on the British semi-armour piercing 6in Common Pointed Ballistic Capped (CPCB) projectile. Some British officers at the Admiralty questioned its effectiveness, especially considering that the first report held that the *Graf Spee* was hit 50 times, twice their estimate of what was required to sink her. If the value of the CPCB shells was inflated in the USNWCMR, it would explain why the actual damage to *Graf Spee* was so much greater in the simulation.

In the battle there were three 8in hits fired from within armour penetration range, and 17 6in hits that would not penetrate belt or deck. In the USNWCMR these hits would have reduced the *Graf Spee* to 68 per cent Life remaining, a damage level that almost perfectly matches *Graf Spee*'s condition at the end of the battle. The values attributed to the 8in and 6in shells appear reasonable.

This passes suspicion to the accuracy of British gunfire. The simulation predicted ten 8in hits and 26 6in hits, versus three and 17 hits actually achieved. *Exeter* lost turrets in the battle earlier than occurred in the simulation. Correcting for this, the simulation predicts seven

Glossary of Decisionmaking Terms

Target Fixation: the tendency to direct attention exclusively on a target, and not processing other incoming information. Harwood could be said to be suffering from Target Fixation when he concentrated on *Graf Spee* while failing to take into account *Exeter's* situation and the state of his own ammunition supply. Target Fixation has been a recognised problem since the First World War.

Situation Awareness: the ability to maintain awareness of the complete picture of the situation. Harwood lost Situation Awareness when he was surprised by the report of low ammunition at the end of the battle. Situation Awareness is a modern term but was recognised during the Second World War, when the phrases 'keeping the picture' and 'maintaining the bubble' (submarines) were often used.

Confirmation Bias: the tendency for a decision-maker to interpret new information as confirming previously-held assumptions or (mis)perceptions.

Status Quo Bias: the tendency for people to adhere to their original plan well beyond the point at which the plan is the best course of action. Harwood suffered from Status Quo Bias by adhering to his original plans for separating his two divisions when flanking fire did not materialise, and when *Exeter* was suffering badly and needed support. He kept his light cruisers at long range well beyond the time when it was evident that their fire was not effective.

8in hits. *Exeter* delivered only 43 per cent and the light cruisers 65 per cent of the number of hits the model predicted. *Graf Spee* scored seven hits in the simulation, while actually getting ten (nine, if the last hit on *Achilles'* mainmast is discounted). Her battle performance exceeded the model by about 30 per cent.

For statistical purposes one wants large data sets before drawing conclusions. River Plate was a long battle, with thousands of rounds fired, nearly emptying the magazines of three ships. Sufficient rounds were fired to support the conclusion that the British shot poorly. Problems with Concentration Fire, the failure to implement Flank Marking, the uninhabitable conditions of the after fire control position in the *Leander*-class cruisers, the damage to *Achilles'* DCT and casualties among her fire control personnel, intermittent firing through smoke, salvo chasing, and the self-inflicted problems with the spotter aircraft all contributed to poor gunnery.

The German gunnery exceeded expectations. The modelling indicates that had Langsdorff not handicapped his gunners with constant changes in course and targeting, *Exeter* and perhaps one of the light cruisers might have been sunk.

The Battle using CB 3011

Using CB 3011, *Graf Spee* is sunk after 21 minutes of combat. Combat in CB 3011 was considerably more lethal than reality. It predicted many more hits at long range than occurred. It expected *Graf Spee* to be disabled after 24 6in hull hits, while she actually absorbed the equivalent of 18 hull hits without significant impairment to her fighting ability. CB 3011 was used at the staff colleges, certainly by Harwood himself, to develop tactics against pocket battleships, and the tactics suggested by CB 3011 match those adopted by Harwood in the battle.

CB 3011 explains why *Exeter* immediately charged to short range. With a better-than-even chance to penetrate an enemy gun turret or magazine before she herself was blown up, the decision to close is understandable. CB 3011 also explains the selection of the range to fight the light cruisers. Expecting a short, decisive battle, and with long-range hit rates predicted at three to 35 times those in the USNWCMR (with far fewer correction factors reducing the accuracy), it is easy to see why Harwood would want to engage from the horizon, and why he would not be concerned about expending his ammunition. Harwood used tactics based on the calculations of all the officers and officials that contributed to formulating CB 3011. He should nevertheless have been alert to the possibility that CB 3011 was in error.

One consequence of the Battle of the River Plate was that expectations in the British fleet changed. *King George V* and *Rodney* engaged *Bismarck* without employing Flank Marking or Concentration Fire.

Summary and Conclusions

Harwood's Tactics

Exeter's immediate charge inside 13,000 yards indicates that Harwood planned for a decisive action, and was willing to risk *Exeter* in an unfavourable match-up to attain it. *Exeter* landed three hits and suffered eight. She survived only through German forbearance.

Harwood initially kept Division 1 at ranges between 17,000–20,500 yards, where his 6in shells could not penetrate. Eventually he realised that his light cruisers were not being hurt. After more than 50 minutes of fruitless fire he discarded the expectation that he could inflict decisive damage from long range and began to close. But he was now in a poor relative position, in a tail chase where the range came down slowly, slower still when Harwood kept turning away to open his 'A' arcs.

When he finally got close enough to pound *Graf Spee* he had run through the bulk of his ammunition. The

Graf Spee ablaze after being scuttled off Montevideo on 18 December 1939. (Conway Picture Library)

Achilles pictured entering harbour after the action. She was the least damaged of the four ships involved; nevertheless the scars of battle are clearly visible along her side. (Conway Picture Library)

report was a shock, and he responded precipitately by breaking off the engagement. He departed with ammunition remaining in his magazines. According to USNWC accuracy tables 440 rounds at 6,000 yards gives 97 hits (48 according to CB 3011), three to six times those inflicted on the *Graf Spee* to that point, and enough to wreck the ship. It is difficult to imagine Nelson, Hawke, or Cunningham withdrawing with ammunition remaining.

Had *Graf Spee* subsequently escaped from Montevideo, there would probably have been a court martial. We can speculate on the charges:

- First, when *Exeter* was *in extremis* Harwood did not support her. He sailed *away*, even after seeing her burning and firing only one gun.
- Second, Harwood maintained Division 1 at long range for far too long. He was beguiled by the illusory benefits of Concentration Fire and Flank Marking, and possibly by a presumption that his 6in shells could penetrate *Graf Spee's* deck armour above 17,000 yards. After 2,000 6in rounds his fire had not affected *Graf Spee's* speed or firepower. He should have grasped that the assumptions drawn from CB 3011 were flawed and his tactics were not working; his epiphany came too late.
- Third, *Graf Spee's* smoke screens hampered Division 1's fire. The tactical situation demanded that Harwood move to clear his gunners' line of sight. He took no action to counter the enemy's use of smoke until the engagement had been proceeding for 75 minutes.
- Fourth, he wasted torpedoes in an impossible salvo, reducing his options for a night action.
- Fifth, he lost Situation Awareness of Division 1's main battery ammunition state at the critical moment in the battle: the point at which he ordered his force to close to decisive range.

After the battle, in the tradition of Nelson's letters to Lady Hamilton explaining his decisions in the battle, Harwood wrote to his wife:

> A raider is thousands of miles from his base. Attack him, make him use his ammunition. Hit him and he can't repair his damage without going in and risking internment. Reduce his efficiency, upset the moral[e] of his crew – he is thus weaker. Some other unit can come later and dispose of him. It is not necessary to sink a raider, lovely of course to do so, lame him is most valuable.

Harwood's conduct of the battle strongly suggests that this was *not* what he had originally intended; rather, this was his apologia for failing to achieve a decisive result. If he was only out to 'make him use his ammunition,' he

The wreck of *Graf Spee* pictured some time after the fires had burnt themslves out. Note how much lower she has settled compared to the earlier photograph. The British intelligence services, intrigued in particular by the radar aerial visible atop the foremast, wasted little time in getting to examine the wreck; using the cover of scrapmetal merchants they spent two months carefully going over everything of interest (see following article). (Conway Picture Library)

would not immediately have sent *Exeter* in to 10,000 yards to kill or be killed, and after 71 minutes of ammunition expenditure he would not have brought his light cruisers in to 8,000 yards, where they might be disabled and rendered incapable of shadowing the pocket battleship. Clearly, he intended to destroy *Graf Spee*. This is reflected in his pre-battle orders: his policy was to 'Attack at once by day or night' and for his captains to act 'without further orders so as to maintain *decisive* gun range' (italics added).[6]

Harwood's tactics, formulated (or at least tested) at the Staff College using CB 3011, were defective, and he recognised this too late; he failed to cripple or sink the raider. Saved by Langsdorff's decision to scuttle, Harwood was knighted and promoted to rear admiral; and the propaganda published following the Battle of the River Plate has coloured the accounts of historians to this day.

Langsdorff's Tactics

The German officers complained about Langsdorff's handling of the battle. They had two criticisms: he mishandled his main battery by continually changing targets, and he manoeuvred the ship constantly so that his guns could not obtain a settled fire control solution. The author estimates seven changes in targeting for one or both of the main battery turrets. The USNWCMR exacts penalties of up to 40 per cent for three minutes for such changes. American professional opinion held that these shifts would greatly reduce the effectiveness of the German fire.

Langsdorff did not close and finish *Exeter*. He had several opportunities, early in the battle and after 0646 when the British divisions were well separated and *Exeter* was down to one main battery turret with all torpedoes expended. He could have completed her destruction at little risk to his own ship. Langsdorff's initial error was his decision to turn the *Graf Spee* to parallel the course of the light cruisers; this kept the latter in the battle. Had Langsdorff turned away, isolated them with smoke, and closed on *Exeter*, he could have destroyed the most powerful British unit before meeting the light cruisers on a more equal basis.

His best chance was to exploit a British tactical error. The British separated their forces, offering Langsdorff his opportunity. Langsdorff laid smoke, split the battlefield, and had multiple opportunities to put *Exeter* beneath the waves, but he did not seize the day. After 20 minutes of battle he turned to the west, and thereafter his decisions were dully reactionary.

Conclusion

The Battle of the River Plate, like most battles, was replete with errors on both sides. It was a gunnery battle commanded by two torpedo specialists, and not well fought. History has misrepresented the battle. The British were not outmatched, did not 'make excellent practice' with their guns, and did not avoid damage by 'the speed and skill with which the ships were handled.'[7]

The Germans won a tactical victory when they escaped being sunk by a superior force; the British won a strategic victory when *Graf Spee* was scuttled. The British seized upon the final result and produced an account that served their immediate propaganda aims but also influenced our perception of the battle to this day. Tactical analysis has helped to see through the smoke.

Sources:

The National Archives, Kew, UK:

ADM 1/9759 'Admiral Graf Spee' Action off the River Plate, report of.

ADM 1/19292 German Pocket Battleship *Admiral Graf Spee* Battle with HM Cruisers *Achilles*, *Ajax* & *Exeter*.

ADM 186/78 CB 3011 War Game Rules 1929.

ADM 223/68 *Graf Spee* 1939 The German Story.

ADM 267/145 The Battle of the River Plate, December 1939.

ADM 239/261 CB 04027, Royal Navy Fighting Instructions 1939.

Archives of the US Naval War College, Newport, RI:

RG 35 Box 2: Maneuver Rules 1937, 1941, 1943.

RG 35 Box 4 folder 12: Red Fire Effect Tables.

RG 35 Box 5a folder 2: Fire Effect Tables 1943.

RG 35 Box 9 folder 6: Construction of the Fire Effect Tables 1922; Notes on Revision of Fire Effect Tables up to March 1930; The 1931 General Revision of Fire Effect Tables, June 1931.

Publications:

Grove, Eric, *The Price of Disobedience: The Battle of the River Plate Reconsidered*, US Naval Institute Press (Annapolis 2000).

Miller, David, *Langsdorff and the Battle of the River Plate*, Pen and Sword (Barnsley 2013).

O'Hara, Vincent, *The German Fleet at War, 1939–1945*, US Naval Institute Press (Annapolis 2004).

Pope, Dudley, *The Battle of the River Plate*. William Kimber and Co (London 1956).

Stern, Robert, *Big Gun Battles: Warship Duels of the Second World War*, Pen and Sword (Barnsley 2015).

Woodman, Richard, *The Battle of the River Plate: A Grand Delusion*, Pen and Sword (Barnsley 2008).

Endnotes:

1. Comments by the First Lord of the Admiralty when welcoming HMS *Exeter*, 15 February 1940, contained in 'The Battle of the River Plate: An Account of Events before, during and after the Action up to the Self-Destruction of the Admiral Graf Spee', HM Stationary Office, 1940, The National Archives, ADM 267/145.
2. 'The Battle of the River Plate: An Account of Events Before, During and After the Action up to the Self-destruction of the *Admiral Graf Spee*', HM Stationery Office, 1940 (TNA, ADM 267/145).
3. *Graf Spee's* speed was much lower than her designed speed; she had a fouled bottom and her engines were in dire need of overhaul. During the engagement *Graf Spee*'s ordered engine speed was 24 knots, but she tracked at only 22 knots according to the British AFCT plot. The British light cruisers' highest pit log reading was 31 knots while executing their weave. This means that the actual speed difference during the engagement was about 9 knots.
4. 'Construction of Fire Effect Tables', Naval War College, March 1922, Naval War College Archives, Record Group 4, Box 70.
5. ADM 223/68, 'Graf Spee 1939: The German Story', The National Archives, Kew, UK, 13.
6. See Grove, 57, and ADM 267/145.
7. See Note 1.

UNDER THE GUNS:

Battle Damage to *Graf Spee* 13 December 1939

In this second of a pair of articles on the Battle of the River Plate, **William J Jurens** looks in detail at the damage sustained by *Graf Spee* using contemporary British and German reports and plans, some of which have only recently seen the light of day.

As technical information is easily located elsewhere, this introduction will be brief. *Graf Spee* was the last, and most technically advanced of three similar German *Panzerschiffe* commissioned between 1933 and 1936. A hybrid type developed to fall within the qualitative restrictions of the Treaty of Versailles, the *Panzerschiffe* incorporated the armament of a small battleship in a large cruiser hull. Although *Graf Spee* and her sisters were, by many accounts, designed to counter French, British and other naval opponents encountered along the western coasts of Europe, they were, in practice, primarily deployed as commerce raiders. Even single units, deployed far from home and threatening British commerce, could tie up an entirely disproportionate number of British warships.

In order to intercept and neutralise raiders of this type, the British deployed eight forces in the South Atlantic (see Table 1). Note that the British appeared to consider that a pair of 8in (203mm) cruisers could control, or at least intimidate a ship like *Graf Spee*. In retrospect, assuming *Ajax* and *Achilles* to be equivalent to a second 8in (203mm) cruiser, it appears they may well have been right.

Graf Spee's officially announced displacement was 10,000 tons, but her design displacement was around 12,340 tons, corresponding to a full-load displacement of around 16,320 tons. *Graf Spee* had a waterline length of 181.7 metres, a beam of 21.65 metres, and a design draft of about 5.8 metres. Nominal frame spacing was 1 metre throughout. The ship was powered by eight diesel engines each with a nominal rating of 6,800bhp, for a total of 54,400bhp. She attained a maximum speed of 28.5 knots with 53,650bhp on trials, although a top speed of 26 knots was more common in service; endurance was estimated at 17,500nm at 15 knots.

Main armament consisted of six 28cm/52 SK C/28 guns, firing a 300kg armour piercing projectile to a range of about 36,500m, plus eight 15cm/52 SK C/28 in simple shielded pivot mounts (MPL), which fired a 45.3kg projectile to a range of about 22,000m. The ships also carried two aircraft (one broken down as a spare), and eight 53cm torpedo tubes located in two quadruple rotating deck mounts. The anti-aircraft armament was typical of German ships of the period. The protective scheme is shown in Figure 1.

After a lengthy and successful commerce-raiding mission in the South Atlantic, *Graf Spee* encountered, and chose to engage the British cruisers *Ajax*, *Achilles*, and *Exeter* at about 0615, 13 December, 1939 off the coast of South America near the entrance to the River Plate.

The action of 13 December lasted from 0618 to 0740, ie one hour and 22 minutes. During that time, *Graf Spee* received something in the order of 25 direct hits or very near misses. About 85 per cent of these appear to have been 6in hits from *Ajax* and *Achilles*. Although the sequence of a few hits can be reconstructed from various narratives, the precise chronology of the hits on *Graf Spee* – and undoubtedly a good deal of detailed information on the damage received – was lost when her records were accidentally destroyed as part of the scuttling process.

Loss and lack of records, plus the inherent errors of measurement, render precise reconstruction of the track chart of the action – as in most cases of naval battles of this era – somewhat problematical. In many cases it is difficult to achieve more than approximate congruence between British and German depictions of the action. It would appear that the best possible solution would be to use the British records throughout for British manoeu-

Table 1: **Allied Deployments in Hunt for *Graf Spee***

Force	Description	Patrol Area
F	*Berwick* & *York* (203mm)	North Atlantic, West Indies
G	*Cumberland* & *Exeter* (203mm)	East Coast of South America
H	*Sussex* & *Shropshire* (203mm)	Cape of Good Hope
I	*Cornwall* & *Dorsetshire* (203mm), *Eagle* (aircraft)	Ceylon
K	*Renown* (381mm), *Ark Royal* (aircraft), *Neptune* (152mm)	Pernambuco-Freetown
L	*Dunkerque* (330mm) , *Béarn* (aircraft), *G Leygues*, *Montcalm* & *Gloire* (152mm)	Brest and Western Approaches
M	*Algérie* & *Dupleix* (203mm)	Dakar
N	*Strasbourg* (330mm) & *Hermes* (aircraft)	West Indies

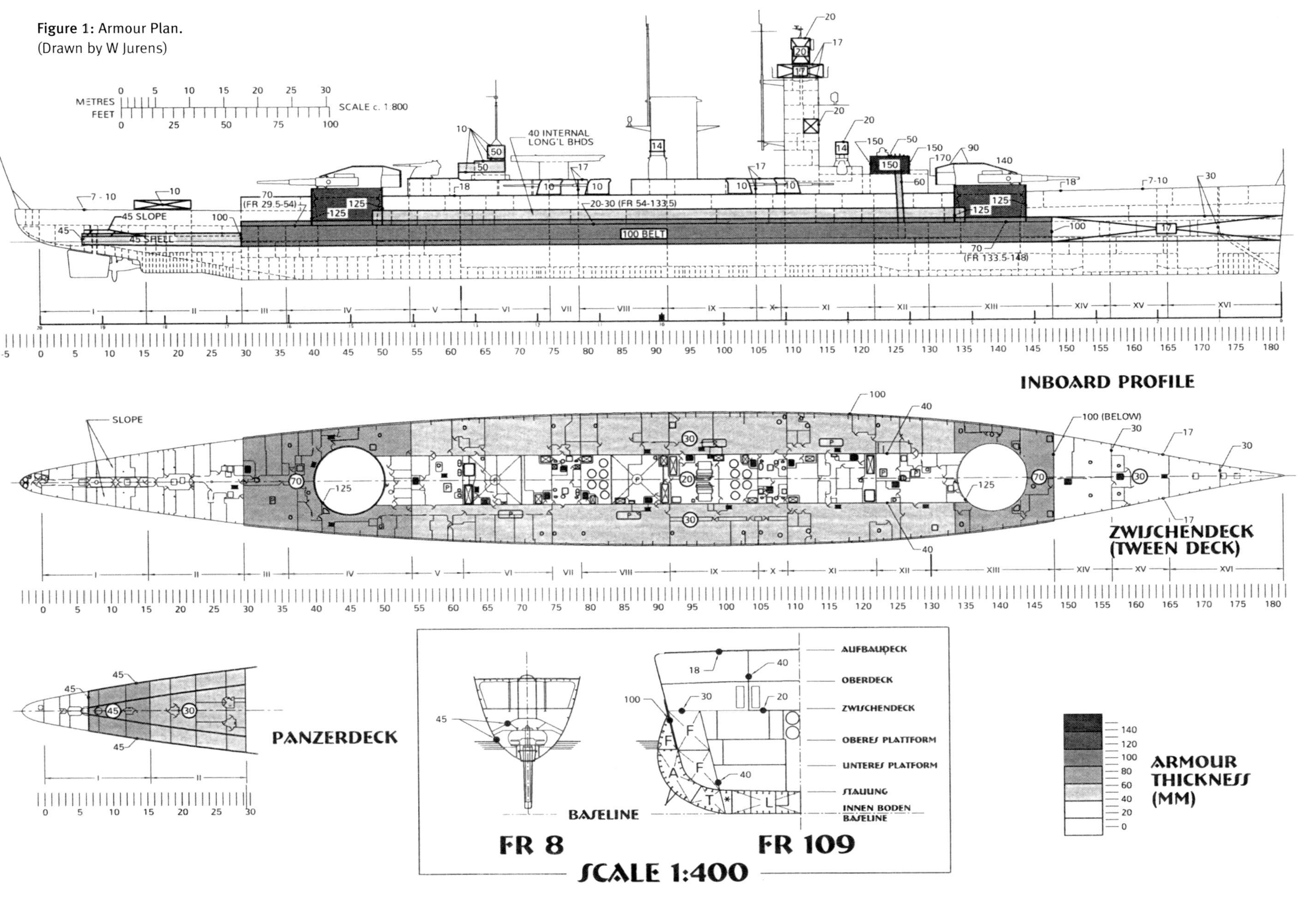

Figure 1: Armour Plan.
(Drawn by W Jurens)

vres, times, and ranges, employing the German track chart – primarily reconstructed from memory – for German movements, particularly early in the action. Comparative track charts are shown in Alan Zimm's accompanying article (see page 34).

The Fire Control Problem

A plot of the relative ranges and bearings from *Graf Spee* to her British opponents is given in Figure 2. The laws of physics suggest that these curves should be relatively smooth, with significant non-linearities associated only with truly dramatic manoeuvres on the part of the participants. When these curves are smoothed, thereby removing the effects of small discrepancies, it becomes possible to deduce approximately when, and at what ranges and bearings the most significant portions of the action took place.

Accurate gunnery requires the ability to predict the position of the target one time-of-flight (TOF) into the future. In practical terms, the fire control system usually predicts future target position by measuring the rate of change of range ('range rate'), and the rate of change of bearing ('bearing rate') and projecting them one TOF ahead. Assume for example an initial range of 20,000 metres with a measured rate of change of range decreasing at the rate of 10 metres per second. Ballistic tables reveal that the time of flight to 20,000 metres is 45 seconds, so during the time of flight the target will have decreased the range by 10 x 45 = 450 metres. This means that in order to intercept the target one TOF into the future the guns should be set to fire to a range of 20,000 - 450 = 19,550 metres. A similar procedure can be used to determine by how much the guns should be offset in azimuth in order to 'lead' the target in train. This example, though highly simplified, nonetheless represents the kernel of the fire control problem.

In general the fire control computer predicts future target position via linear extrapolation. This means that accurate prediction of target position requires that range and bearing rates must remain reasonably constant, ie behave in a linear fashion as well. The problem is most easily and accurately solved when both range and bearing rates are simultaneously zero, corresponding to a situation where both ships are stationary, or when both are proceeding on identical courses at identical speeds.

Regarding the action with *Exeter*, Figure 2 suggests that this ideal condition was most closely met between about 0640 and 0700 after which *Exeter*, heavily damaged, disengaged. The bearings from *Graf Spee* to *Exeter* during the main part of the action seem to have ranged between 240 and 280 degrees, ie with *Exeter* bearing slightly aft of *Graf Spee*'s port beam, at ranges averaging between 11,000 and 12,000 metres. After *Exeter* disengaged, it appears that *Graf Spee* chose to disengage as well. The remainder of the engagement thus evolved into a pursuit action with the British light cruisers gaining at a rate of about 8.4 knots with most incoming British rounds approaching *Graf Spee* – which was at times manoeuvring quite dramatically – from off

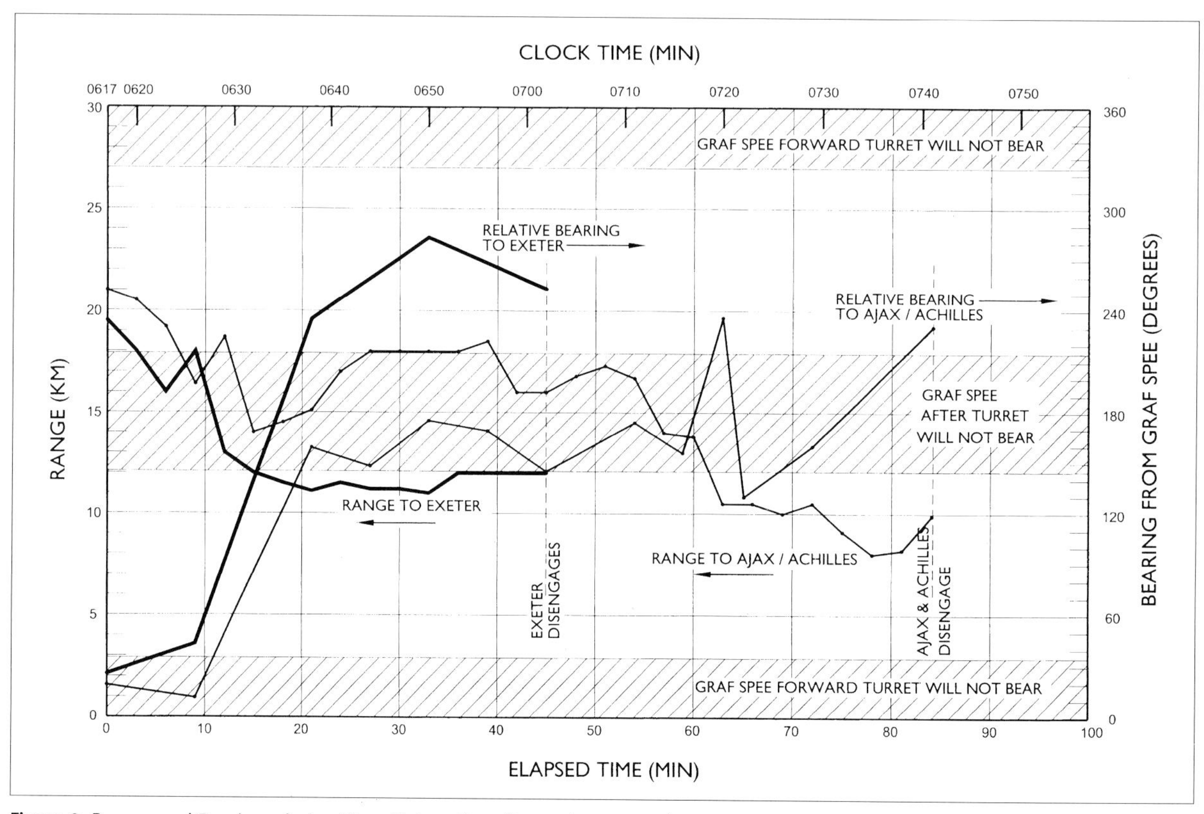

Figure 2: Ranges and Bearings during River Plate action. (Drawn by W Jurens)

her starboard quarter. The difficulties associated with shooting nearly off the bow suggest that although ranges might have been correct, British misses in deflection would have been common. During the majority of the action, relative bearings from *Graf Spee* to *Ajax* and *Achilles* ranged from about 130 degrees to 160 degrees, with ranges varying from 8,000 to 15,000 metres.

The design of *Graf Spee* left a 70-degree arc fore and aft where only one turret (ie only three main battery guns) could engage, leaving a 110-degree arc on each beam where both turrets could train simultaneously. It should be noted that *Exeter* was knocked out relatively early while engaged on bearings where the Germans could employ both main battery turrets simultaneously – albeit at the cost of ignoring *Ajax* and *Achilles* altogether. *Graf Spee*'s decision to retreat from the action early resulted in a stern chase which essentially 'wooded' her forward turret and, because her secondary battery was not designed to engage on bearings well off the beam, rendered it ineffective for much of the action, from 0635 to 0725. During the latter part of the action, Langsdorff's decision to run meant that only one of *Graf Spee*'s main battery turrets could bear at any one time; provided *Ajax* and *Achilles* maintained a reasonable separation, at least one of them could always remain out of major calibre danger. Had *Graf Spee* chosen to turn and fight, the action of 13 December might have turned out differently, though the final result from a longer-term perspective would probably have been similar. Indeed, once *Graf Spee* began to manoeuvre more aggressively – ie to turn so as to bring full broadsides to bear – the British, running short of ammunition, with *Exeter* vulnerable and *Ajax* heavily damaged, chose to break off the chase. This occurred at about 0736. Thereafter, *Graf Spee*, more heavily damaged than the British thought she was, and also running short of ammunition for her after 28cm turret, commenced a more organised retreat into a friendly, or at least neutral harbour, incidentally (and probably accidentally) intercepting – but only lightly harassing – a British merchantman encountered on the way.

During the main part of the action, *Exeter* maintained a range between 11,000 and 12,000 metres. In ballistic terms this corresponds to a striking velocity of about 450m/s at an angle of fall of 10-12 degrees for her 8in guns. *Ajax* and *Achilles* spent most of their time at ranges between 14,000 and 18,000 metres, corresponding to a striking velocity of about 330m/s and an angle of fall varying between 28 and 33 degrees for their 6in guns. Late in the action, ranges to *Ajax* and *Achilles* fell below 8,500 metres.

British Weapons

British projectiles of the period fell into several categories; only the most relevant will be detailed here. The Semi-Armour Piercing Capped (SAPC) variety used in 8in guns was a lighter version of the full Armour Piercing Capped (APC) design intended for use against battleships, specifically designed for use against lightly armoured ships. The projectiles were filled either with 'Shellite' with a picric acid exploder, or TNT/BWX (beeswax) with a TNT or CE exploder; they were usually fitted with a No 345 or No 346 base fuse. The Common Pointed Ballistic Capped (CPBC) design was supplied for both 8in and 6in guns. It was considered to be a semi-armour piercing type, and was typically filled with TNT or TNT/BWX, and the 6in projectile was fitted with a No 479 or No 480 base fuse. The latter could be adjusted before firing to act almost instantaneously, or set for a short delay in order to allow the projectile to penetrate armour before detonating. Delays varied from design to design, and were often somewhat unpredictable in service, but most fuses were designed to provide delays in the order of 0.03 seconds. In any case detailed records of British fuse settings during the River Plate action appear not to have survived and, even if they varied, could probably not be effectively correlated with damage except in unusual situations.

The majority of the British 6in projectiles fired from *Ajax* and *Achilles* would have been of the SAPC type, a 4-calibres long 7.5 CRH tangent ogive design with an all-up weight of about 50.8kg, including a 1.7kg TNT burster. An HE variant with the same weight and geometry carrying a 3.6kg burster was also available in much smaller numbers. The normal 8in projectile fired by *Exeter* would have been a CPBC design, 4.5 calibres long with an ogive approximating a 5/10 CRH and with an all-up weight of about 116.1kg and a 5.2kg TNT burster. As with the 6in projectile, an HE variant of similar weight and geometry with a 10kg burster was also available in small numbers. A graphic range table for each of these projectiles is given in Figure 3. (Although the British launched several torpedoes during the action, insofar as no hits were achieved, technical details have been omitted.)

Graf Spee's protection, shown on Figure 1, was generally similar to that provided to contemporary heavy (ie 8in) cruisers. Although the thickness of the belt is often given as 80mm, detailed official drawings of the ship show a 100mm belt, almost certainly of homogeneous armour. Deck protection over the engineering spaces was either 20mm or 30mm, with 70mm over the magazines and plating in the region of 45mm around the steering

Table 2: **Estimated Penetration Limits For British Guns**

Calibre	30mm deck (Engineering)	70mm deck (Magazines)	100mm belt (90 degrees Inclination)	100mm belt (60 degrees inclination)
6in	14,500yds	26,000yds	9,600yds	8,230yds
8in	14,000 yds	23,000yds	19,200yds	16,500yds

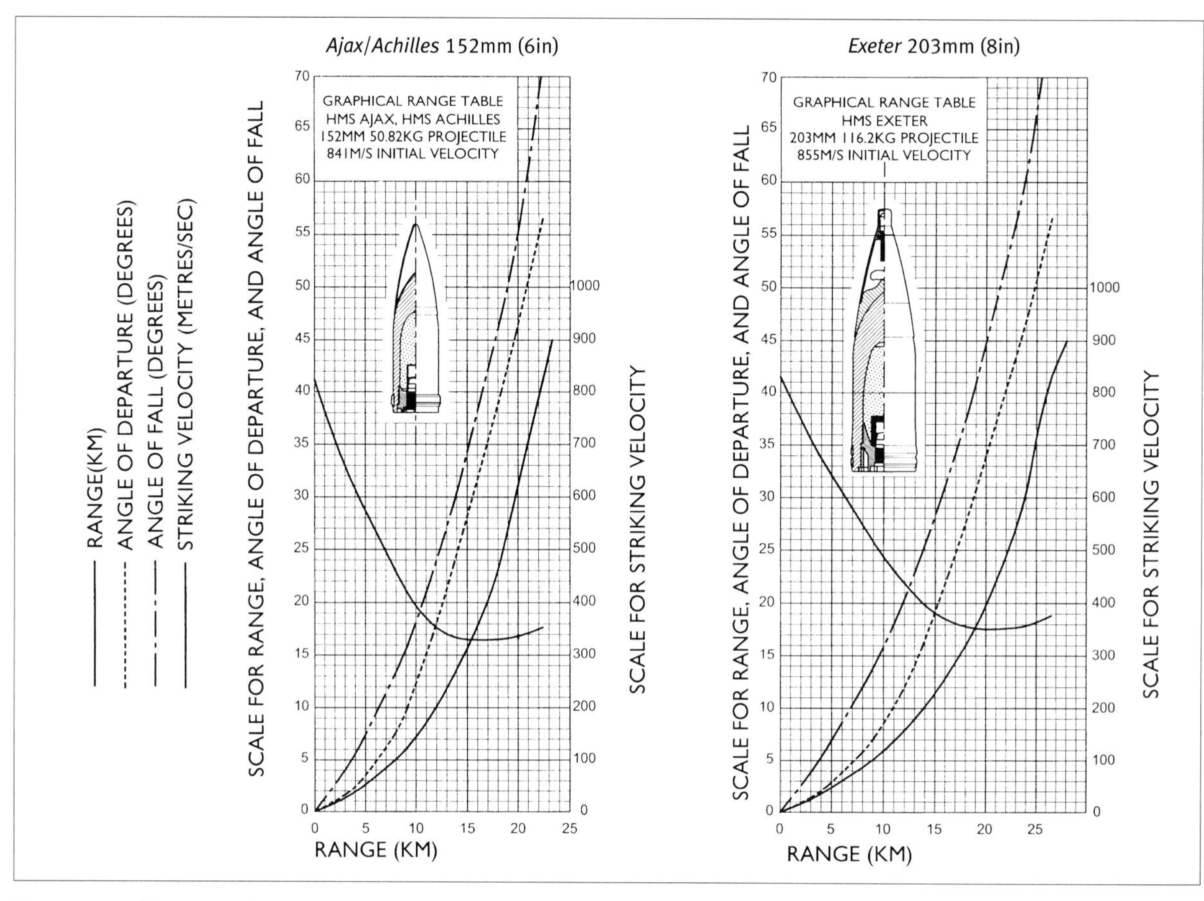

Figure 3: Graphic Range Tables for British 6in and 8in Projectiles. (Drawn by W Jurens)

gear. This proved to be adequate to resist most British projectiles during this relatively short-ranged action. There is not enough primary-source information available to predict the probability of penetration of *Graf Spee*'s belt by *Exeter*'s 8in projectiles with any certainty, except to state that penetration would be unlikely unless ranges were relatively short and the attack was received from nearly off the beam, ie if the striking obliquity was low. The chances of penetration by projectiles from *Ajax* and *Achilles* is considered minimal, particularly as they engaged from off the port and starboard quarters where striking obliquities against the belt would have been very high. Material extracted from contemporary Confidential British documents (eg CB 04039) suggests the penetration data shown in Table 2.

It should be noted that Commander Rasenack, the German gunnery officer aboard *Graf Spee*, observed that most of the British 6in projectiles were armour-piercing, and felt that much greater damage would have been inflicted had contact-fused common projectiles been employed instead. On the other hand, he was very impressed with the effects caused by Exeter's 8in armour-piercing rounds, which proved capable of penetrating German belt and side armour.

The drawings and textual descriptions which follow are based on an amalgamation and reconciliation of all readily-available primary-source material. At least three primary hit surveys were conducted. The most important of these appears to have been inadvertently incinerated during the destruction of German records prior to the scuttling. Reports include the German MDV 550 *Heft 1*, produced largely from memory, a British report primarily based on observations made in Montevideo shortly after the action, and a collection of British reports composed during and after a thorough examination of the still accessible portions of the ship after scuttling. These are listed under Principal Sources.

As might be expected, German and British sources disagree somewhat regarding the impact point and subsequent trajectory of each hit, but in most cases a reasonable correlation can be obtained. (The German diagram from MDV 550 is somewhat confusing insofar as it colours hits as being either from port or starboard, but these define the angles of approach, not the side of the ship on which the projectile may have exploded and/or exited.) For convenience, hits recorded by the Germans have been enumerated in the 'G' series, whilst the major hits recorded and examined by the British and recorded in ADM 281/85 have been numbered as the 'B' series, with a suffix 'P' or S' denoting a port or starboard approach respectively. ADM 281/85 was preceded by ADM 267/145, produced shortly after the action and based upon observation before *Graf Spee* was actually boarded, with hits recorded on profiles clearly redrawn from a profile of *Admiral Scheer* in *Jane's Fighting Ships*. Hits in this series have been recorded as PB ('Preliminary British') 1–50. This latter report tends to somewhat exaggerate the actual damage received; in many cases multiple holes caused by fragments appear to have been recorded as separate hits. There are a few anecdotal reports, and at least one photograph, depicting one or more additional British projectiles embedded in *Graf Spee*'s belt armour, not described in *Graf Spee*'s reports. Additional background information has been extracted from Sir Eugen Millington-Drake's *The Drama of* Graf Spee *and the Battle of the Plate*, published in 1965 (see Sources), an excellent treatment which collates and reproduces material derived from a wide variety of secondary sources.

For the purposes of this article, the German enumeration will be primarily retained, with hits numbered and designated as G1–G21. Diagrams describing the hits are appended as Figures 4–13, and shown in the photographs.

We tend to visualise trajectories in oncoming coordinates until impact occurs, then mentally rotate our frame of reference to look along the trajectory line thereafter. Thus, a projectile coming in at an azimuth of 135 degrees would first be seen as coming from the starboard quarter, but after impact the coordinate system would be reversed and the projectile would be seen as travelling along a 315-degree flight path thereafter. This convention has been retained in the narrative.

It should be noted that analysis of the damage is complicated by the fact that the Germans and the British assigned different designations to decks in the superstructure. This results in a particularly confusing situation insofar as the British and German decks are both listed from the top of the superstructure downward, ie 'A' deck is higher than 'B' deck, but with both sequences offset one letter (eg the British 'D' deck was designated 'E' deck by the Germans). For the purposes of this treatment, the German numbering system has been retained. The signal deck, located between decks G and F, has been arbitrarily assigned the designation 'G1'.[1]

Shell Hits

The German accounts enumerate the following hits:

G1: This projectile, of uncertain calibre, struck Section X. The shell came from the starboard quarter, passed through the starboard boat deck, the crew's galley, the upper deck on the port side in the forward magazine protection chamber, and detonated between the splinter bulkheads in the upper deck of Section X, destroying Ammunition Hoist II, the power supply for all 15cm hoists of the forward group, fire main riser IV, supply and exhaust steam piping for the laundry, and the centreline pressure cooker in the crew's galley. Insofar as this hit took place in sections of the ship that were completely destroyed during the scuttling process, it cannot be specifically identified in British accounts. Emergency power was restored within five hours. Damage to the galley equipment proved to be irreparable.

G2: This 15cm (6in) projectile hit Section VIII starboard. The shell (considered to be of uncertain calibre by the Germans) came from the starboard bow, passed through the protective shield of the starboard 10.5cm HA gun, the after boat deck and the bakery, and detonated in the searchlight workshop. Explosive effects were directed at the upper deck of Section VIII starboard. The shell destroyed the starboard 10.5cm HA gun, the starboard 10.5cm chain hoist, the searchlight workshop, and the drinking and wash water supply lines, which were pressurised in action in order to supply the main engine installation and the battle dressing stations. The lines drained out, with flooding from the upper deck downward into the fore and aft passage, Section VIII, between decks, and from there outboard. Damage was repaired by 16 December, except for the hoist, which was considered beyond salvage.

G3: This 20.3cm (8in) projectile hit Section XI between

A fine image of *Graf Spee*'s 28cm turret Anton, taken taken ca 1939. This image was published in the contemporary booklet *Deutsche Seemacht*. (US Naval History and Heritage Command, NH 104024)

Figure 4: Starboard and Port Profiles Showing Location of Hits.
(Drawn by W Jurens)

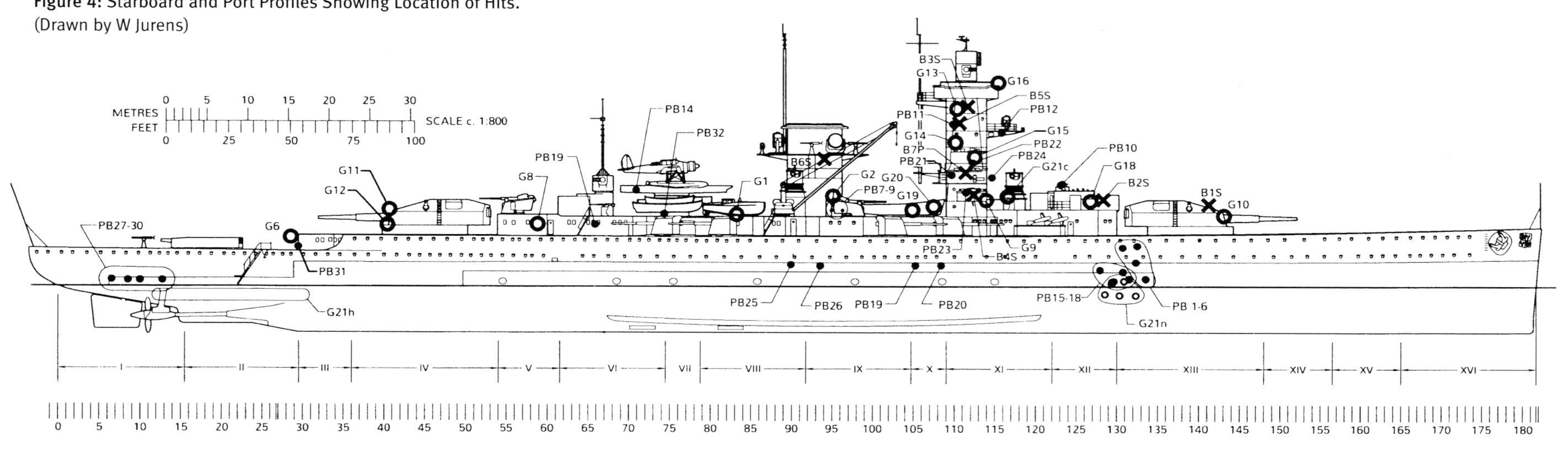

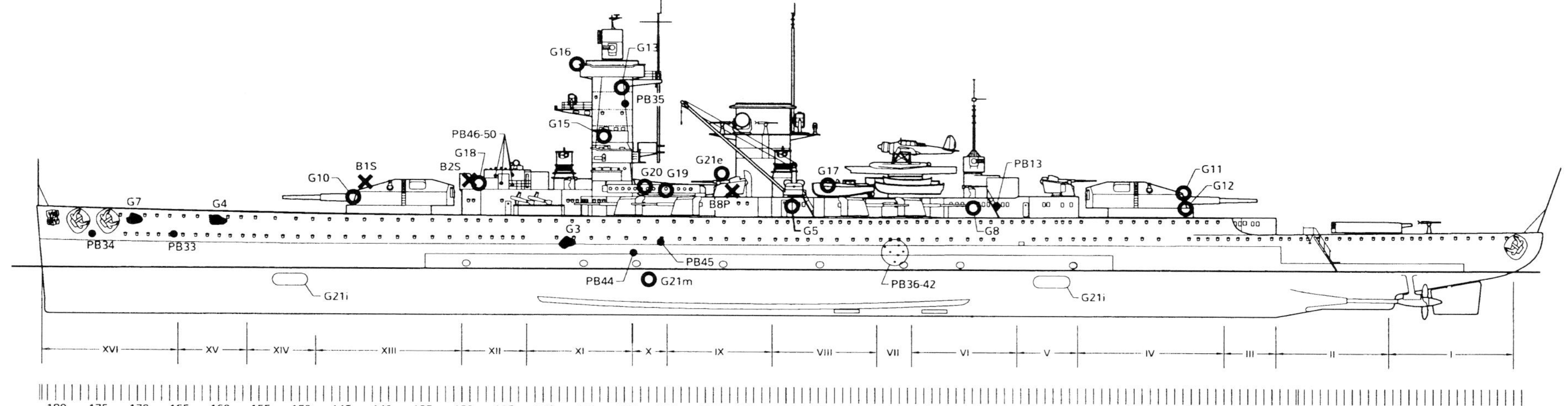

decks. The shell passed through the 100mm armour belt and the 40mm thick splinter bulkhead above the armoured deck travelling from port to starboard roughly perpendicular to the centreline, detonating on the armoured deck between the splinter bulkheads. Various stores between the splinter bulkheads were destroyed and the armoured deck was indented by about 25cm over an area of about one square metre, with several small cracks about 10cm long. Fire main riser V and the 'Ardexin' (methyl bromide fire-extinguishing) installation was destroyed. (Methyl Bromide was also used as a pesticide, but it is believed that this was not the intended use aboard *Graf Spee*.) This material caused chemical burns which led some German crew members to believe that the British were employing gas-filled projectiles, which of course they were not. Fresh and sea water piping along the starboard side was destroyed. (Ardexin was normally stored in the engineering spaces, but in this instance reserves were stored outside as well.) The bulkhead door in the starboard splinter bulkhead jammed and could not be closed. Water entering from the hole in the belt armour and from destroyed piping caused problems which were exacerbated by smoke and gas from the detonation, Ardexin, and nitrogen. Fragment damage associated with fire, smoke and gas caused problems for damage control crews in the area, especially in the upper deck Section XI. The sick bay area and the operating theatre were lightly flooded by the sea and/or water escaping from damaged fire mains. Damage was repaired by noon, 14 December.

G4: This 15cm projectile struck Section XV. The shell came from the starboard quarter, passed through the forecastle deck on the starboard side, and detonated near the upper deck at the shell plating on the port side of the Petty Officers' washroom, creating a hole about 1m x 2m in size. (Uruguayan technical inspectors described it as being two metres square.)

G5: This 15cm projectile came from the port quarter, passed through the port cutter and the port 10.5mm chain hoist, and detonated close to the ventilation hatch for Engine Room III. The chain hoist bearing brass was destroyed and there was heavy fragment damage to the ventilation hatch for Engine Room III, the forward exhaust installation, and the auxiliary boiler uptakes. The hoist was promptly repaired. Wreckage from this hit temporarily restricted intake air to one of the diesel engines, which had to be throttled back until flow was restored, briefly reducing *Graf Spee*'s speed from 24 to 18.5 knots.

Graf Spee's port bow, taken in Montevideo harbour. The crewman between the anchors appears to be spreading canvas to cover the entrance hole for Hit 7. The exit hole for Hit 4 is prominent on the right-hand side of the photo. (NHHC, NH 59660)

Graf Spee anchored off Montevideo following the battle. The camouflage pattern is interesting, but its effectiveness has to be regarded as questionable. (NHHC, NH 59658)

G6: This 15cm projectile hit Section III between decks. The shell, apparently an inert practice round, came in from the starboard quarter, passed through the starboard hawser reel behind the officers' mess, the starboard torpedo loading station, the warrant officers' mess and several warrant officers' lockers on the port side, interrupting a variety of communication circuits.

An internal view of Hit 4 looking forward and to port. Note the apparent fragment damage to bulkhead 163.5, which suggests an explosion just inboard of the shell plating. (Courtesy HO Cochi and AM Haulpczok, from a photograph taken by crewmember *Maschinengefreiter* Erich Haulpczok, via maritimequest.com)

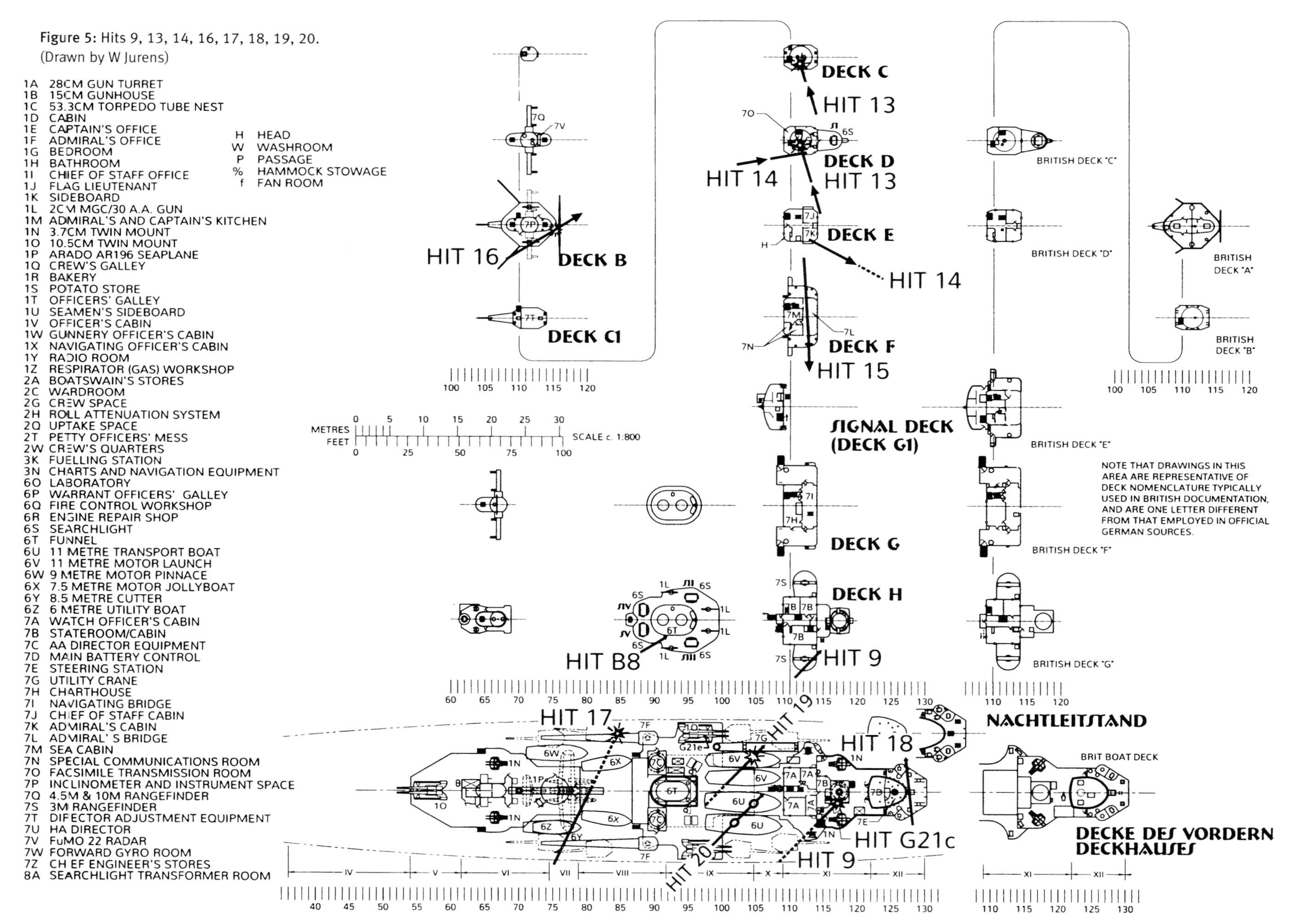

Figure 5: Hits 9, 13, 14, 16, 17, 18, 19, 20.
(Drawn by W Jurens)

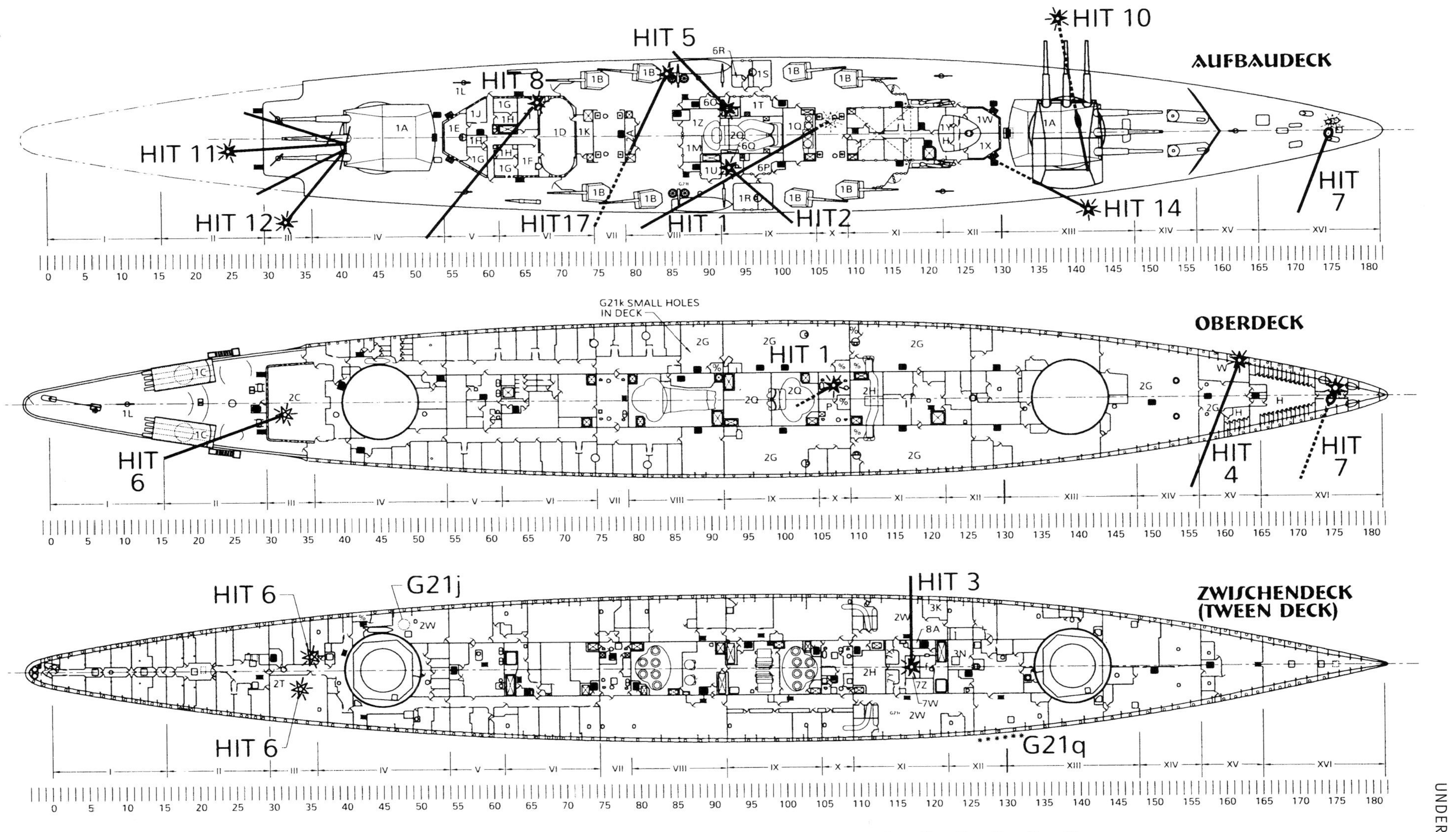

Figure 6: Hits 1, 2, 3, 4, 5, 6, 7,8, 10, 11, 12, 14, 17.
(Drawn by W Jurens)

G7: This 15cm projectile came from the starboard quarter in Section XVI on the upper deck. The shell passed through the forecastle amidships and exploded at the shell plating on the port side in the damage control material storeroom, causing a fire in the stored damage control materials. Uruguayan technical inspectors noted the hole in the shell plating to be about 60cm x 70cm in size.

G8: This 15cm projectile came from the starboard quarter on to the shielded port of the admiral's cabin. The shell detonated in the pantry on the port side, causing little significant damage.

G9: This 15cm projectile came in from the starboard quarter and struck in Section VII, about 1.5 metres above H deck. The projectile passed through the bulwark of the platform holding the starboard rangefinder, intercepted part of the forward control tower, passed through the ready locker of the starboard forward 3.7cm gun mount and detonated in the foundation of the forward HA command post. The starboard forward gun director was put out of action by splinters, the ready ammunition of the 3.7cm gun burned, and the gyroscopic mechanism and cable connections in the director were damaged. The 3.7cm gun was repaired within six hours; repairs to the gyroscopic section of the gun director and associated cabling took seven hours to complete.

This hit was described in detail by the British as well, enumerated 4S. Their observations noted that the projectile, which they also believed to be of 6in calibre,

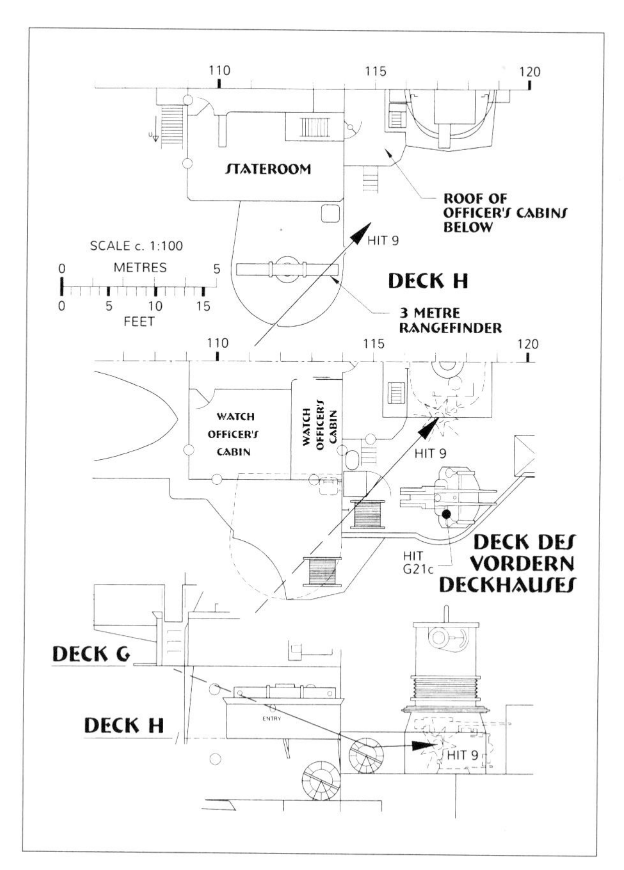

Figure 7: Hits 9 & G21C. (Drawn by W Jurens)

View of the starboard side of the forward superstructure, in Montevideo. The entrance hole for Hit 9 can be seen above and to the left of the large cable reel. The entrance hole for Hit 14 can be seen near the aft edge of the tower superstructure approximately level with the searchlight, just inside the dogleg of the prominent grey vertical camouflage stripe. The exit hole for Hit 14 appears just below and to starboard of the searchlight platform. The exit hole for Hit 15, as yet unpatched, can be seen just above and to the left of the three rectangular windows on the bridge. The scars in the shell plating are from hits PB1–6 and PB15–18, most or all from the explosion of the projectile responsible for Hit 14. (NHHC, NH 59662)

approached from an angle of 128 degrees at about a 6-degree angle of fall. The projectile struck halfway up the 5mm bulwark of the rangefinder platform, transecting a 60 x 38 x 5mm stiffener and leaving a rectangular hole about 230mm long x 165mm high. It then penetrated the 5mm-thick H deck at the lower junction of the forward bulwark, leaving a hole 405mm long and 460mm wide in the deck, and an oval hole in the mild steel bulwark 560mm long x 320mm wide. The projectile then glanced off the top of a hawser reel, and passed through the outboard side of a 37mm ready service locker, the door of which was open at the time, creating a 305mm x 165mm hole in the 11mm plating. The locker was torn from the deck.

After passing across the 37mm thick gun deck (German: *Des Vorderen Deckhauses*, or Forward Deckhouse), the projectile hit the starboard side of the Gyro and Director control room, impacting at the intersection of a 100 x 64 x 10mm vertical external tee and a 75 x 100 x 10mm internal horizontal tee, making an irregular hole in the 13mm plating. The British were of the opinion that the shell did not explode, noting that there was no detectable splinter damage inside the compartment and that equipment along the predicted flight path, including the swinging balance of the director, appeared to have been undamaged. The British concluded that since the projectile could not be found, it must have lodged in the side of the director cabin, later

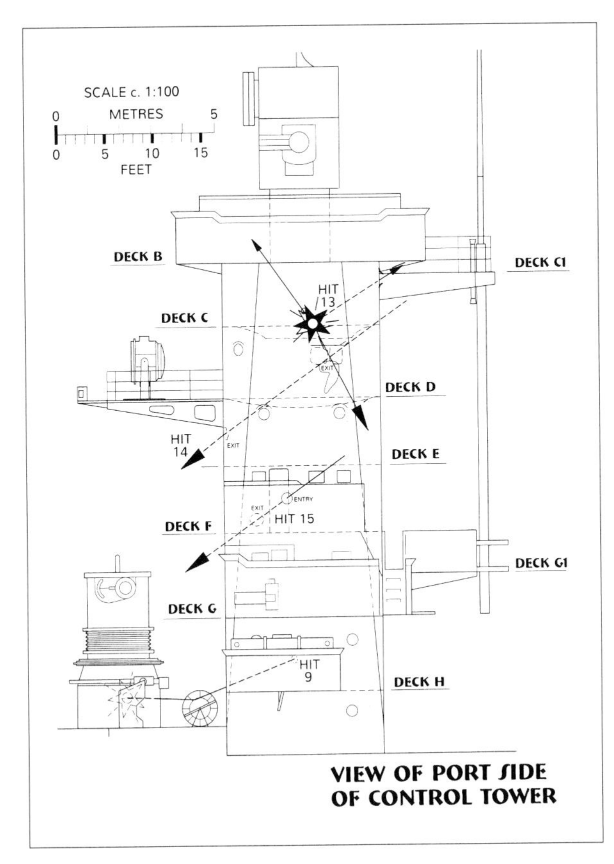

Figure 8: Hits 9, 13, 14 & 18. (Drawn by W Jurens)

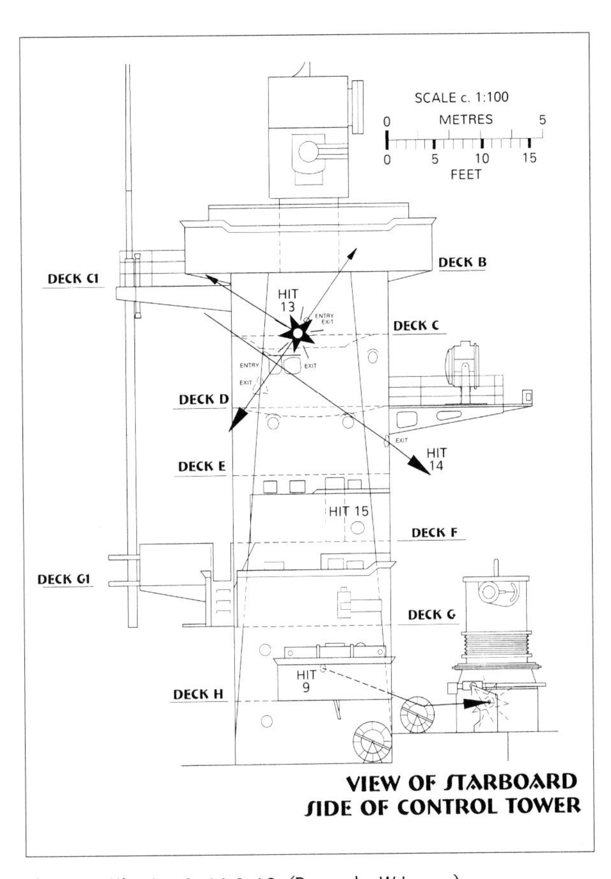

Figure 9: Hits 9, 13, 14 & 18. (Drawn by W Jurens)

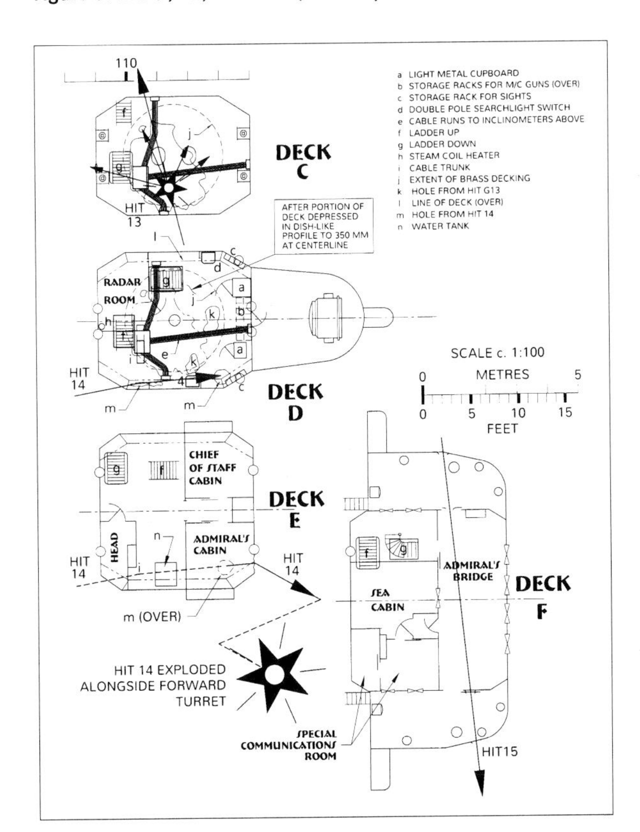

Figure 10: Hits 13, 14 & 15. (Drawn by W Jurens)

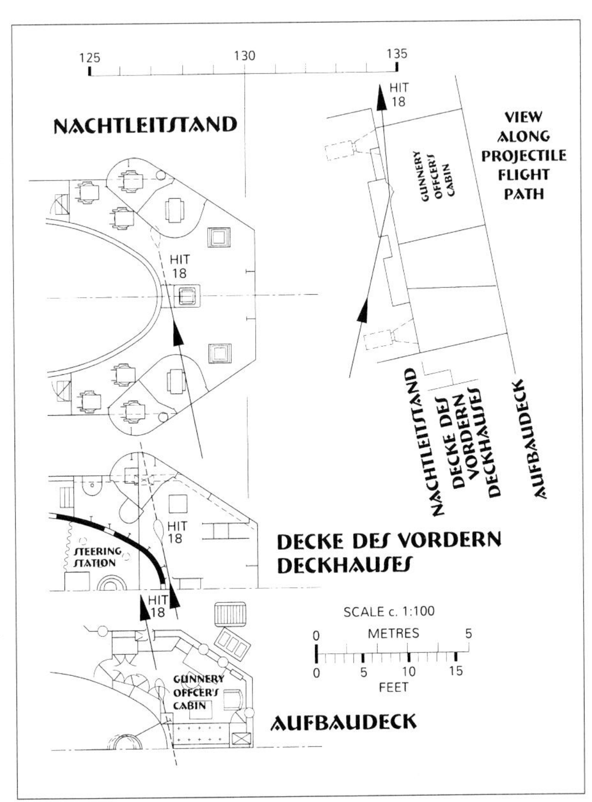

Figure 11: Hit 18. (Drawn by W Jurens)

to be removed by the Germans. The total flight path after impact was approximately 7.5 metres.

G10: This 15cm projectile struck on the oblique front armour of turret Anton; the shell deflected and detonated outboard. British accounts estimate that it was a 6in bullet coming in from 90 degrees at an angle of fall of about 20 degrees. The hit occurred on the sloping face of the gunhouse, making a 'slight regular groove' 125mm long by 70mm wide by 19mm deep, oriented about 10 degrees to the centreline. The British concluded that this projectile must have been from *Ajax* or *Achilles*, and was received while turret Anton was itself trained on *Exeter*.

G11: This 15cm projectile hit on the oblique front armour of turret Bruno; the shell deflected and detonated outboard.

G12: This 15cm projectile hit on the oblique front armour of turret Bruno; the shell deflected and detonated outboard.

G13: This 15cm projectile passed through the starboard side of the foremast underneath the flying bridge, through the deck directly underneath and through the port side wall of the mast without detonating, cutting the cables leading to searchlight II.

The British investigation after the sinking described this hit in great detail, concluding that it was caused by a 6in projectile approaching from a relative bearing of about 75 degrees at an angle of fall of 27.5 degrees.

The shell struck the 5mm thick plating of the control tower on the starboard side between Decks B and C, creating a 180mm diameter regular entry hole and severing or damaging a very large number of cables mounted to the inside wall upon entry. The projectile detonated just above Deck C and broke into four large pieces and one small one. The first large piece, which the British considered was probably the nose, cut through three quarters of the cables under C deck leading forward from the cable trunk, and exited through the port side plating just below C deck, breaking a vertical 210 x 75 x 10mm tee bar and leaving an irregular triangular hole roughly 600mm wide and 600mm deep in the portside plating.

The other projectile fragments were apparently of relatively low velocity. The second and third large fragments passed through the hole in C deck, and pierced the 5mm brass plating of D deck below. Shell fragment No 4, apparently travelling at a lower velocity, moved upward, and after barely penetrating the aft 5mm mild steel shell just under C deck, caused slight damage to the underside of the mast platform. The remaining small fragment penetrated the residual portions of D deck, made of 5mm brass, and the outer shell, which was of 5mm thick mild steel.

The British were not able to recover any fragments from this projectile, and concluded that they had been removed by the Germans. The blast damage, attributed primarily to explosive overpressure and confined by the relatively heavy plating of the superstructure, was primarily directed downwards. The 5mm thick roof of the compartment (Deck B) was buckled and distorted.

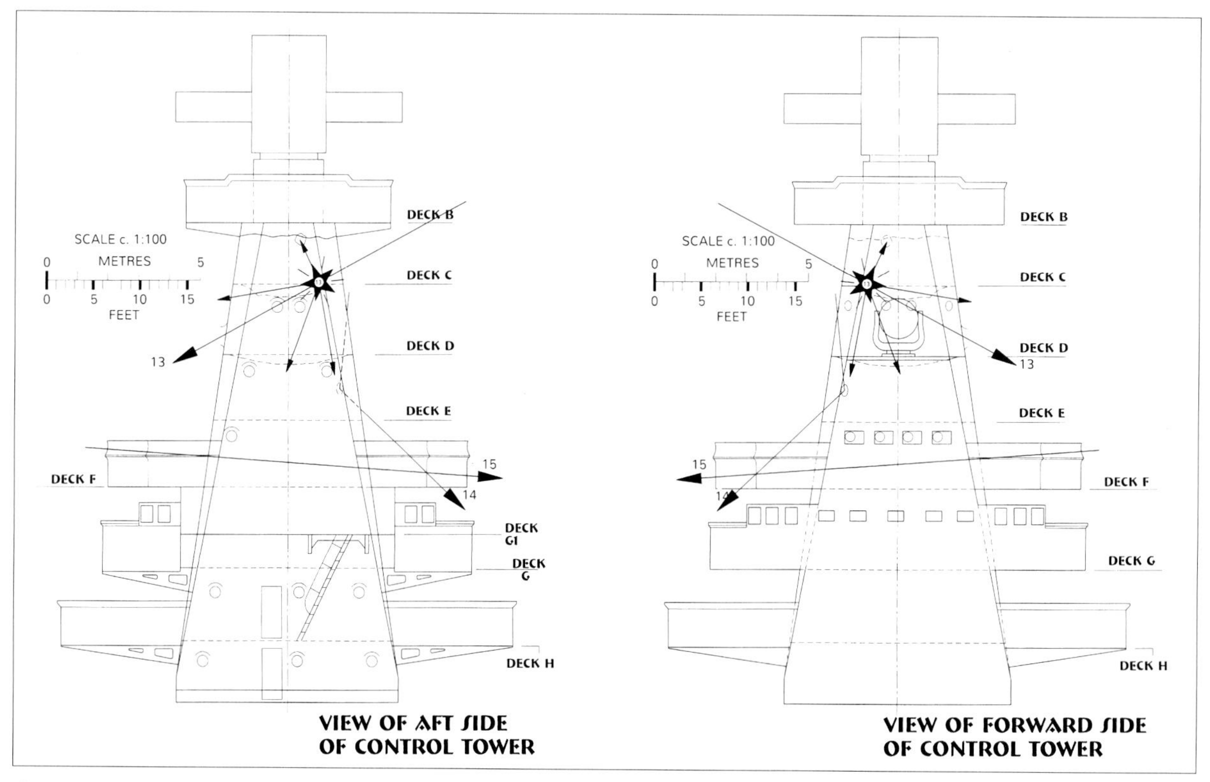

Figure 12: Hits 13, 14 & 15. (Drawn by W Jurens)

Close-up views of the two sides of the tower, taken in Montevideo. In the port-side view, a close examination reveals the exit hole from Hit 13. The slightly elongated entrance hole for Hit 15 is much more prominent. The bulwarks show several small fragment holes.
In the view from the starboard side, the entrance hole for Hit 9 can be seen above and to the left of the prominent cable reel. The rectangular dark patch covering the exit hole from Hit 15 can be seen slightly below the searchlight platform. The exit hole for Hit 13 appears below the left-hand scuttle in the forward shell of the superstructure. (NHHC, NH 59659, NH 59663)

A large part of the brass circular portion of Deck C was blown away, apparently demolishing most of the instruments that were attached to it. This deck was 5mm thick and stiffened by two transverse and two longitudinal beams, 160 x 75 x 10mm, made of brass in way of the circular brass section of deck. The part of this blown away was about 2.6m x 1.5m in size and roughly triangular in shape.

Parts of C deck still remaining were blown down about 600mm around the hole, while the after part of D deck below was depressed about 300mm, though not broken. The distance from impact to explosion of the projectile was estimated at about 4.5 metres.

The British noted that although the Germans had carefully removed all of the instruments from the affected compartments, the connecting cables had been left in place, and little or no effort had been made to restore service using the spare cables located on C deck. Noting that this hit '... must have put the Aloft Control Position Inclinometers [and equipment on decks C and D] whatever their use completely out of action ...', the British concluded that the effect of this hit '... was possibly so great that it was one of the main reasons which led to the German Captain's decision not to take part in any further action.' Insofar as German reports seem to treat this hit as relatively unimportant, it is possible that the Germans' failure to deploy spare cabling was due to the fact that the original cabling remained intact and that the British confused damage occurring during the battle with damage done during the process of scuttling and destruction.

G14: This 15cm projectile, approaching from about 170 degrees and falling at a 45-degree angle, struck the starboard shell plating of the superstructure tower between decks C and D at about Frame 111. The projectile passed through the starboard side wall of the mast, removed plating along the starboard side, punched a roughly-circular hole 500mm in diameter in the starboard forward corner of D deck, and exited the forward bulkhead of the superstructure between D and E decks on the starboard side of the forward bulkhead of the superstructure tower. British analysis concluded that the effects of this hit were essentially nil, while German accounts state that the projectile entered from starboard and exited on the port side of the superstructure, putting the radar room out of action. German diagrams show the point of entry somewhat farther forward than shown in British diagrams and suggest a more transverse trajectory through the superstructure. It is likely that those attempting reconstructions were unsuccessful in clearly discriminating one trajectory from another.

G15: This 20.3cm projectile (recorded by the British as Hit 7), falling at an angle of about 6 degrees at an approach angle of about 276 degrees, passed through two layers of 3mm plating, forming two round holes about 250mm in diameter in the wind deflector of the Admiral's Bridge. The projectile passed just aft of the port wing gyrocompass, crossed the bridge wing and entered the main superstructure through the 10mm thick sliding door of the Admiral's Bridge, which was closed at the time, making a slightly oval hole about 300mm x 250mm about 1.1 metres above the deck. Travelling nearly athwartships and at a mean height of about 750mm the projectile traversed the Admiral's Bridge, piercing the 6mm plating and a 125 x 60 x 8mm stiffener on the starboard side and also the starboard-side sliding door, which was open, about 450mm above the deck, leaving a 250mm x 430mm hole. The projectile then traversed the starboard wing of the bridge nearly horizontally, passed just forward of the starboard gyrocompass, and exited through the 6mm thick starboard bulwark. The Germans cut away the damaged plating in the bulwark after the action, leaving a hole about 1 metre long and 600mm high.

The projectile was fuse-delayed, and travelled about 9.5 metres from bulwark to bulwark before exiting, in the process piercing about 32mm of structure along its path. It may have detonated just after passing through the starboard bulkhead, sometime thereafter, or not at all.

G16: This 15cm projectile, coming from the starboard quarter, grazed the upper edge of the bulwark at the leading edge of the foretop gallery without exploding or causing any damage. This damage may show on NHC photograph NH59663 (qv). If so the projectile apparently made only one point of contact, on the forward port quarter of the bulwark.

G17: This 15cm projectile, coming from the starboard quarter, passed through the starboard boat skid and detonated on the left side of the protective shield of No 3 port 15cm gun mount. The gun shield was damaged, killing or wounding several members of the gun crew and effectively putting the mount out of action. The official German diagrams of the damage show the explosion occurring on the aft leg of the boat davit on the port side immediately forward of gun PIII, but post-action photos show this item to be intact, rendering this precise location implausible. The same photos show little or no apparent damage to gun PIII, but insofar as the mount is then shown trained fore-and-aft, details of the port shield and the area immediately behind the gun would have been hidden from the camera. It seems likely that this hit exploded adjacent to the mount itself. If this took place while the mount was trained fore-and-aft, or nearly so, damage would not be apparent. If the mount was trained outboard for action when the hit occurred, it appears that the damage did not prevent the mount from being trained back into the fore-and-aft position after damage. Photographs reveal fairly extensive splinter damage to the hull in the immediate vicinity, and it is possible that what appears to be damage inflicted from the outside may actually reflect the passage of fragments from the inside of the hull outward. This area was more-or-less completely demolished during the scuttling process and was not extensively visited by the British. The damaged portion of the gun shield was cut away, cabling repaired, and the elevation mechanism jury-rigged, rendering the mount 'conditionally operative' on 15 December.

G18: This 15cm projectile, recorded by the British as Hit No 2, approached from 85 degrees at an angle of fall of about 25 degrees. It passed over the top of the starboard bulwark protecting the *Nachtleitstand* deck in Section XII, narrowly missed a 'calculating instrument', and made contact with the port bulkhead of the Central Selector Position Well at about Frame 128, leaving a 260mm x 200mm hole through the 5mm plating. The projectile then traversed at a height of 650mm the space between the *Nachtleitstand* and the 20mm-thick *Decke Des Vordern Deckhauses*, cutting the cables to the port torpedo training mechanism and searchlights II and III. The projectile then made a 300mm gouge in the deck itself before penetration, leaving a 300mm x 600mm irregular hole. Deflecting upward, the projectile punched a regular round 300mm hole in a vertical 20mm-thick bulkhead, entered a ventilation duct and thereafter passed out over the side at about Frame 126.5 without exploding, the residual angle of fall then being about 8 degrees. Total travel through the ship was about 2.75

Graf Spee's port side, amidships in Montevideo harbour. Note the burned-out aircraft and the apparent fragment or shell hole in the horizontal crane boom, perhaps from Hit 17, just above the aftermost 15cm mount. The fragment damage along the side near the centre of the photograph is from British hits PB36–42. (NHHC, NH 59661)

Graf Spee's No 4 15cm gun mount (second gun in the forward port-side group), photographed on 2 February 1940. Note the position of the waterline, and the angle iron attached to the perimeter of the roof. The British suspected the angle was mounted to present the illusion of extra roof armour when viewed from dockside. (NHHC, NH 50958)

metres, during which the projectile defeated a 5mm bulkhead plus two layers of 20mm protective plating. German descriptions list this hit as coming from the starboard quarter instead of from slightly ahead of the starboard beam. The cabling for one of the searchlights was repaired after action.

G19: This 15cm projectile approached from the starboard quarter, passed through the port officers' boat and a part of the boat skid without detonating inside the ship. This hit does not appear on British diagrams and, although the boats were still in place when *Graf Spee* was photographed after entering Montevideo, it remains difficult or impossible to locate on photographs. By the time the post-scuttling examination had been done, the boats were gone and this area of the ship had been heavily damaged by German sabotage.

G20: This 15cm projectile almost duplicated the effects of hit G19. It approached from the starboard quarter and passed through the starboard and port officers' boats without detonating.

Splinter and Residual Damage

Minor splinters and fragments caused the following damage, enumerated for reference purposes as 'G21a' to 'G21s':

G21a: Splinters damaged the helix and segment for the tilting mirror and dirtied the lenses for the starboard target training and azimuth indicator, putting it out of action.

G21b: The right prism assembly of the flying bridge rangefinder was destroyed by splinters, putting the unit out of action.

G21c: Fragments damaged the elevation mechanism and the left sighting gear on the starboard forward 3.7cm gun.

G21d: Fragments damaged searchlight II and rendered it unusable.

G21e: A large fragment bent the right barrel of HA mount No II, rendering it unusable. The British reported a hit, 'probably 8-in' in this area, enumerated as '8P', which approached from an azimuth of about 265 degrees, passed just under the gun mount just forward of the base ring, in the process tearing a 1,220mm x 610mm hole in the roof of the deckhouse and afterwards disappearing. Photographs of the area do not seem to support this damage, and it is possible that this was confused with Hit No 2, which had a roughly similar trajectory reversed port to starboard.

G21f: A splinter destroyed the mirror of searchlight V. This was repaired after the action in three hours.

G21g: Splinters destroyed the catapult turning gear. Although not specifically mentioned in German sources, a hit in this area caused a fire which destroyed the Arado aircraft – which was in any case non-operational due to a cracked engine block – mounted on the catapult.

G21h: Four small leaks were caused in Sections I–III on the starboard side. These were stopped using wedges and caulking.

G21i: Two small leaks in the port side in Sections V and XIV were blocked off with wedges and caulking.

G21j: A crack 15cm long was discovered in the between deck Section IV port.

G21k: Several 15cm holes were found in the upper deck, Section VIII port.

G21l: A dent 15cm long was found in the upper deck, Section VIII starboard.

G21m: A hole 10cm in diameter was made in the trim tank at the waterline of Section X on the port side.

G21n: Several holes about 15cm in diameter were opened up at and beneath the waterline on the starboard side, Section XIII; trim tanks XII 4.6 and XII 4.8 were full of water.

G21o: Riser pipes IV and V were damaged and out of action.

G21p: The damage control telegraph and telephone between decks Section III were unusable.

G21q: The trim tank Section X port side was full of water.

G21r: Drinking and wash water piping was damaged Section XI between decks, starboard side.

G21s: Service conduits were damaged in Section VII.

Captain Langsdorff's After-Action Report

Although Captain Langsdorff was twice temporarily incapacitated by shock and fragments, he apparently retained effective command throughout the action, albeit at times with significant assistance from junior officers. After the action, he surveyed the ship with the executive officer and concluded that the ship was no longer seaworthy and must therefore retreat to a local port. Buenos Aires was rejected as a potential refuge due to concerns that shallow water and mud in Indio Channel might clog the engine cooling system; Montevideo was therefore chosen instead.

In his first report after the action, the captain prepared the following summary (in translation, with this author's commentary in square brackets):

General: The survey shows that all galleys are out of action with the exception of the Admiral's Galley. The possibility of repairing them with the ship's own resources is doubtful. Penetration of water into the flour store makes the continued supply of bread questionable, while hits in the forward part of the ship render her unseaworthy for North Atlantic winter conditions. One shell has penetrated the armour belt and the armoured deck has also been torn up in one place. There is also damage in the after part of the ship.
Main Battery: Foretop rangefinder destroyed by fragments, otherwise nothing out of action; no derangement of fire control or turrets. Main battery ammunition expended: 324 rounds; main battery ammunition remaining: 378 rounds.
Secondary Battery: Starboard target indicator in conning tower destroyed. Electrical supply for forward magazine hoist disrupted. Ammunition hoist No 1 forward destroyed. 15cm gun No 3: gun shield heavily damaged, elevation mechanism jammed. Ammunition remaining: 423 rounds.
Anti-aircraft Battery: Forward HA director out of action due to hit at base. HA mount No I damaged on left side. HA mount No II: right gun barrel damaged by large shell fragment. Elevating mechanism and left-sighting mechanisms damaged on starboard forward 3.7cm mount. Starboard chain hoist for 10.5cm ammunition destroyed. Chain hoist bushing for port 10.5cm ammunition hoist

Graf Spee's No 2 10.5cm twin HA gun mount (port side, amidships), photographed on board her wreck on 2 February 1940. Reported damage to this mount, as described in Hit G21e, is difficult to substantiate. (NHHC, NH 50959)

Table 3: **Summary of Hits on *Graf Spee***

Hit No	Description and Path	G No	PB No	B No	Comments
1	Stbd Sect VIII	G1	–	–	Not recorded by British
2	Stbd Sect XI	G2	PB7–9	–	–
3	Port Hull Sect XI	G3	PB44(?)	PB45(?)	–
4	Port Hull Sect XV	G4	PB33	–	–
5	Port Superstructure – Section VIII	G5	–	–	Not recorded by British
6	Stbd Superstructure – Sections II–III	G6	PB31	–	–
7	Port Hull Section XVI	G7	PB34	–	–
8	Stbd Superstructure – Sections V–VI	G8	PB13	–	–
9	Stbd Superstructure – Section XI	G9	PB 46–50(?)	B4S	–
10	Faceplate Turret A	G10	–	B1S	–
11	Faceplate Turret B	G11	–	–	Not recorded by British
12	Faceplate Turret B	G12	–	–	Not recorded by British
13	Superstructure Tower	G13	PB11(?) PB35(?)	B3S	–
14	Superstructure Tower	G14	PB22(?)	B5S	–
15	Superstructure Tower	G15	PB22(?)	B7	–
16	Superstructure Tower	G16	–	–	Not recorded by British
17	Hull Section VIII	G17	–	–	Not recorded by British
18	Superstructure Sect XII	G18	–	B2S	–
19	Superstructure Sect X	G19	–	–	Not recorded by British
20	Superstructure Sect IX	G20	–	–	Not recorded by British
21a	Unknown	–	–	–	Location unidentifiable
21b	Main Rangefinder	–	–	–	Splinter damage to optics
21c	3.7 cm Gun Mount	G9	–	B4S	–
21d	Searchlight II (Funnel Platform Port Fwd)	–	–	–	Fragment damage
21e	AA Gun II	–	–	B8P(?)	–
21f	Searchlight V (Funnel Platform Starboard Aft)	–	–	–	Broken mirror
21g	Superstructure Sect VII	–	PB14(?) PB32(?)	–	–
21h	Hull Sect I–III Stbd	–	PB27–30	–	–
21i	Hull Sects V & XIV	–	PB36–42(?)	–	–
21j	Hull Sect IV Port	–	–	–	Crack in *Zwischendeck*
21k	Upper Deck Sect VIII Port	–	–	–	Small holes
21l	Upper Deck Sect VIII Stbd	–	–	–	Small dent
21m	Hull Sect X Port	–	PB44(?) PB45(?)	–	Small hole in shell
21n	Hull Sect XIII Stbd	–	PB1–6	–	–
21o	Firemain Damage	–	–	–	Not shown on diagrams
21p	Communication Damage – Section III	–	–	–	Not shown on diagrams
21q	Hull Section X Port	–	–	–	Apparently duplicates G21m
21r	*Zwischendeck* Section XI Stbd	–	–	–	Fresh water supply damaged
21s	Service Conduits – Section VII	–	–	–	Not shown on diagrams
B6	Funnel	–	–	B6	Not recorded by Germans
B8	Superstructure – Section IX	21e(?)	–	B8P	German hit 21e?

Notes:

G No: hits as recorded in German sources.

PB No: hits as described by British observers while ship was anchored in Montevideo harbour.

B No: hits as described by the British during their inspections of the wreck after scuttling.

Assuming that the German sequence is not sequential, it is possible, though admittedly in a very crude way, to reconstruct the hit chronology by correlating ranges, derived from angle of fall values, and bearings. In some cases, the azimuths and angles of fall were recorded in British surveys of the wreck. In most other cases, the azimuths and angles of fall must be approximated from German descriptions of the trajectory, which are often rather generalised.

damaged. Ammunition remaining: 2,470 rounds 10.5cm, full supply of 3.7cm and 2cm.

Torpedo Installation: Port torpedo train angle indicator destroyed by direct hit. Starboard foremast director damaged by shell fragments and shock. Torpedo distance converter damaged by 20.3cm projectile in transverse passage Section X on the *Zwischendeck*. Main telephone from torpedo reporting station damaged by fragments. Starboard torpedo loading station damaged by direct hit. Starboard spread firing apparatus damaged by shock, but usable. Power to torpedo director and training mechanism for port tubes interrupted, probably through bad connection. Torpedo tube shields damaged from gun blast from Turret II. Telephone in torpedo workshop inoperative. One torpedo tube out of action, six torpedoes remaining.

Aircraft: Catapult training mechanism destroyed by fragments. Aircraft destroyed by fire.

Communications: Dispatch tubes from sending and receiving room shot away. Reserve Radio Room transmitter damaged. Both Direction Finding rangefinders damaged. One long-wave and three short-wave aerials, four receiving aerials and four aerials on funnel damaged. Radio Photo Post [Radar?] room destroyed by direct hit. Radar gear damaged by shock.

Hull: Large hole in side Section XV port. 15cm crack on *Zwischendeck*, Section IV port. Several holes about 15cm diameter in upper deck Section VII port. Dent 15cm long in upper deck Section VIII starboard. Hole 10cm in diameter at waterline of Section X port flooded trim tank. Several holes about 15cm in diameter at or beneath waterline Section XIII starboard. Trim tanks XII 4.6 and XII 4.8 flooded.

Engineering Plant: The engines are available for maximum speed with the exception of defects of long standing in the auxiliary engines. [Note that this omits any mention of problems with fuel oil filtration.]

Casualties: 37 dead, 57 wounded. [Assuming a complement of 30 officers and 1,000 men, and that ⅓ of those listed as wounded might be considered permanently disabled, this represents about 5.4% of the crew.]

Graf Spee After the Action

Although the Germans reported little damage to the fire control equipment, is appears clear that the hits on Decks C and D in the upper superstructure, plus fragment damage to the rangefinder itself, would have destroyed the effectiveness of the forward director. While *Graf Spee* could certainly use the after director, the latter was mounted fairly low in the ship, only about 13.5 metres above the design waterline, lacking the command of the forward director, which was mounted 14 metres higher. The nominal visual range of the forward director on a 35 metre-high target would have been about 40km vs about 34km for the after unit.

It appears that the main battery itself remained essentially undamaged by British fire. It had developed a variety of technical problems during the action but these were, in the overall context of things, unremarkable – similar problems would have developed aboard British ships as well. The secondary battery appears to have had one gun of eight damaged, an effective loss of 25% in firepower on one beam only. The destruction and damage suffered by the ammunition hoists represented a more serious problem; after ammunition in the ready-use lockers had been expended, and long-term manual resupply of the 15cm projectiles, which had an assembled round weight of 69kg, at the rate required to support a high rate of fire would have been difficult, perhaps impossible.

The damage to the hull forward presented a serious concern regarding seaworthiness. The holes were large, and located in areas where dynamic loading from wave action, particularly in heavy seas, would have been substantial. Repair facilities in Montevideo were very limited, and the failure of makeshift patches at sea might have led to catastrophic results. Complete repair would have been essentially impossible in the time granted for *Graf Spee*'s sanctuary in Uruguay. Moreover, *Graf Spee* (and her near-sisters) appear to have been at times susceptible to wave damage. The British noted that the portside crane alongside *Graf Spee*'s forward superstructure had been disabled due to heavy seas prior to the action. Groener notes that all of the *Panzerschiffe* suffered from regular loss of all equipment near the stern, and that the lack of an 'Atlantic Bow', which *Graf Spee* never received, would have rendered her wet in head seas.[2]

There were certainly serious concerns regarding the condition of the ship's galleys and food supplies. A great deal of the flour on board had been contaminated by seawater flooding, and a single projectile, Hit No 1, had passed through the petty officers' and crew's galleys, rendering both unusable.

The engineering plant appears to have been essentially undamaged, although one hit (Hit No 5) had caused heavy fragmentation damage to the uptakes of the auxiliary boiler installation, the funnel, and to air intake shaft No 3. Some postwar commentary has suggested that Hit No 5 rendered the main engines unusable because it destroyed the ability to pre-heat – and reduce the viscosity of – fuel oil prior to injection into the diesels, and/or because it destroyed the oil purifiers. This does not strike the author as credible. First, it is believed that a workaround to this problem would have been fairly simple, especially as the ambient temperature in Montevideo at that time would have averaged around 20 degrees Celsius. Although the situation may have been somewhat less than optimal for injection, it is likely that fuel would not have needed heating at all.

Commander Klepp, *Graf Spee*'s Engineering Officer, noted that damage to the boiler – he mentions only one – which supplied steam to the distilling plant and the purifying plant for the lubrication and fuel oil systems was limited to numerous fragment holes in the uptakes, which were repaired while in Montevideo. Clearly, *Graf Spee* operated for some time during the action itself without at least one of the auxiliary boilers in operation, and was able to start engines and depart for scuttling, which

Table 4: **Reconstructed Timeline of Hits on *Graf Spee***
(Figures in bold considered most reliable)

German Hit No	British Hit No	Estimated Calibre	Estimated Bearing	Estimated Angle of Fall	Derived Range (m)	Derived Striking Velocity	Derived Time
G1	–	?	150°	18°	?	?	?
G2	–	6in?	42°	?	?	?	?
G3	–	8in	270°	?	?	?	?
G4	–	6in	111°	30°	14,000m	333m/s	0635
G5	–	6in	228°	15°	8,800m	425m/s	0735?
G6	–	6in	160°	15°	8,800m	425m/s	0732?
G7	–	6in	111°	45°	18,000m	331m/s	0643+/-
G8	–	6in	129°	11°	7,000m	500m/s	0735?
G9	B4	6in	128°	17°	11,800m	360m/s	0722
G10	B1	6in	270°	20°	12,600m	345m/s	*
G11	–	6in	*	?	?	?	?
G12	–	6in	*	?	?	?	?
G13	B3	6in	75°	27.5°	14,900m	330m/s	0631
G14	B5	6in	170°	45°	19,600m	337m/s	0717+/-
G15	B7	8in	264°	6°	8,000m	540m/s	0650+/-
G16	–	6in	?	?	?	?	?
G17	–	6in	115°	15°	8,800m	425m/s	0728?
G18	B2	6in	85°	25°	14,100m	335m/s	0632
G19	–	6in	134°	?	?	?	?
G20	–	6in	134°	?	?	?	?

* No estimate possible; hit was on rotating structure.

suggests that there were no serious problems with the engines. Fuel oil purifiers, although certainly installed, would probably have been mounted deep in the ship well below the waterline. Meticulous examination of primary damage reports reveals no references alluding to problems with the fuel supply to the main engines at all. It appears most likely that these accounts stem from misunderstood (mistranslated?) commentaries from crew members who were actually referring to difficulties obtaining sufficient cooking oil for the galleys.[3]

Precise details of the ship's distilling arrangements have not been located, and it is possible that the destruction of the ship's auxiliary boilers – assuming both were rendered inoperative – may have rendered ongoing replacement of fresh water for the use of the crew somewhat questionable. While there certainly were concerns that *Graf Spee*'s main engines, which were then in need of major overhaul, might not have stood up to an attempt at lengthy high-speed transit back to Germany, these problems were indicative of ongoing maintenance issues not directly related to the action off the River Plate.

Casualties to personnel would certainly have depleted morale and reduced capabilities; however, the crew should have retained about 95% availability, suggesting that personnel issues did not represent a serious problem. Although other accounts differ, *Graf Spee*'s surgeon reported that the medical services aboard ship performed well throughout the action, and confirmed that the remainder of the crew was well up to operating the ship efficiently.

It appears that only one hit, probably from *Exeter*, penetrated the main armour belt. Although this hit, which exploded near the centreline at about Frame 117 on the *Zwischendeck*, caused a lot of internal damage, it does not appear to have affected the overall fighting capacity of the ship to any great extent.

Graf Spee was scuttled on 17 December, 1939. The hull, heavily damaged, ended up sitting nearly upright on the mud bottom in about ten metres of water which, allowing for some settlement, suggests that she was scuttled with about three metres of water below the keel. Discounting tidal effects, this placed the waterline slightly above the top of the main armour belt. Propellant bags had been distributed throughout the ship. These, when ignited, caused severe fire damage, particularly in the superstructure. Accounts seem to indicate that the vessel burned out, at naturally decreasing levels as fuel was exhausted, for two or three days.

By 2 April 1940 the hull had sunk about 4 metres into the mud and had attained a starboard list of about 12 degrees, leaving only about one third of the weather deck above the water. On 14–15 April 1940 a huge gale severely battered the wreck. After the storm, the ship ended up buried about 11 metres in the mud with a starboard list of about 50 degrees. At that point, very little of the main hull would have remained above the water. Over the years, the wreck continued what amounted to a slow capsize to port, and appears now to be lying on its side, more or less completely submerged in the mud. A few major artefacts, including one of the main gun directors, were salvaged a few years ago during abortive attempts to raise the wreck for restoration. These are now on display in local museums and on street corners.

Conclusions

It would appear clear that *Graf Spee*'s initial retreat into Montevideo harbour represented a wise, if not entirely necessary conclusion. The condition of the ship after the action effectively ended (ie after 0740 on 13 December) would suggest that continued operation, even in the absence of British pursuit and harassment, would have been questionable. The large holes in the bow on the port side would, as the captain suggested, have rendered the ship – which had already suffered severe storm damage – essentially unseaworthy over any extended period. Damage from British gunfire, although in no instance instantly conclusive, was nonetheless incremental and cumulative, severely restricting the ship's ability to protect herself from further attacks, either from the surface or by air. Damage to equipment and supplies rendered long-term support of the crew problematical.

Graf Spee's survival revolved around her ability to intercept and destroy lightly-armed targets and to rapidly disappear thereafter. Once she encountered warships capable of tracking her on a more-or-less continuous basis, her effective lifetime, operating alone far from friendly bases, would probably have been short. Once the British knew where she was, it was a fairly simple matter to concentrate forces so as to eliminate her, either via decisive action or attrition, within a week or so. Later, this also happened to *Bismarck*.

In that regard, the flight into Montevideo represented, at best, a temporary reprieve. To begin with, an extended stay in that harbour was diplomatically forbidden. Even in the absence of such a restriction, as long as she remained outside an area controlled by the Germans the ship would have been to all intents and purposes permanently interned wherever she went, insofar as the British certainly had the resources to prevent any attempt at escape on a more-or-less permanent basis. Even if Uruguay had agreed to allow *Graf Spee* to remain indefinitely, fully repairing the ship in such a remote location would have been problematical at best, as German ships carrying repair materials would have been intercepted via a British blockade. *Graf Spee*, victim of 'the Death of a Thousand Cuts', was effectively 'mission-killed'.

It is unfortunate that German records describing the damage have not survived in detail, insofar as one of the most interesting aspects of this sort of investigation would have otherwise lain in the postwar comparisons that could have been made between German and British assessments of the damage. It should be noted that in some cases the assessed severity of damage from any given hit differs, sometimes quite dramatically, when seen through British or German eyes. It should also be noted that a significant portion of the 'mission kill' inflicted upon *Graf Spee*, for example the critical damage to the galleys, can be attributed to damage to structures and facilities not normally considered to be worthy of protection behind armour. Few nations considered galleys as being of sufficient importance to place behind armour, presumably on the assumption that after damage other friendly ships would normally be in close proximity, or that safe and well-equipped anchorages were only a few days away. This might well have been true had *Graf Spee* been damaged in the North Sea, but proved to be of essentially fatal consequence when the ship was operating alone and far from home.

Ironically, had *Graf Spee* been operating, as originally designed, in the North Sea or the Baltic, the lack of protection provided to her galleys and her lack of overall seaworthiness might have been seen as inconsequential. But that is another story ...

Principal Sources:

MDV 550 Heft 1, vorläufige Zusammenstellung über das Seegefecht vor dem La Plata am.13.12.193.

ADM 267/145 'The Battle of the River Plate, December 1939', an unclassified narrative published by His Majesty's Stationary Office in 1940.

ADM 269/46 '*Graf Spee* Battle with HM Cruisers *Achilles*, *Ajax*, and *Exeter*'.

ADM 1-9759 '*Admiral Graf Spee* – Action off the River Plate'.

ADM 223/68 '*Graf Spee* 1939 – The German Story'.

ADM 281/84 Report on visit to *Graf Spee* Wreck, April, 1940 Part I (Technical Report) and Part II (Shell Damage to Superstructure). These reports were generally classified SECRET when produced.

Millington-Drake, Sir Eugen: *The Drama of Graf Spee and the Battle of the River Plate: A Documentary Anthology, 1914–1964*, Peter Davies (London 1964).

Endnotes:

1 This convention renders reference to British documents confusing, insofar as discussions in British sources referring to, for example, 'Deck D', have necessarily been rewritten to refer to 'Deck E', which may at first glance appear to represent a transcription error. For convenience, the drawings attached have been augmented with an additional set of superstructure layouts showing corresponding British deck nomenclature.

2 Groener, *German Warships 1815–1945*, US Naval Institute Press (Annapolis 1990), Vol 1, 60ff.

3 It appears most likely that subsequent narratives along this line spring from an article by B Ruff, 'A Mystery Solved: The Scuttling of the *Admiral Graf Spee*', published in the December 1999 edition of *World and I*, a religiously-funded pseudo-scientific magazine published by New World Communications via the Washington Times. The magazine ceased print publication in 2004. The article states in part: 'The diesel engines that powered the ship used a thick oil that had to be thinned with heat before it could be injected into the motors. This was accomplished with a large boiler located at the base of the smokestack. This boiler was, uncharacteristically for a German design, unprotected and so was destroyed by British gunfire during the Battle ...'. It is worth noting that *Graf Spee* was equipped with two auxiliary boilers rather than one, and that these appear to have been primarily designed to operate steam-powered fire and auxiliary pumps in action, and to supply steam to the galleys, etc, otherwise.

THE ARMOURED CRUISER *JEANNE D'ARC*

Jeanne d'Arc was the first of a series of large, fast French armoured cruisers designed by the celebrated naval architect Emile Bertin. In this article **Luc Feron**, with the help of **Jean Roche**, traces the history of the ship from her earliest beginnings to the end of her career in 1934.

The first French armoured cruiser *Dupuy-de-Lôme*, designed by Louis de Bussy, was followed by four smaller 4,700-tonne armoured cruisers (*Amiral-Charner*, *Latouche-Tréville*, *Chanzy* and *Bruix*)[1] and the *Pothuau* of 5,400 tonnes, all designed by Jules Thibaudier. The next armoured cruiser was a much larger ship displacing in excess of 11,000 tonnes designed by Emile Bertin and named *Jeanne d'Arc*. However, she acquired her name more or less by default, as it had already been allocated to an earlier projected ship which did not get beyond the design stage.

The 2nd class armoured cruiser *Amiral Charner* in Toulon roads, 1900. (Author's collection)

The Protected Cruiser Project

In the early 1890s the design of a 'first class cruiser for foreign stations' had been under consideration, and had

The cruiser *Pothuau* in Toulon roads. (Author's collection)

evolved to the point of specifying that the ship in question should be fitted to act as flagship for a rear admiral, with a main armament of 24cm guns and a displacement of 8,000 tonnes. Even so the *Conseil des travaux*, in its session of 7 July 1891, was of the opinion that:

> It does not seem possible, with a ship of suitable length, to achieve the requirements before the Council without considerably exceeding the specified 8,000-tonne displacement. Even by reducing the endurance at 10 knots from 6,000nm to 5,500nm and adopting water-tube boilers whilst retaining the proposed armament it does not seem possible to resolve the problem without an increase in displacement to 8,600 tonnes and in length to 118 metres.

The following characteristics were proposed:

Armament: 4 x 24cm guns (Mle 1887) with 110 rounds per gun mounted in enclosed turrets, of which two were to be on the centreline forward and aft and the other two on the beam; 8 x 14cm QF on the upper deck, and five torpedo tubes.

Protection: A protective deck of curved or angular profile over the lower decks and hold spaces, with its outer edge 1.40m below the waterline and raised in the centre; the deck to be 50mm thick (including plating), increasing to 100mm on the slopes over all the spaces occupied by the engines and boilers; access hatches to be protected by armoured coamings; an 8mm splinter deck to be fitted below the protective deck over the machinery.

Propulsion: Machinery to be two triple expansion engines each driving a single screw propeller and located in separate engine rooms; the engines to be of robust construction; boilers to be of either the Lagraffel–d'Allest or Belleville type and of large dimensions; a speed of 19 knots to be attained with moderately forced draught but without closed stokeholds (at least 18 knots to be attainable without forced draught).

Various: Hull to be of steel and to have a wooden cladding with copper sheathing below the waterline; the accommodation to include facilities for a rear admiral commanding a squadron; the ship to be entirely lit by electricity; length and draught not to exceed 118m and 7.5m respectively.

Accepting that the project could probably not be realised without a considerable increase in displacement, the *Conseil* would give preference to this solution over one which reduced either the protection or the number/calibre of the main armament in order to maintain the displacement at 8,000 tonnes. If, however, this latter solution were to be adopted, the maximum length referred to above should be reduced to 115m.

The Minister did not, however, act on the Council's advice, and it was later decided to limit the number of 24cm guns to two. A despatch of 2 March 1892 designated the ship as 'First class cruiser' and also prescribed discarding sail(!). Four naval engineers, Mssrs Tréboul, Schwartz, Besson and Raymond, were requested to submit preliminary designs, as were three private shipyards (Forges et Chantiers de la Méditerranée, Ateliers et

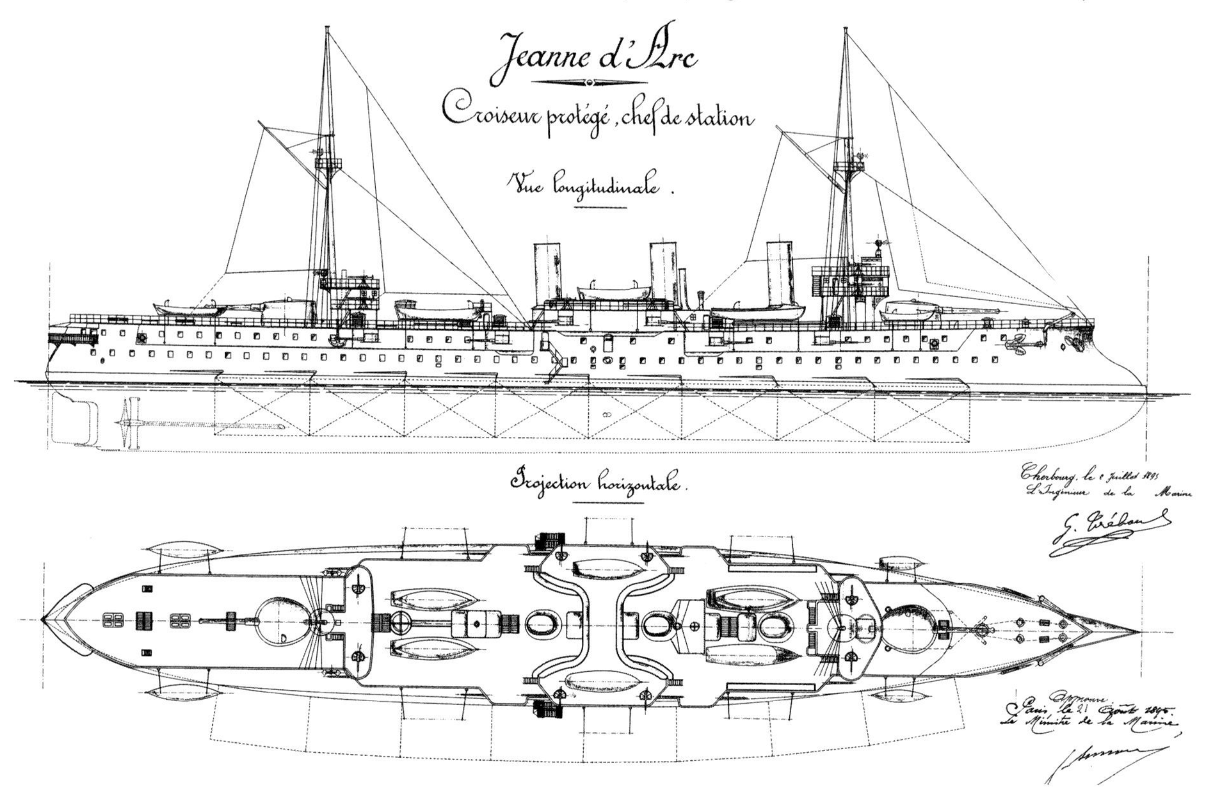

Tréboul's design for a protected 'station cruiser' to be named *Jeanne d'Arc*. The plans are dated 2 July 1895. (Courtesy of the author)

Chantiers de la Loire and Chantiers et Ateliers de la Gironde). Having examined these designs on 29 March 1892, the Council's decision was to decline those submitted by engineers Schwartz and Besson and those of the Loire and Gironde yards, but to request Tréboul, Raymond and F C Méditerranée to submit definitive designs taking into account remarks emanating from the various departments.

On 31 January 1893 the *Conseil des travaux* examined the revised designs for a first class protected cruiser and approved those from Tréboul and F C Méditerranée, subject to some modifications, and concluded that if only one of these designs were to be retained, preference should be given to that of Tréboul. On 3 August 1893 the *Directeur du matériel*, engineer Bienaymé, informed the Minister that, in accordance with a report that he had sent to the Minister on 8 February and which he had approved, Tréboul had been requested to revise his design, taking into account the observations of the *Conseil des travaux* at its session of 31 January 1893 as well as matters indicated in despatches dated 23 February and 10 March of the same year. A further study of the plans had shown that it would be desirable to complete the dossier and also to make a few minor alterations. Tréboul was also to review the design of the 24cm turrets in the light of the observations made. If the Minister were to signify his approval of the attached plans, the Commission for Machinery and Large-scale Equipment could be instructed to request tenders for the construction of this cruiser, which would be shown on the fleet list as *Jeanne d'Arc*.

On 8 November, however, a contract was signed with F C Méditerranée (La Seyne) for the construction of a cruiser of 8,309 tonnes full load to a design by engineer Lagane for a price of 16,693,477 francs, to be named *D'Entrecasteaux*.

Tréboul was not immediately informed that his design had been abandoned, and the accompanying drawings were approved by the Minister on 21 August 1895. No details have been found either of the decision not to proceed with the construction of this ship or of the manner in which Tréboul was informed of this; in the meantime the fleet list continued to include a protected cruiser with the name of *Jeanne d'Arc*.

On 28 December 1895, with the protected cruiser still languishing unbuilt on the fleet list, the then-Minister, Edouard Lockroy, instructed the Toulon dockyard to lay down a first class armoured cruiser to be named *Jeanne d'Arc*, 'to replace the one of the same name whose construction was suspended following a decision by the Budgetary Commission on 24 October.'

The Armoured Cruiser

No trace has been found in the records of the *Conseil des travaux* of any discussion of a programme for this ship, which seems to indicate that it was proposed directly to the Minister by the then-director of naval construction, Emile Bertin.

In 1890 Bertin had returned from a five-year secondment to Japan where he had overseen the development of the Sasebo and Kure dockyards, and was appointed deputy director of construction at Toulon. In 1892 he took over as director at Rochefort, following this with a period as director of the school of applied engineering. In 1895 he was appointed *Directeur du matériel* and then director of the newly-created *Section technique*, in which the role of the *Conseil des travaux* gradually diminished until it was finally abolished in 1905.

According to the ministerial announcement, the machinery of the new ship was to comprise three vertical triple-expansion engines each of 9,500CV with steam supplied by 24 double-ended Sigaudy–Normand boilers. Displacement was to be about 9,000 tonnes and the armament comprised two 194mm guns in centreline turrets forward and aft and fourteen 138.6mm guns in casemates.

However the specification seems to have evolved as, on completion, the displacement had increased to 11,264 tonnes and the length between perpendiculars to 145.40m. At an early stage it was proposed that the secondary armament should consist of eight 138.6 mm and ten/twelve 100mm QF guns, but the latter were eventually replaced by six more 138.6 mm in shielded upperdeck mountings, thus reverting to the original proposal of fourteen 138.6mm. Designed for an output of 28,500CV, the engines finally installed attained in excess of 30,000CV during trials, with steam provided by 36 Guyot–du Temple small-tube boilers exhausting through six funnels, an arrangement which gave the ship a decidedly unusual appearance for the time. The engines and boilers were manufactured at the Navy's Indret works.

Armament

Main turrets

From the outset, Bertin had decided to include a novel element in his design of the 194mm turrets in that he abandoned the pivot-shaft, which had been the standard system since the battleship *Brennus*, in favour of a turret supported by a horizontal roller path at upper-deck level with a vertical roller-path for guidance, the roller paths and the training mechanism being protected by a fixed

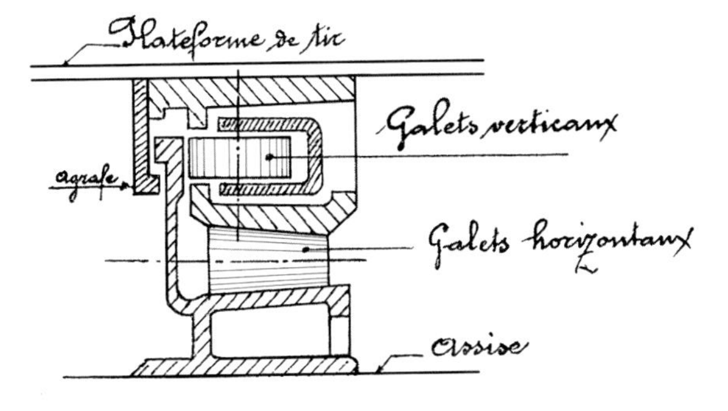

A sketch of the roller paths adopted for the new turret, showing the vertical guidance rollers above the horizontal rollers on which the turret revolved. (Courtesy of the author)

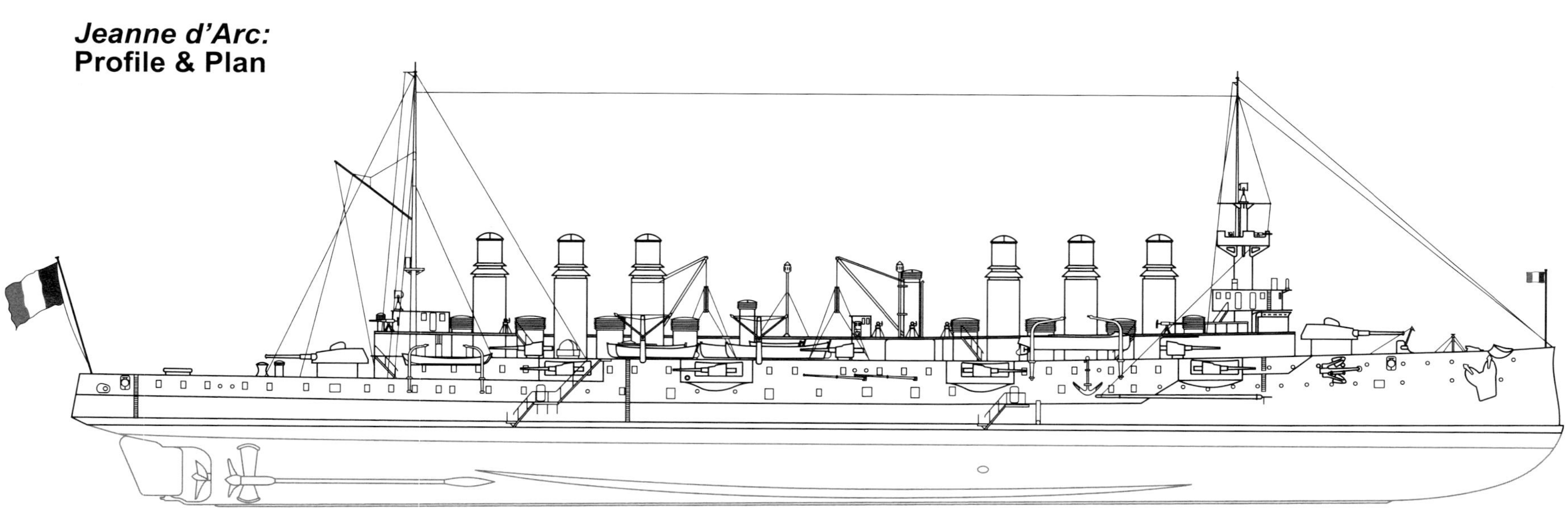

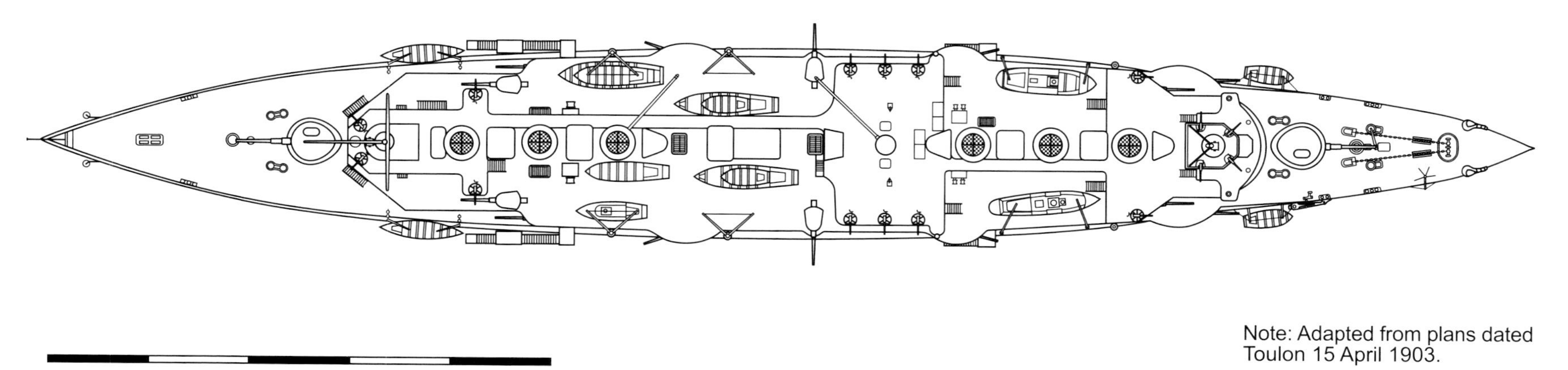

Note: Adapted from plans dated Toulon 15 April 1903.

© John Jordan 2016

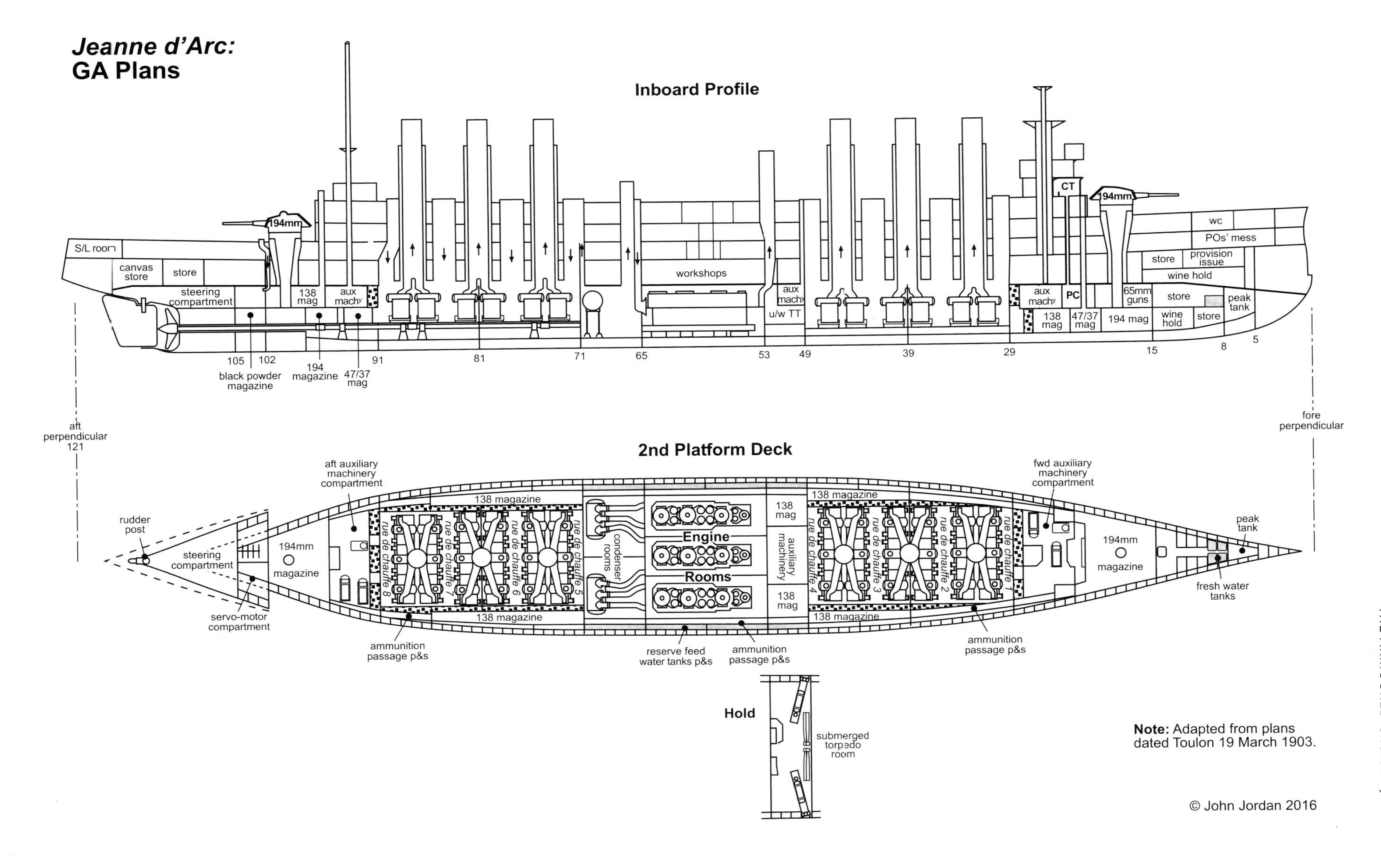

Jeanne d'Arc:
GA Plans
Inboard Profile
S/L room
canvas store
store
steering compartment
194mm
138 mag
aux mach^y
workshops
aux mach^y
u/w TT
aux mach^y
138 mag
PC
47/37 mag
CT
194mm
65mm guns
194 mag
wc
POs' mess
store
provision issue
wine hold
store
wine hold
store
peak tank
105
102
194
91
81
71
65
53
49
39
29
15
8
5
black powder magazine
194 magazine
47/37 mag
aft perpendicular 121
fore perpendicular
2nd Platform Deck
aft auxiliary machinery compartment
rudder post
steering compartment
servo-motor compartment
194mm magazine
138 magazine
rue de chauffe 8
rue de chauffe 7
rue de chauffe 6
rue de chauffe 5
condenser rooms
Engine Rooms
138 mag
auxiliary machinery
rue de chauffe 4
rue de chauffe 3
rue de chauffe 2
rue de chauffe 1
fwd auxiliary machinery compartment
peak tank
fresh water tanks
ammunition passage p&s
reserve feed water tanks p&s
Hold
submerged torpedo room
Note: Adapted from plans dated Toulon 19 March 1903.
© John Jordan 2016

ring of 176mm armour. In order to differentiate between the two systems he adopted the term 'barbette–turret' (*tourelle–barbette*), which has given rise to a certain amount of confusion. The accompanying sketch shows the roller-path arrangement more clearly than would a lengthy explanation.

This proposal had been submitted, as was custom and practice, to the ordnance department. In a note of 9 March 1896 the head of that department, General Desbordes, had indicated his approval, remarking that eliminating the pivot-shaft which normally served to support the rotating platform, and locating both the supporting and directional roller paths immediately below the turret provided a gain in simplicity and economy of weight, which was particularly advantageous in a cruiser such as the *Jeanne d'Arc*. Given the height of command of the fore turret, it was calculated that the height of a pivot-shaft would have been 12.24 metres. However, with the system proposed by Bertin the axial tube would become no more than an ammunition passage whose protection could be achieved with a considerable saving in weight. On 24 June the construction department informed the ordnance department of its final approval of the new arrangement.

Five companies (Schneider, Société des Batignolles, Châtillon, Commentry, Saint-Chamond and F C Méditerranée) were invited to tender for the supply of '... a balanced barbette–turret consisting of a rotating platform guided and supported by two paths of vertical and horizontal rollers respectively and protected by an armoured cupola rotating with it as well as a fixed parapet above which the gun would be trained.' The aim of these conditions was:

- to enable the turret to be trained manually with the same degree of protection and the minimum of weight and dimensions
- to eliminate the need for a pivot-shaft.

After a study of the various proposals it was considered that Saint-Chamond's design was the closest to requirements and also the least expensive. However, a note to

***Jeanne d'Arc:* After Single Turret for 194mm Mle 1893 Gun**

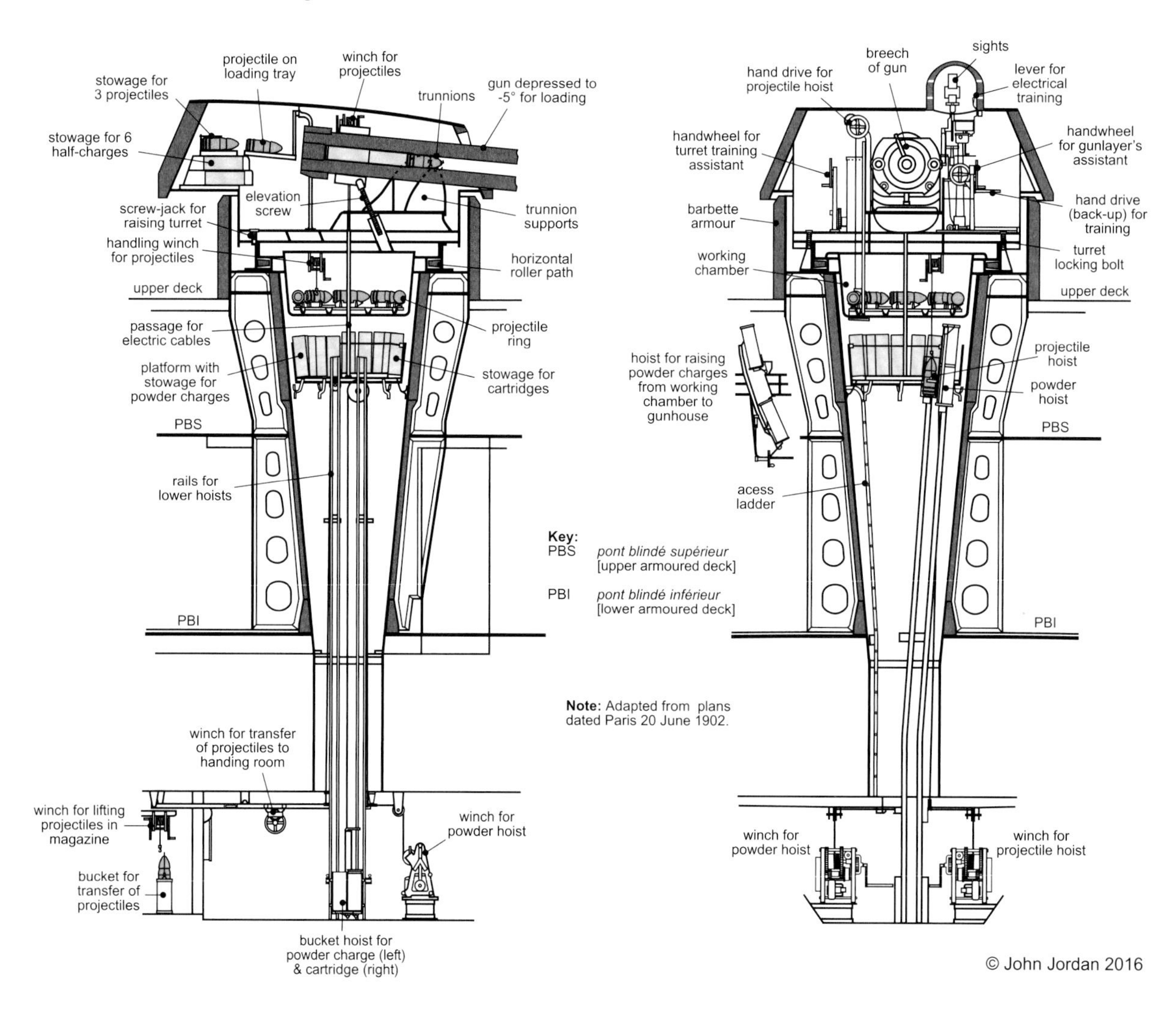

the Minister dated 10 May 1897 indicated that the ordnance department had raised a number of objections to the design submitted by Saint-Chamond and concluded that, despite the higher price quoted by the Société des Batignolles, it was the latter company's design that should be accepted. After further technical considerations it was, in the end, the Société des Batignolles proposal that was approved and the contract confirmed on the 18 March 1898.

From the drawings of the turret found in the ship's atlas of plans (*atlas de coque*) it is evident that the tube linking the turret to the magazines was constructed of light plating which had to support only the ammunition hoists; the latter terminated just below a shell chamber which was suspended beneath the turret and revolved with it. The 194mm projectiles were then lifted onto the circular shell carrier in the chamber using a hand winch, and the cartridges and bagged charges were lifted by hand and stowed around a fixed platform beneath the revolving shell chamber. There was a hoist with hand drive to raise the heavy 194mm projectiles from the shell carrier to the gunhouse, and a separate manual hoist for the propellant charges. Between the lower edge of the barbette and the armoured deck this combination was enclosed in an inverted cone of plating, the upper section of which, covering the working chamber, was 140mm thick and the lower section 60mm. with armoured coamings of 130–70mm at the base where it joined the armoured deck (see drawings).

QF guns

The arrangements finally adopted for the secondary armament of fourteen 138.6mm QF guns were unusual. The eight hull-mounted guns (four per side) were mounted on projecting sponsons rather than in conventional casemates. All the guns, including the six weather-deck mountings, had full shields of 72mm steel, and the 40mm screens at the top, bottom and sides of the sponsons were simply to provide a degree of splinter protection for the guncrews from shells striking the upper part of the hull. The unusually wide openings, which enabled the guns to be fired over broad arcs, were covered by drop-down weatherproof hatches when the ship was at sea (see drawing).

Magazine arrangements

With almost three-quarters of the ship's length taken up by the machinery and boiler spaces, the location of *Jeanne d'Arc*'s magazines was something of a problem. Those for the 194mm guns were situated directly below the turrets at both ends of the machinery spaces. Those for the medium and small-calibre guns were located in long-narrow spaces below the armoured deck, outboard of the boiler rooms and the coal bunkers and also above the torpedo compartment, between the forward group of boilers and the engine room. All the magazines (including the 194mm magazines fore and aft) were connected by ammunition passages which ran outboard of the internal bulkheads enclosing the machinery spaces (see GA plan and midship half-section drawings). Rails allowed the movement of projectiles and charges from one end of the ship to the other so that any gun could be supplied from any of the magazines. All the ammunition hoists were electrically operated.

ATB guns

For torpedo boat defence *Jeanne d'Arc* was fitted with sixteen of the then-standard Hotchkiss 47mm Mle 1885 guns: four in the military foremast and twelve on the shelter deck (two forward, four aft and six amidships). This was a superior arrangement to contemporary battleships, in which a number of the ATB guns fired through ports in the hull and had restricted arcs in consequence; it also made it easier to coordinate fire against attacking torpedo boats.

Sponson for 138.6mm Mle 1893

Face: Forward Starboard Mounting

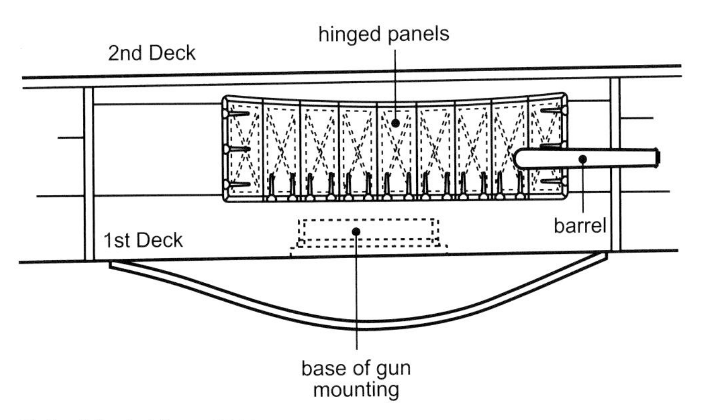

Side View

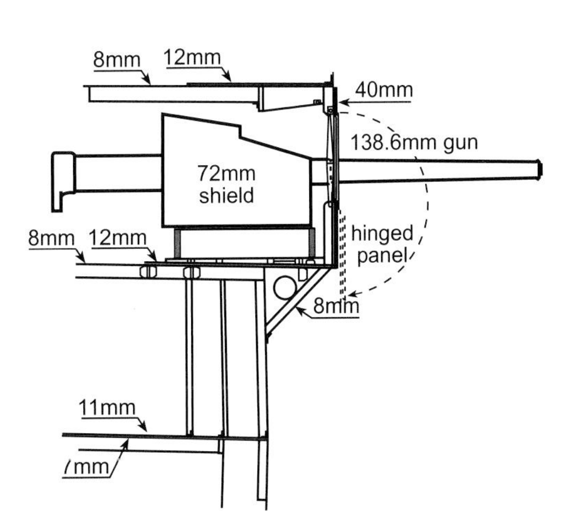

Note: Adapted from plans dated Toulon 15 May 1903.

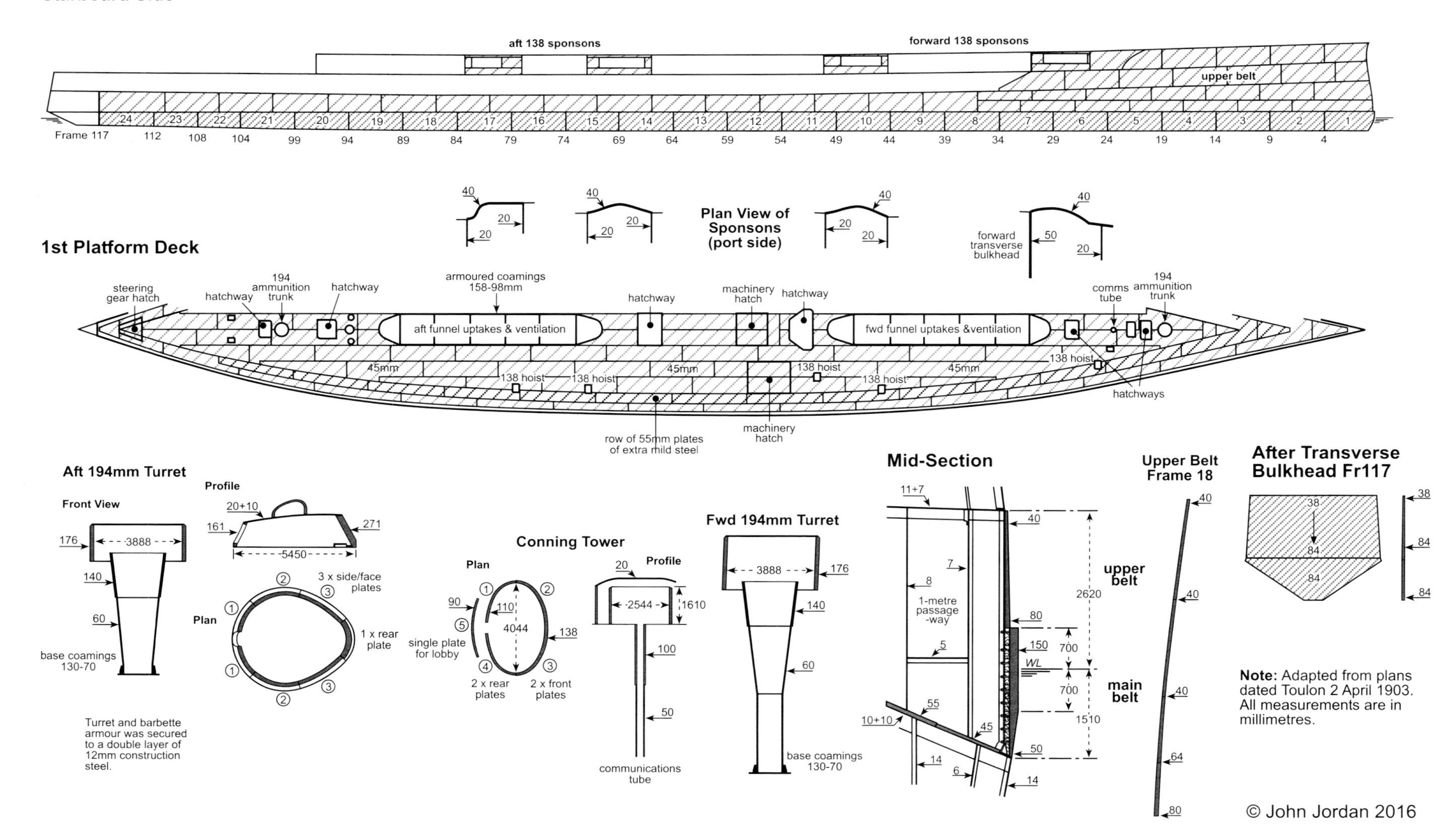
Jeanne d'Arc: Protection
Starboard Side
aft 138 sponsons
forward 138 sponsons
upper belt
Frame 117
112
108
104
99
94
89
84
79
74
69
64
59
54
49
44
39
34
29
24
19
14
9
4
1st Platform Deck
Plan View of Sponsons (port side)
forward transverse bulkhead
steering gear hatch
hatchway
194 ammunition trunk
armoured coamings 158-98mm
aft funnel uptakes & ventilation
machinery hatch
fwd funnel uptakes &ventilation
comms tube
138 hoist
45mm
hatchways
row of 55mm plates of extra mild steel
Aft 194mm Turret
Front View
Profile
Plan
3 x side/face plates
1 x rear plate
base coamings 130-70
Turret and barbette armour was secured to a double layer of 12mm construction steel.
Conning Tower
single plate for lobby
2 x rear plates
2 x front plates
communications tube
Fwd 194mm Turret
Mid-Section
1-metre passage -way
upper belt
main belt
WL
Upper Belt Frame 18
After Transverse Bulkhead Fr117
Note: Adapted from plans dated Toulon 2 April 1903. All measurements are in millimetres.
© John Jordan 2016

Torpedo tubes

Jeanne d'Arc was fitted with only two submerged torpedo tubes. These were in a special compartment amidships, between the forward boiler rooms and the engine room (see GA Plans), and were angled at 13 degrees forward of the beam. Six 450mm torpedoes were stowed on tiered racks against the forward bulkhead; these were of the Mle 1892 type, which weighed 530kg and had a length of 5.05m. The Mle 1892, which was built under licence from Whitehead at Toulon, was powered by a piston engine using compressed air, carried a 75kg warhead and had a maximum range of 800m at 27.5 knots.

Hull Protection

The side armour was arranged in two strakes. The waterline belt (*cuirasse épaisse*) was 2.2m high, with the upper edge 0.70m above the waterline. It was 150mm thick amidships, reducing to 100mm forward and 80mm aft, tapering to 50 mm at its lower edge, and was secured to a 70mm teak backing. Above this was an upper belt (*cuirasse mince*) which was 1.92m high amidships; the upper belt was 80mm thick at its lower edge reducing to 40mm at its upper edge. Forward of the foremost 138.6mm gun sponson three more strakes of 40mm armour covered the entire bow section to the height of the forecastle. The sponsons themselves were protected by 40mm plating and had 20mm internal splinter screens

The ship had two protective decks. The upper protective deck was at main deck level; it comprised 11mm armour on 7mm plating and rested on the upper edge of

Jeanne d'Arc: Midship Half-section

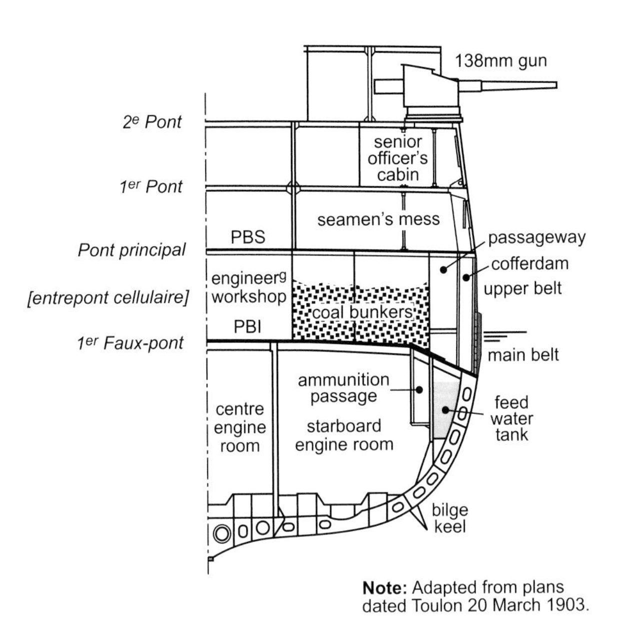

the upper armour belt. The lower protective deck was at the level of the first platform deck: the plates over the horizontal central section, which was just above the load waterline, were of 45mm mild steel on two layers of 10mm plating; the outboard plating of this deck sloped downwards at an angle of 23° and joined the lower edge of the main belt 1.50m below the waterline; the inboard strake of the slopes had 55mm extra-mild steel armour in place of the 45mm plates of mild steel, for a total thickness of 75mm.

The plating for the armour belt was to be supplied by Marrel Frères of Toulon, the plating for the armoured deck by the Forges de la Chaussade at Guérigny. On 26 May 1896, engineer Korn had undertaken to supply all this plating by 1 July 1897.

The space enclosed by the two protective decks and the side armour contained the cellular layer which formed a key part of the ship's protective system. Behind the side armour was a cofferdam, inboard of which was an 8mm longitudinal watertight bulkhead. The space between the bulkhead and the cofferdam was 1m wide and provided access to allow inspection and repairs to be made to any action damage to the cofferdam. Scuppers in the floor allowed any flooding to drain directly into the bilges.

The space between the port and starboard longitudinal watertight bulkheads was closely subdivided: some of the compartments were left empty, others were used as coal bunkers or for boiler feed water. Where the ammunition hoists, boiler uptakes, ventilation shafts and access hatches passed vertically through this section they were enclosed in watertight casings with armoured coamings at their lower ends.

There were 15 watertight bulkheads running from the double bottom to the main armoured deck (first platform

Jeanne d'Arc: Armour Plate Thickness Main Belt

Plate	Single Strake *upper/lower edge*
1-4	100/50
5	113/50
6	126/50
7	138/50
8-17	150/50
18	136/50
19	122/50
20	108/50
21	94/50
22-24	80/50

Notes:

1. All the protective side plating was of special steel; the decks were of mild steel, while the 55mm reinforcing plates on the slopes were of 'extra mild' steel.
2. The upper belt was in four strakes, of which three covered the bow section up to the foremost casemate. The single-strake upper belt amidships tapered from 40mm at its upper edge to 80mm at its lower edge. Forward, the upper two strakes were of 40mm special steel, the lower two were tapered from 40mm to 80mm through 63–65mm at the joint.

Jeanne d'Arc: Table

Characteristics

General	
Length of keel (bearing on ground)	112.90m
Length between perpendiculars	145.40m
Length at load waterline	145.40m
Length overall	147.00m
Beam (maximum at load wl)	19.42m
Beam (maximum at 0.70m above load wl)	19.42m
Depth of hull at centre of wl zero	7.40m
Draught (fwd/mean/aft)	7.08m/7.54m/8.00m
Displacement (during stability testing)	11,264 tonnes
Metacentric height:	
as designed	1.184m
as measured 09/05/02	1.458m
Protection	
Main armour belt (special steel):	
height fwd/amidships/aft	2.20m
thickness at upper edge	100mm/150mm/80mm
thickness at lower edge	50mm
Upper armour belt (special steel):	
height	8.00m fwd/1.92m amidships & aft
thickness at upper edge	40mm on 8+8mm
thickness at lower edge	80mm
Main armoured deck(mild steel):	
flat	45mm mild steel on 10+10mm
inboard slopes inclined at 23°	55mm extra-mild steel on 10+10mm
outboard slopes inclined at 23°	45mm on 10+10mm
Upper armoured deck (mild steel)	11mm on 7mm
Conning tower (special steel):	
front/rear	138/110mm on 11+11mm
rear lobby	90mm
communication tube	100mm upper/50mm lower
roof	20mm on 10mm
194mm turrets:	
fixed armour (barbette)	176mm on 12+12mm
gunhouse (cemented armour	161mm on 12+12mm face & sides
except rear plate)	271mm rear
roof	20mm on 10mm
138.6mm sponsons	40mm on 8+8mm
Armament	
Main guns	2 x 194mm/40 Mle 1893 in single turrets
QF guns	8 x 138.6mm/45 Mle 1893 on sponsons (1st Deck)
	6 x 138.6mm/45 Mle 1893 on upper deck (2nd Deck)
ATB guns	16 x 47mm/40 Mle 1885 QF
Torpedo tubes	2 x 450mm submerged
Machinery (Indret)	
Engines	Three 4-cylinder triple expansion
designed horsepower:	28,500ihp
max rpm	138.75 (port engine 23/01/03)
designed speed	23 knots
max speed	21.72 knots with 29,690CV (23/01/03)
Propellers	Three 3-bladed
diameter	4.70m (centre)/5.00m (wing)
pitch	5.54m (centre)/ 5.55m (wing)
Boilers	36 Guyot–du Temple small-tube
Coal	1,400 tonnes normal (9,000nm at 10 knots)
	2,100 tonnes full load (13,500nm at 10 knots)

deck). These were at Frames 5 (fwd end peak tank), 8 (stores), 15 (194mm magazine), 23 (138mm magazine), 29 (boiler room no 1), 39 (boiler room no 2), 49 (138mm magazine/torpedo flat), 53 (engine rooms), 65 (condenser rooms), 71 (boiler room no 3), 81 (boiler room no 4), 91 (47mm magazine), 96 (194mm/138mm magazines), 102 (steering servo-motors) and 105 (steering compartment).

Construction Delays

Laid down at Toulon in October 1896, *Jeanne d'Arc* was launched on 8 June 1899 but would not enter service until 19 May 1903. The Toulon dockyard was not renowned for speed of construction (the battleship *Carnot*, laid down in July 1891, did not join the fleet until July 1897), but as far as *Jeanne d'Arc* was concerned this was not the only factor that delayed her completion.

On 27 August 1896 assistant engineer Maugas who had been supervising the construction of the battleship *Carnot* was transferred to Brest by the then-Minister, Admiral Besnard, for reasons that are not entirely clear but probably originated in friction between the construction department and the naval administration over matters related to errors in the stability calculations for *Carnot*. Subsequently the most senior of the dockyard's naval constructors, including the director, were also officially reprimanded for having openly demonstrated their support for Maugas, The simmering discontent in the construction department which resulted affected the activities of the dockyard to such an extent that work on *Jeanne d'Arc* made very little progress between September 1896 and June 1898.

It was not until 13 December 1898 that the head of the *Section technique*, Emile Bertin, was able to report to the Minister (by now again Edouard Lockroy) that, following the aforementioned lengthy interruption, work on the hull of the *Jeanne d'Arc* had recently recommenced in earnest, the number of men assigned to it being between 1,200 and 1,300. Provided there were no further serious interruptions in the work of the dockyard, it could be assumed that the cruiser would be ready for launch in March 1899 with completion a year later. The Indret works should therefore be asked to proceed as a matter of urgency with the completion of the boilers, as it was considered preferable that the weightier and bulkier components of the machinery should be installed before launch.

However, on 4 February 1899 engineer Lhomme, who was supervising the construction of the engines, stated that:

> The whole of the propulsion machinery cannot be ready before January 1900; the first (starboard) engine is to be completed in November but there is a serious problem with the port engine whose moulded steel high-pressure cylinder had to be rejected during machining in July 1898. The Firminy works, contracted to supply this item, estimates that it will require three and a half months' work before it can be fitted and adjusted. The final assembly of this engine, like that of the starboard engine, will therefore not be completed until November ...

As a result, *Jeanne d'Arc* was launched on 8 June 1899 without her engines. The operation went off smoothly but it may be noted that only one more large warship, the armoured cruiser *Dupetit-Thouars*, would be allocated to the Toulon dockyard.

A note dated 27 October 1899 advised that the transport *Drôme* then loading at Saint-Nazaire would bring *Jeanne d'Arc*'s starboard engine and the remaining two boilers of the steam plant. The port engine had been

Jeanne d'Arc shortly after her launch at Toulon on 8 June 1899. (Author's collection)

Jeanne d'Arc at the fitting out quay at Toulon, probably in 1900–01. (Author's collection)

completed, was in the process of being dismantled and would be ready in 25 days.

There was to be a further incident affecting the machinery. Indret had sub-contracted the manufacture of the crankshafts of all three engines to Schneider at Le Creusot. On 21 July engineer Simonot had raised objections to the way in which the supplier was making these shafts, and proposed sacrificing one of them to provide test bars. The supplier refused to agree and it was only on 8 December that the shafts were finally accepted with serious reservations. In fact, Simonot's fears proved to be unfounded, as *Jeanne d'Arc* was to come to the end of her long career with her original crankshafts still in place.

Machinery Trials

The ship's records show that the installation of the engines started on 2 February 1900, and the boilers were first lit on 20 February of the following year. The ship was manned for trials on 1 March 1901, and three days later CV Boisse was appointed in command. The initial

Jeanne d'Arc during her full power speed trials on 23 January 1903. (Constructions Navales)

static trials took place 23–30 March. A preliminary trial of the central engine was carried out on 5 July 1901; it proved satisfactory with 104rpm. A trial for smooth running on the 20th was also successful. On 28 August the trials commission had the boilers stoked so that the outboard engines ran at a little over 100rpm. With 24 boilers fired there were only minor problems with the engines. The commission nevertheless remarked that inadequate ventilation and poor insulation of the boiler lagging resulted in an excessively high temperature (65°C) on the stokehold platforms, and that frequent failures of the feed pumps were caused by the temperature of the feed water being too high due to overheating of the condensers.

Following this trial a despatch from the Minister requested Indret to make arrangements with the Toulon dockyard to carry out such alterations to *Jeanne d'Arc*'s boilers as were specified by the *Section technique*. The main improvements were completed in March 1902; in an initial static trial on 26 March with fuel consumption up to 175kg per square metre of grate the temperature in the stokehold did not exceed 25–30°C and the boiler feed functioned well, suggesting that the alterations had been effective.

A preliminary trial at sea with 12 boilers lit took place on 12 April 1902, the aim of this trial being to run the central engine at its maximum speed. During the whole of the morning of the 12th the boiler feed proved inadequate, and the Belleville boiler feed pumps failed or were not delivering. The reason for this was soon discovered: an inspection of the pistons following these trials revealed the possibility of an incident similar to one which had previously occurred on the protected cruiser *Jurien-de-la-Gravière*. Indret therefore undertook the replacement of the rings retaining the packing of the pistons in the high-pressure cylinders of all three engines and of the after low-pressure cylinder of the starboard engine; the rings of the other pistons were strengthened. After the trials had been completed the rings of all the pistons were modified.

The sea trials were resumed after a static trial on 29 September 1902 and took place on 2, 25 and 31 October and 8 November 1902, the latter being at maximum power; 33,000CV was obtained at the specified revolutions but this produced a speed of less than 22 knots. Without further trials it could not be assumed that the propellers were too resistant and should be modified. An inspection by divers following this trial revealed that they were in fact thickly encrusted with marine growth, which could explain the very high power needed to attain the required rate of revolution.

During the time at maximum power, coal consumption was up to 190kg per square metre of grate without the gratings having been completely cleared of slag. Combustion was good and the temperature on the stokehold platforms did not exceed 36°C; there were no problems with boiler feed.

The naval construction department's summary report of 13 November 1902 advised the authorities that *Jeanne d'Arc*'s engines were ready for acceptance trials. These official trials took place without any interruptions other than those required for the usual inspections after each trial and for such docking as was necessary:

- A reduced speed trial on 5 December 1902 was rated very satisfactory, although the coal used was from Anzin with 7% ash instead of coal with 5% ash; consumption was 695g per CV per hour.
- A consumption trial at maximum power with natural draught on 16 December 1902 was also deemed very satisfactory, the coal being the same as for the previous trial; consumption per CV per hour was 702g.
- A maximum power trial on 23 January 1903 found the functioning of engines and steam plant to be completely satisfactory; consumption was 854g per CV per hour. With 29,690.8CV the speed attained was 21.724 knots rather than the 23 knots expected. In order to improve on this performance it was suggested that the following measures should be implemented: replace the three propellers by others with increased propulsive area; shorten the bilge keels (which ran for half the length of the ship) from 72.7m to 43.5 m; replace the hollow moulded steel propeller-shaft brackets by thinner solid forged steel brackets. (It appears that the proposed modifications were, indeed, carried out but with unconvincing results.)
- A 24-hour trial was first attempted on 17 February, but had to be abandoned due to bad weather; when a second attempt was made 20–21 February 1903 the results were very satisfactory, with a rate of consumption of 670g/CV/hour for the first six hours and 694g over the 24 hours.

The trials of the armament which had started on 28 December 1901 had given rise to no problems worthy of note. The ship's general characteristics on completion can be found in the accompanying table.

The trials commission's final report was dated 12 October 1903. However, by that time *Jeanne d'Arc* had already undertaken a special mission. Commissioned on 10 March, she sailed on 12 April from Toulon to Marseille, leaving that port on the 14th for Algiers with the President of the Republic, Emile Loubet, on board. She remained on the North African coast until the 29th,

A contemporary postcard commemorating President Emile Loubet's visit to Algiers in April 1903. (Author's collection)

on which date she sailed from Bizerta to take the President back to Marseille before returning to Toulon to resume her trials.

On 20 May she left Toulon, again manned as for trials, for Brest where she arrived on the 25th, replacing the armoured cruiser *Bruix* in the Northern Squadron from 1 June; CV Pivet replaced CV Boisse in command. The Northern Squadron then consisted of the battleships *Masséna* (flag of VA Caillard), *Formidable* and *Dévastation* forming the 1st Battle Division. A second division comprised the coast-defence ships *Bouvines* (CA Stéphan), *Amiral-Tréhouart* and *Valmy* as well as a division of cruisers comprising *Jeanne d'Arc* (flag of CA Bugard), *Guichen*, *Dupuy-de-Lôme* and the despatch vessel *Cassini*. There was also an Atlantic Naval Division commanded by CA Rivet comprising the cruisers *Tage*, *Jurien-de-la-Gravière*, *Troude* and *D'Estrees*. The new cruiser took part in the division's various exercises on the Brittany coast and in Quiberon Bay.

On 14 September, following serious problems with her boilers, *Jeanne d'Arc* was placed in normal reserve at Brest. Manned for further trials on 8 October she was decommissioned for repairs on 15 November. In May 1905 she was again manned for trials, only to return to the reserve on 6 August of that year.

Service 1906–1914

On 26 May 1906 *Jeanne d'Arc* was assigned to the Mediterranean Light Squadron to replace *Marseillaise* as flagship of CA Campion. On the same date CV Pivet was replaced in command by CV Guépratte, the future admiral. The cruiser left Brest on 5 June for Tangier, arriving there on the 8th. On 2 July she sailed from Tangier for Gibraltar, returning to Tangier on the 9th. On the 12th she sailed to join the Mediterranean Squadron.

The Grand Manoeuvres having been reinstated, the Mediterranean Squadron, commanded by VA Touchard, had sailed from Toulon on 28 June for Algiers where it was joined two days later by the Northern Squadron. *Jeanne d'Arc*, coming from Tangier, joined them at Bône on 12 July. The manoeuvres continued until 28 July when all the ships returned to Toulon, having made calls at Bougie, Cap Blanc and Bizerta.

Apart from the usual exercises at Les Salins d'Hyères, it would be 13 September before the cruiser again sailed with the Squadron to Endoumes roads at Marseille where a grand naval review was held on the 16th to mark the laying of the first stone of the Rove canal, attended by President Armand Fallières on the torpedo gunboat *La Hire*; HMS *Cumberland*, the Italian cruisers *Giuseppe Garibaldi* and *Varese* and the Spanish cruiser *Carlos Quinto* were also present. The ships returned to Toulon on the 18th.

On 18 October *Jeanne d'Arc* sailed for Bizerta where she remained for 48 hours, returning to Toulon on the 25th, but she made no further movements worthy of note for the remainder of the year.

On the 2 February 1907 the cruiser anchored in Villefranche roads with the Squadron. On the 23rd CV Ramey de Sugny replaced CV Guépratte who, in his report on his time in command noted:

> As a seagoing unit *Jeanne d'Arc* is an excellent and handy ship which makes life easy for her captain ... As a fighting unit her value lies mainly in her speed, range and manoeuvrability ... The secondary armament is ridiculously weak in terms of both calibre and ammunition supply ...

On 24 March 1907 the cruiser sailed from Toulon for Tangier. After various movements including calls at Casablanca, Mogador and Mers el-Kebir she sailed from Tangier on 7 June for Cherbourg, arriving there on the 11th and then proceeding to Brest three days later. She remained there until 3 July for routine hull-cleaning and returned to Toulon on the 20th, having called en route at Bône and Les Salins where she had rejoined the Mediterranean Squadron on the 19th. On 1 June CA Krantz had replaced CA Campion in command of the Light Squadron. On 3 August '*La Jeanne*' (as she was

Jeanne d'Arc moored in Toulon roads shortly after she entered service. (Author's collection)

Jeanne d'Arc moored in Toulon roads during the early 1900s. Note the black-painted funnel caps. (Author's collection)

familiarly known) left Toulon for Morocco; she arrived at Tangier on the 9th having called in at Mers el-Kebir and Casablanca. On 5 October she went to Rabat (Morocco), remaining there until the 10th. On 23 January 1908 she again departed for Rabat and sailed from there on 7 February for Brest, where she arrived on the 15th having made a call at Gibraltar from the 8th to the 12th. Shortly after she put to sea a serious accident to boiler no 8 had caused the deaths of five stokers; three others had been badly burned.

In reserve from 15 April 1908, *Jeanne d'Arc* was initially attached to the Northern Squadron's 2nd Cruiser Division but was taken out of commission for modifications which would enable her to be used as a training cruiser for officer cadets from the naval college. At this time CV de Sugny completed his term in command and noted in his final report: 'the designer has sacrificed a great deal to speed and range ... but the cruiser has never exceeded 21.7 knots, despite replacement of the propellers and the shaft brackets and shortening of the bilge keels ...'.

Recommissioned on 20 May 1911, the cruiser was attached to the 3rd Division of the Reserve Squadron but remained at Brest. On 1 May 1912 she was assigned to

Jeanne d'Arc in the mouth of the River Penfeld at Brest on 15 February 1908, following the boiler explosion off Gibraltar which killed five stokers and injured three more. She is still in her early livery of black hull and buff upperworks. (Author's collection)

the Atlantic Schools Division which was then commanded by CA Bouxin (replaced by CA Le Cannellier on 20 October 1913) and included *Duguay-Trouin*, *Châteaurenault*, *D'Estrées*, *Guichen*, *Chamois* and *Borda*. On 10 October 1912, commanded by CV Grasset, she sailed on her first 'world cruise' which took her to Bahia (Brazil), Fort-de-France, Dakar, Toulon, Naples, Bizerta, La Pallice, Kronstadt, Bergen and Trondheim, returning to Brest on 29 July 1913. Her second cruise, starting on 10 October, took her to the Indian Ocean, returning via the Suez Canal and arriving at Brest on 27 July 1914.

War Service

When mobilisation was declared on 1 August 1914, *Jeanne d'Arc* was assigned to the Northern Squadron which was, essentially, the 2nd Light Squadron commanded by CA Rouyer. Rouyer's flagship *Marseillaise*, together with *Jeanne d'Arc* and *Amiral-Aube* constituted the 1st Division while the 2nd Division consisted of *Gloire*, *Gueydon* and *Dupetit-Thouars*. Following the German declaration of war on 3 August, this less-than-imposing 2nd Light Squadron was faced with the unenviable prospect of having to oppose a possible incursion into the Channel by Germany's powerful High Seas Fleet, an eventuality which was averted by Britain's entry into the war on the following day. With the start of hostilities *Jeanne d'Arc*, together with the other ships of the 2nd Light Squadron and British naval units, was engaged in patrol duties in the western Channel, intercepting possible blockade runners or ships carrying wartime contraband. On 21 December 1914 *Jeanne d'Arc* was docked at Brest for general maintenance and repairs to her quarterdeck and the tube for her central stern shaft. Further repairs were needed on 1 March 1915.

After the Allies' failed attempt to force the Dardanelles on 18 March 1915 there was a major reorganisation of French naval forces in the Middle East, and *Jeanne d'Arc* was assigned to the former Syrian Division, which became the 3rd Squadron commanded by VA Dartige du Fournet. The squadron comprised a 1st Division with *Saint-Louis* (flag), *D'Entrecasteaux* and *D'Estrees* and a 2nd Division consisting of *Jauréguiberry* (flag of CA Darrieus), *Henri IV* and *Jeanne d'Arc*, which was still commanded by CV Grasset. *Jeanne d'Arc*'s first task was to escort a convoy of French and British troopships to Trebouki and Mudros and then, with her division, to support the forces landing on the Gallipoli peninsula during 24–26 April, firing on targets at Yenisheir, Besika Bay and Kum Kaleh. On the 26th she was hit by two 150mm shells from a shore battery, one of which damaged a 138.6mm sponson, causing a number of casualties and starting a small fire which was quickly extinguished. The other penetrated an officer's cabin, two coal bunkers and ended up in a stokehold passageway without exploding; it was taken on deck and thrown overboard. Having dealt with the damage, *Jeanne d'Arc* returned and silenced the battery.

From this time until 30 March 1916 *Jeanne d'Arc* would be the flagship of three vice admirals: Dartige du Fournet (6 May–12 September 1915), Gauchet (5–24 October 1915) and Moreau (8 November 1915–20 March 1916).[2] She was based at Port Said, patrolling the Syrian coast with other members of the squadron and enforcing the Allied blockade of Turkish ports, frequently intercepting dhows and other small craft suspected of carrying contraband.[3] Other notable activities in which she took part during this period included the occupation of the island of Ruad (Arwad) on 30–31 August and Castellorizo (Kasteilorizon) on 28 December 1915. There were also frequent bombardments of shore targets held to be of strategic value including:

- a Turkish barracks and oil store at Makry on 9 May 1915
- an oil depot at Jaffa on 30 May and another at Caiffa (now Haifa) the following day
- an oil depot at Tripoli and a railway bridge at Homs on 2 June
- a precision attack in company with the battleship *Jauréguiberry* on a railway junction east of Caiffa and the German Wagner munition works at Jaffa on 12 August
- bombardment of the German metallurgical works at Caiffa and return to Jaffa to complete the destruction of the Wagner works on 16 December
- bombardment of the railway stations at Medina and Caiffa on 7 and 9 January 1916 respectively.

There were also two retaliatory actions against German consulates. The German consul at Alexandretta (modern Iskenderun) having persistently defied earlier attempts to dissuade him from flying the German ensign when French ships appeared, the consulate's flagmast was brought down on 30 May 1915 by a well-aimed shot from the cruiser *D'Estrées*. *Jeanne d'Arc* returned there on 4 July to find the German flag again flying on a new metal mast; it was promptly destroyed and the consulate building demolished with ten well-placed rounds of 138.6mm shell, an operation which CV Thomazi claims, in his account of the naval war in the Mediterranean, was observed with interest by the local people sitting in nearby cafes. At Caiffa, where the consul had incited Turkish troops to fire on a French emissary approaching under a flag of truce[4] and also to desecrate French war graves from the Egyptian campaign of 1799, the German consulate was demolished in a similar operation on 31 May.

During 8–10 September *Jeanne d'Arc* was briefly involved in the organisation of a humanitarian operation. On 5 September the cruiser *Guichen* signalled that she was in contact with a group of 700 Armenians who, with dwindling stocks of ammunition, were attempting to defend some 3,000 of their compatriots, men, women and children, who had taken refuge on the seaward side of Djebel Moussa (Musa Dagh) on the gulf of Alexandretta and were in danger of being massacred by Turkish forces surrounding them. *Jeanne d'Arc* arrived on the 8th with VA Dartige du Fournet, who obtained the necessary authorisations for evacuation but was urgently recalled to Port Said on the 10th, having been appointed in command of the Dardanelles Squadron, leaving the rescue operation in the capable hands of CA Darrieus. By 12 September *Guichen* had been joined by *Desaix*, *Amiral-Charner*, *D'Estrées* and *Foudre*, which used their guns to hold the Turks in check and on 12–13 September embarked some 4,100 refugees who were taken to Port Said.

Having taken on coal, water and provisions at Port Said, *Jeanne d'Arc* sailed on the 12th for Mudros, where VA Dartige transferred to his new flagship, the battleship

At Port Said in December 1915, when *Jeanne d'Arc* was serving with the 3rd Squadron (formerly the Syrian Division). Note the Barr & Stroud 2-metre rangefinders sided atop the bridge. (Author's collection)

Jeanne d'Arc at Toulon during the late war or postwar period. The main turrets have had the original paint scraped from them and have been coated with a mixture of soot and grease informally known as *bouchon gras*. Note the aerial spreaders on the first and fourth funnels. (Author's collection)

Saint-Louis, on 16 September. Ordered to Malta where she arrived on 3 October, the cruiser underwent routine docking and then brought Dartige's successor, VA Gauchet, to Port Said and from there on to Mudros and Salonika. Gauchet then transferred his flag to the battleship *Patrie* as commander of the Dardanelles Squadron (4th Squadron) on 24 October, Dartige having by then taken up his new appointment as C-in-C of the French naval forces (*Armée Navale*) in the Mediterranean.

Following the departure of VA Dartige du Fournet on 12 September the 3rd Squadron was temporarily commanded by CA Darrieus with his flag in *Jauréguiberry* until 8 November when VA Moreau was appointed, hoisting his own flag in *Jeanne d'Arc*.

Jeanne d'Arc as a school ship during the early 1920s; the photo was taken during a visit to Antwerp. (Leo van Ginderen collection)

Jeanne d'Arc in the 'ships' graveyard' at Landévennec in Britanny in 1931, awaiting disposal. (Author's collection)

Worn out by her constant patrols, *Jeanne d'Arc* was withdrawn from the 3rd Squadron on 20 March 1916, and after handing over surplus stores to the 4th Squadron at Salonika she sailed on the 31st for a period in reserve and a major refit.

In January 1917 she was assigned to the West Indies station (4th Light Division, comprising the armoured cruisers *Gueydon, Montcalm* and *Dupetit-Thouars*) which was commanded by CA Grasset, her former captain. On 28 May 1918 CA Grasset was replaced by CA Grout with his flag in *Gloire*. We have no precise details of the cruiser's movements up to the end of hostilities, but she is recorded as having called at many ports on the coasts of South and North America and the West Indies, including La Guayra (Venezuela), Kingston (Jamaica), Puerto Limón (Costa Rica), Colón (Panama), and San Juan (Puerto Rico), sometimes in company with British units. On one occasion the British authorities requested that she call at Montserrat, where her arrival seems to have had a calming influence on some civil unrest in the colony.

Jeanne d'Arc returned to Brest in 1918, escorting a convoy, and was placed in reserve with a reduced crew. In 1919 she was refitted to enable her to resume her pre-war role as a training cruiser and was commissioned as such in August 1919. Thereafter she made nine annual cruises, departing from Brest in September or October and returning the following summer, during the last of which (1927–1928) she was commanded by CV François Darlan (the future admiral). Replaced thereafter by the armoured cruiser *Edgar-Quinet*, the aged *Jeanne* was placed in reserve. In 1930 she was renamed *Jeanne d'Arc II* to release her original name for the new training cruiser then under construction, and she was stricken from the fleet list on 15 February 1933. Laid up at Landévennec, she was condemned on 21 March of that year. She was sold on 9 July 1934 and was towed to the La Seyne shipbreaking yard.

In conclusion, it may be noted that after the Second World War it was claimed by some authors that during the war years *Jeanne d'Arc* had been used by the Germans at Brest as a decoy intended to divert air attacks from the *Scharnhorst*. This is an error: the ship used for that purpose was the old armoured cruiser *Gueydon*, with the sloops *Aisne* and *Oise* moored alongside to make a more imposing target.

Acknowledgements:

Luc Feron's original article was published in the French naval journal *Marines* (No 157, June/July 2015). It has been translated by Jean Roche who, in conjunction with the Editor, has provided additional material on the technical characteristics and the service history of the ship. The plans for this article have been specially drawn by John Jordan.

Endnotes:

1 The author's articles on *Dupuy-de-Lôme* and the armoured cruisers of the *Charner* class were published in *Warship* 2011 and *Warship* 2014 respectively.

2 The second half of 1915 saw a rapid succession of changes in the command of the French naval forces in the Mediterranean. These began when VA Nicol, commanding the Dardanelles Squadron, resigned due to ill-health. He was replaced by VA Dartige du Fournet, who was to have been replaced in command of the 3rd Squadron by VA Gauchet. However, within days of the latter arriving in Port Said, Dartige was appointed to replace VA Boué de Lapeyrère as C-in-C of French naval forces in the Mediterranean. Gauchet was then appointed in command of the Dardanelles Squadron with CA Darrieus temporarily commanding the 3rd Squadron until VA Moreau was able to take over on 8 November.

3 The area referred to as Syria, then under Turkish domination, extended from the South of Anatolia to the Egyptian border; it included the territories which have become the present-day states of Syria, Lebanon, Jordan and Israel as well as the Gaza strip.

4 In order to avoid antagonising the local population, who were no friends of the Turks, it was customary, whenever possible, to send an emissary ashore under a flag of truce before a bombardment in order to warn civilians to keep clear of the ship's intended target.

BREAKING 'ULTRA':

The Cryptologic and Intelligence War between Britain and Italy, 1931–1943

Enrico Cernuschi debunks some of the myths that have grown up around the success of 'Ultra', and provides an alternative Italian perspective.

The wartime British decryption organisation, the Government Code and Cypher School (GC&CS) whose output is known commonly as 'Ultra', has become a symbol of the Second World War on a par with the Battle of Britain, Churchill's cigar, and HMS *Ark Royal*.

Enthusiasm for the achievements of 'Ultra' has helped re-write the history of war, but the price paid for the plethora of new documentation and information released since 1974 has been high. 'Ultra' has become a sort of key to unlock the secrets of Allied success in the European conflict. However, this is by no means universally accepted by historians. Alexander S Cochran Jr wrote in 1982:

> The 'Ultra' material now available at the PRO [now TNA] is overwhelming in quantity and consists almost exclusively of flimsies written in intelligence jargon, the digests of raw intercepts as sent from the intelligence 'huts' at Bletchley to the military operation centers [sic] in London. There are significant chronological gaps in these records, and the critical indexes plus many original intercepts have been destroyed.[1]

Because so many of the original intercepts have been destroyed their digests, which come from many different sources including human intelligence (HUMINT – intelligence gathered by people), do not allow the researcher easily to identify the cypher or the code decrypted for any given interception. The gaps create additional problems. Christopher Andrew, considered a leading historian of the British intelligence community, has described the British public archives as 'laundered'.[2] Once a more critical approach is adopted there is thus ample space for new insights and discoveries.

Banners and Bugles

The four well-known volumes – five if one takes into account the division of Volume 3 into two parts – of Hinsley's *British Intelligence in the Second World War: Its Influence on Strategy and Operations* and its companion *Codebreakers: The Inside Story of Bletchley Park* (see Sources) are of debatable value. Hinsley was an analyst, not a decryption specialist, and there are a number of occasions, particularly when he writes about

Sailors on board an Italian torpedo boat working on a crossword puzzle during the summer of 1941. The activities of the *Regia Marina*'s codebreakers were sometimes not too different. (Signal 1942, Deutsche Verlag)

Italian signals, where he relies on his own generic recollections and opinions without verifying wartime decryptions and appreciations against available sources. In the process he commits errors which undermine the value and objectivity of his account.

For example, on page 210 of Volume 1 (*British Intelligence*), it is stated:

> The Italians had used a version of the Enigma machine carelessly during the Spanish Civil War. In 1940 they brought an improved version of it into use for the Navy, and it was this which was broken on September 1940.

> Unfortunately, it carried only one or two messages a day up to the summer of 1941, when it was withdrawn from naval use and confined to the traffic of the Italian SIS (the Italian Navy's intelligence unit).

This is incorrect. In December 1936 the Germans gave their Spanish and Italian allies 20 commercial Enigma mechanical cypher machines to allow the three navies to communicate with each other. In March 1937 British intelligence bribed an Italian operator, inducing him to pass, between 1937 and Spring 1939, the internal and external keys of the Enigma machine to GC&CS so that wireless traffic could be easily read. However, after their Italian agent was transferred to other duties, the British lost their ability to decrypt that machine, and the often-related 'rodding' technique invented in 1937 by the British veteran cryptographer Dillwyn 'Dilly' Knox to break messages encyphered on Enigma machines that did not have a plug board proved to be worthless until September 1940 when, by chance, the 19-year-old Mavis Lever, working at Bletchley Park, found a crib. The *Regia Marina* never purchased an improved version, as claimed by Hinsley, and the machines were used only by the Italian SIS, between 1940 and 1941, and never on warships, as there were only ten machines available.

On page 209 of Volume 1 Hinsley credits the identification of call signs and direction-finding intelligence, as well as signals in the Italian Fleet code and in plain language, for revealing Italian intentions prior to the July 1940 fleet engagement off of Calabria:

> [It] enabled [Admiral Cunningham] to establish the intention of the Italian C-in-C, which was to lure his opponent into a submarine and aircraft trap by waiting back off the Italian coast.

The reference for this assertion is to the Royal Navy Staff Summary of the Battle of Calabria, SO Playfair's official history of the Second World War in the Mediterranean published by HMSO, and a report by the Mediterranean Operational Intelligence Centre. However, there were no broadcasts in Italian Fleet code or plain language that were intercepted about the alleged submarine/aircraft trap, nor were there any such messages in the original 'Action off Calabria 9th July 1940 – List of Enemy

I gruppi da battaglia in Mediterraneo Orientale da A fino a D per le operazioni M.A.5 ricevono dal C.C. Mediterraneo l'incarico di arrivare fino a 75 mg. a ovest della Sicilia.
Per questa operazione sono indicati i giorni da 1 a 4.
I gruppi B e C (il gruppo C ha lasciato Alessandria il 7/7) devono trovarsi il giorno "3" alle ore 0600 in: lat.35°40' N. – long.20°30' E. Alle ore 1400 dello stesso giorno i gruppi "A" e "B" devono trovarsi in: lat.36°30' N. – long. 17°40' E. ed alle ore 1800 trovarsi come segue: Il gruppo "A" in lat.37°20' N. – long. 16°45' E. – gruppo "B": lat.37°00' N long.17°00' E. – gruppo "C": in lat.36°20' N. – long. 17°00' E.
Il giorno "1" è probabilmente il 7 luglio (partenza del gruppo da battaglia da Alessandria. Idrov. partiranno per pattuglia da Malta ogni 50 minuti.

SUPERMARINA

C. S. M. 3 - Capo di Gabinetto 5 - O. A. 7 - I. A. 9 - S. I. M. 11 - S. I. G. 13 -
S 4 - Ufficio Trattati 6 - M. D. S. 8 - V. N. 10 - S. I. A. 12 - Stamage 14 - M.N.T.

The Cunningham signal decrypted on 5 July 1940 by the Italian Navy which set in motion, four days later, the action off Calabria. (Ufficio Storico Marina Militare)

The cypher office of the battleship *Littorio* in 1942. On the table are an Enigma machine to read the Luftwaffe signals and a C 38m cyphering machine. (Enrico Cernuschi, courtesy of Cdr Vittorio di Sambuy)

signals Intercepted' record available at The National Archives (TNA), ADM 223/121. The five Italian submarines present that day were not deployed to trap the British fleet; they were about 60 miles south of the area where the Action off Calabria was fought. As this information was readily available in *I sommergibili nel Mediterraneo*, Vol I, published by the *Ufficio Storico della Marina Militare* (USMM – the Italian Navy Historical Branch) in 1967, it would not have been difficult to check this detail and verify that the whole story of the submarine trap which Admiral Cunningham adroitly avoided was simply a myth.

In Volume 2 there are further examples of incorrect or misinterpreted information regarding Italian intercepts. On page 22 Hinsley writes:

> The Italians had adopted [the C 38] system[3] under pressure from the Germans, who claimed that the existing Italian cyphers were vulnerable.

This is again incorrect: the Italians did not adopt C 38

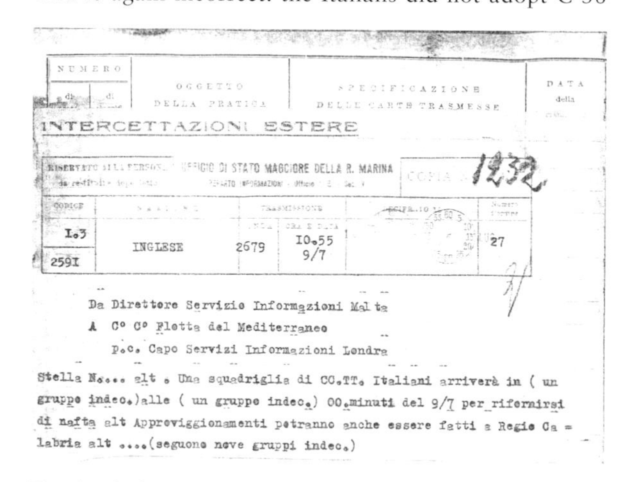
NUMERO | OGGETTO DELLA PRATICA | SPECIFICAZIONE DELLE CARTE TRASMESSE | DATA della

INTERCETTAZIONI ESTERE

UFFICIO DI STATO MAGGIORE DELLA R. MARINA
REPARTO INFORMAZIONI

COPIA 1232

CODICE		TRASMISSIONE			Numero
I.3 / 2591	INGLESE	2679	10.55 9/7		27

Da Direttore Servizio Informazioni Malta
A C° C° Flotta del Mediterraneo
P.c. Capo Servizi Informazioni Londra

Stella N.... alt . Una squadriglia di CC.TT. Italiani arriverà in (un gruppo indec.) alle (un gruppo indec.) 00.minuti del 9/7 per rifornirsi di nafta alt Approviggionamenti potranno anche essere fatti a Reggio Ca = labria alt ....(seguono nove gruppi indec.)

The signal of 14 July 1940 which betrayed the fact the British were reading one of the most important Italian naval cyphers. (Ufficio Storico Marina Militare)

because of German pressure. The British knew from their *Kriegsmarine* sources that the Germans had proposed in December 1940 that the Italians adopt their Enigma system, and that the Italian Navy had refused. The decision made by the *Regia Marina* in the summer of 1940 to buy ten Swedish C38 machines off-the-shelf was dictated by the cutting of the telegraph cable between Syracuse and Tripoli by two British submarines on 8 August 1940. This left Italy with only the cable between Syracuse and Benghazi. Laid in 1912, that surviving telegraphic link was much deeper and thus secure, but it could not handle all communications traffic and had to be supplemented, for naval administrative signals, by wireless traffic. This, for reasons of speed, was machine cyphered, a practice the *Regia Marina* had long resisted. That last cable, cut and concealed on February 1941 when Benghazi was occupied by the British, was recovered on April 1941. However, it became worn out in spite of renewed maintenance and ceased transmitting once and for all in November 1941, exposing all communications to interception by the British. From late December 1942, however, the Italian Navy reactivated and relaid in the direction of Italy the two cables between Tunis and Malta which the Italian cable layer *Giasone* had cut on 14 June 1940. This meant that communications between Tunisia-Sicily and the Italian peninsula could no longer be intercepted over the airwaves. From that point until the armistice of 8 September 1943 the importance of 'Ultra' intercepts of Italian signals declined dramatically.

Page 324 has a minor error of the type commonly found in histories, but which provides an interesting example of the mistakes which can result a wrong analysis of a correct decryption. Hinsley writes:

> C 38m had disclosed at the end of October that the Italians had carried out a large-scale minelaying operation (operation B) off Benghazi ...

In fact, although the operation had been planned and three cruisers, accompanied by six destroyers, had sortied from Taranto on 12 October 1941, the following day, after two British battleships were sighted 130 miles away on a converging course, the Italian squadron returned to base with the mines still on board. This turn of events had been a complete accident, as the message had been decrypted (or analysed) only at the end of October, while the mission was scheduled for the night of 13/14 October. USMM's book *La guerra di mine* (Eng: 'The Mine War') tells the whole story on page 239. The book was published in 1966, so it would have been easy for Hinsley to discover that this minefield was never laid. In the event, the decrypted signal spared Benghazi any threat of bombardment from the sea until a minesweeping operation undertaken by the Royal Navy on January 1942 off that harbour demonstrated that the minefield did not exist.

On page 348 Hinsley states:

> Early on May 1942 C 38m was immediately useful in helping to avert a further disaster at Alexandria. On 5 May it gave a precise warning that there was to be another attack on that same base by Italian human torpedoes, the chief target being the floating dock in which the *Queen Elizabeth* was being made seaworthy. The attack was made on 14-15 May and was duly frustrated.

No DEFE ('Ultra') file is cited to identify the decryption that was the source of this warning and the source Hinsley quotes is Playfair, which was written pre-'Ultra'. In fact, a C 38m signal had been decrypted but was evaluated correctly only on 3 August 1942, helping to avoid a similar, planned attack against the submarine base at Haifa. The X MAS attack craft action against Alexandria of May 1942 had been frustrated by a navigational error by the three human torpedoes crews; the effort caught the British by surprise.[4]

Arguably, Hinsley's biggest error from a cryptology point of view is his statement on page 283:

> The C 38m was a most invaluable acquisition. Almost all the traffic could be intercepted in the United Kingdom. In the year after it was broken the Italians made steadily increasing use of it: the number of signals rose from 600 in August[5] to nearly 4,000 in July 1942, the peak month. As it was often used by Italian naval shore authorities and fleet units, it often provided information about the intentions of the enemy's main fleet. It was also used by non-Service organizations such as the Italo-French Armistice Commission for general purposes.

In January 1983 Admiral Luigi Donini, head of the Italian Navy's British decryption section[6] between 1939 and 1979, published an article in *Rivista Marittima* (The Italian Navy Staff monthly) in which the wartime records of the external C 38m keys were reproduced. Between June and October 1942 the naval commands of Libya, plus Tripoli and Benghazi, had a total of just 3,100 exterior keys available for messages over 122 days; each of these keys could be used only once, either to broadcast or to receive a single signal. The figure of 3,100 was less than the total of C 38m decrypts Hinsley claims for 'Ultra' for July 1942 alone.[7]

Once the material available at TNA is examined, it is easy to appreciate that more than 70% of the so-called 'Ultra' C 38m decrypts involving Italian Navy activities (labelled ZTPI) are, in fact, German Enigma machine signals or messages emanating from the C 35 equipment operated by the *Regia Aeronautica*.

In 1941 the Italian Air Force purchased from the Germans hundreds of ex-French Army Hagelin C 35 cypher machines to speed its wireless communications. Judged insecure by the Italian Navy codebreakers in December 1935, the C 35 was a 5-rotor system with fixed lugs and a maximum period of 3,900,225. The C 38, having six rotors and 131 mobile pins for the internal key, had a theoretical total of 101,405,850 and a practical total of about 100,000, as it was necessary to avoid both excessively easy solutions and the possibility that the initial part of a message would be casually encyphered with a portion of keystream which duplicated the tail of a past or future message. Since the total of possible daily keys for the *Luftwaffe* electromechanical Enigma – which, from May 1940, was routinely broken by the first electromechanical 'bombes' within a few hours after midnight, allowing the British to read in real time all signals of the day by breakfast – was about 159 million million million,[8] the C 35 was clearly no match for the British Government Code and Cypher School at Bletchley Park.

The British labelled C 38m a 'medium-grade' cypher. There were never enough bombes at Bletchley Park, and German traffic had higher priority; British decryption efforts focused on that system from July 1941 initially required weeks before the discovery of the month's internal key and the consequent entire keystream. Once this goal had been achieved, the following step was to do all the necessary trials to discover the internal key. The interruptions at the beginning of any month were later reduced. By the end of 1941 three days were sufficient, and by the end of 1942 it took less than 24 hours for the generation of timely decryptions.

A working model of the bombe used to crack the Enigma cypher at Bletchley Park Museum. Each of the rotating drums simulates the action of an Enigma rotor. There are 36 Enigma-equivalents and, on the right-hand end of the middle row, three 'indicator' drums. (Antoine Tavenaux)

Two serious crises in this process are quoted in *British Intelligence*: the first in March 1942 and the second in October 1942, both resulting from the introduction by the Italian Navy's *Servizio Telecomunicazioni*[9] of the masking of the indicators of the external keys in the signals. In the meantime the *Regia Aeronautica* used its C 35 excessively, not only for tactical messages (as originally planned) but for everyday communications, because of its speed and ease of use. This continued despite the regular protests of the *Regia Marina*.[10] The Italian Navy used the C 38m for administrative traffic only; the machine was not used by individual warships nor by the commands at sea, which could only receive and read signals sent to them (generally circular letters) encrypted with that system (but not broadcast by that cypher); for transmissions they used only their classic naval codes, which the British could not read after 14 July 1940. The average number of messages encyphered by the single commands with the C 38 was about two in any 24 hours (and not every day), and they were usually administrative in nature.[11] In December 1941, 'Ultra' conducted a lengthy investigation following the intercept of a series of C 38 signals signed by Admiral Inigo Campioni, then Governor of Rhodes, and sent to the Germans and the Bulgarians. Bletchley Park was convinced that the signals related to a pending Axis combined operation before discovering, after a week, that the Admiral's valet had lost a sea chest of the governor's best silk shirts somewhere between Sofia and Athens.[12]

The inflated importance Hinsley assigned to C 38m is the product of confusing Italian naval SIS's commercial Enigma machine-cyphered messages generated from March 1941 with German messages, Italian Air Force C 35 messages and the few genuine Italian naval C 38m messages. Hinsley's work provides a distinctly shaky foundation for appreciating the Mediterranean intelligence situation, but the understanding of 'Ultra' and Italian signal traffic deteriorated even more in the wake of the publication of Hinsley's *British Intelligence*. In lectures subsequently delivered by the author, statements such as the following were made:

> ...the Germans and the Italians assumed that we had 400 submarines whereas we had 25. And they assumed that we had a huge reconnaissance air force at Malta, whereas we had three aeroplanes!

and:

> ...we instructed our people when interrogated by Germans – our pilots for example – to propagate the view that we had absolutely miraculous radar which could detect a U-Boat even if it was submerged from hundreds of miles. And the Germans believed it.[13]

Clearly Hinsley is exaggerating the impact and importance of 'Ultra' beyond its actual achievements and is journeying into the realm of fantasy. When both sides of the story are considered the picture becomes more complex.

The Other Side of the Hill

The decrypted messages (every origin and nationality) relating to Italian naval affairs between 1940 and 1943 in The National Archives number 42,163 from mechanical and electromechanical machines (these last German only) and 3,699 hand cyphers. These messages were intercepted between 1 April 1940 and 9 September 1943, but only around 50% were actually decrypted during that period; the remainder were decoded in 1944–1945 mainly for historical and statistical purposes. The times of codebreaking varied between real time (once the machine cypher keys had been broken) to several days for the much harder nuts to crack: messages composed using the classic cypher books. Although a comparison between signals and the different sets of data is always difficult, as routine radio interceptions made in lower-grade codes are included, between 10 June 1940 and 1 September 1943 Italian Navy codebreakers intercepted and decrypted 36,262 British, 209 French up to 24 June 1940, 4,002 French post-armistice, 989 Yugoslavian, and several hundred American, Greek, Free French, Russian, and Turkish signals.[14] The times varied from less than 15 minutes for the RAF reconnaissance signals (reduced to under two minutes from January 1942 when the first IBM machines modified in Italy became operative for decrypting purposes) to hours or days in the worst cases.[15]

Lacking the appreciable assets the British government dedicated to the GC&CS, the Italian Navy had relied from February 1922 on the ingenuity of two very young lieutenants, Giorgio Verità Poeta and Luigi Donini, who had demonstrated – quite by chance – a special talent for this kind of work. It was a classic, Italian-style artistic improvisation. On occasions this excessively economical method paid dividends, at other times not. Backed by a small budget and supported by a small team of seven people including archivists, the two officers were able, during the 1920s, to achieve some occasional results against diplomatic cyphers. Their greatest achievement during that decade, however, centred around a very dangerous double mission in 1928: one in the Arctic on the cableship *Città di Milano* and the other in the Black Sea. A large amount of Soviet cypher traffic was intercepted and broken. This success allowed Mussolini to read at breakfast, over a period of almost ten years, the coded instructions Moscow broadcast to its agents in Italy, France, Germany, Austria and in the Balkans, enabling him to destroy the Italian communist clandestine network in 1930, 1932 and 1938, and to evade successfully two attempts to his life.[16]

Admiral Alberto Lais in January 1940. He was the head of the intelligence service (SIS) of the *Regia Marina*, becoming naval attaché in Washington in 1940–41. (Ufficio Storico Marina Militare)

The then-Lieutenant Giorgio Verità Poeta in the early 1930s. (Courtesy Verità Poeta family)

The activity of these two officers was initially on a part-time basis, as they had to serve periodically at sea in order to advance their respective careers. However, in 1931 the new head of the *Reparto Informazioni dello Stato Maggiore della Marina* (later known as the *Servizio Informazioni Segrete* or SIS), Rear Admiral Alberto Lais, decided to create a professional cryptological section led by Verità Poeta and Donini. Two years later the new secretary of the Italian Navy (and, in the following year Chief of Staff), Admiral Domenico Cavagnari, put the organisation on a war footing, having appreciated that Mussolini's political strategy would inevitably lead to conflict with Britain, and possibly France, within a few

The Italian Naval Ministry in Rome in 1940. The codebreakers' offices were in the left corner of the Palazzo Marina. (Ufficio Storico Marina Militare)

years. The first target was the main French naval cypher, the TMB. It was considered impregnable, but it was cracked within a few weeks. A lengthy game of bridge between the two officers and two ladies suggested to the male couple the secret of that cypher, the key to which was based on the classic combinations of diamonds, hearts, clubs and spades. The new TBM 2 and 3 versions adopted the following year were quickly broken, and until June 1940 that cypher granted high dividends to the Italian Navy.

The next goal was, of course, the British. In November 1934 the Royal Navy administrative Naval Code was first broken. Twenty-four hours were originally needed to reconstruct the keys, but from 1935 this delay was reduced to a few hours.

The much more complex and more important Naval Cypher proved, however, to be impossible to break. It was necessary for Verità Poeta and Donini to spend three long seasons at sea, between 1936 and early 1938, in the Mediterranean and in the Atlantic Ocean on two spy ships, the civilian motor fishing vessels *Pegaso* and *Procione*, to intercept and read a large amount of Royal Navy wireless traffic and to accumulate a large enough body of signals to work with. The solution was found by Verità Poeta in January 1938, after a full night spent in his office on the top floor of the Naval Ministry, just a flight of stairs up from Cavagnari's office. The next morning a very shabby Verità Poeta, with a two-day growth of beard and bloodshot eyes, was sitting drinking his tenth cup of coffee in ten hours while the tiny and always elegant Cavagnari and the large Lais were reading in amazement a series of messages transmitted the previous day between Admiral Pound, then C-in-C of the Mediterranean Fleet, and the Admiralty. Verità Poetà had discovered the 'impossible' key word used by the British to superencypher the letters in numbers for any signal they sent: Trafalgar!

There were gaps, of course, but on September 1938 Lt-Cdr Francesco Camicia, commander of the sloop *Lepanto* stationed in China, was able to rent, and photograph over the course of several hours the first of the two books of the Naval Cypher (the so-called 'dictionary') from a NCO of the destroyer *Decoy*.[17] This he sent home on the liner *Conte Biancamano*.

That cypher remained in use until 20 August 1940. It was necessary to reconstruct periodically, by cryptological systems, the new supercyphers introduced every month, but it was a task the mathematical methods conceived by Donini and Verità Poeta could always handle in less than a week. The cryptology section was also increased between 1938 and 1939 with the appointment of a dozen codebreakers, including a sort of magician of tactical signals, Lt Eliso Porta, nicknamed '*Il principe azzurro*' (Prince Charming) for his successes

Lt-Cdr Luigi Donini (right) and Captain Mario De Monte, head of the Italian Navy's cryptographic bureau, visiting a German WT interception station in the Spring of 1942. (Ufficio Storico Marina Militare)

Lt- Cdr Donini and Captain De Monte entering the German interception center. (Ufficio Storico Marina Militare)

with both young women and foreign tactical codes classified with female names – 'Boadicea' and 'Rowena' for the British.

On 22 October 1939 Verità Poeta, who had received the Gold medal from the King – the reason was kept secret – died in Rome, poisoned with antimony. Two days later Donini had a miraculous escape when his car plunged off the Parioli Hill in Rome; an inquiry discovered the brakes had been sabotaged. On January 1940 it was the turn of Captain Eligio Giacopini, the head of the Planning Office of the Navy and Cavagnari's right-hand man: antimony again. By this time Italy had so many potential enemies – the British, the Russians, the French and even the Germans – that responsibility was never established. However, it was clear that war was coming.[18]

Tit for Tat

Among the most important and best documented successes gained by the *Regia Marina*'s codebreakers was the message, decrypted on 5 July 1940, from Admiral Cunningham to the Admiralty announcing his imminent sortie to attack the Sicilian coast on 9 July and listing in detail his forces. It had been an ill-advised idea to broadcast such a comprehensive warning, especially since the cables between Alexandria, Malta and Gibraltar (and from there to Britain) remained operational until 16 August 1940, when the last link between Malta and Gibraltar was finally located and cut by the MFV *Orata* (a twin of *Pegaso* and *Procione*). It is possible that Cunningham had been provoked by some earlier Churchillian request for 'Action Now'!

On 14 July 1940, after just 100 hours of effort, a Royal Navy message revealed that the British had been able to decrypt a signal broadcast using one of the two new, main Italian Navy cyphers introduced on 2 July 1940, the SM 19S. An emergency order was issued to the effect that, for the superencyphering of the SM 19S, the initials of the captain of any vessel was to be used pending delivery of a new (currently reserve) cypher to all ships before the end of that month. As Hinsley acknowledged,

The office of the Italian Navy's Chief of Staff at *Supermarina*, December 1941. The armchair and the office remain the same to this day. (Ufficio Storico Marina Militare)

'...the cyphers used by [the Italian] fleet for most of its important communications were never read again after July 1940 except for a few brief intervals as a result of captures after the middle of 1941'.[19]

The action off Gavdos Island on 28 March 1941, in which the light cruisers of the British Mediterranean Fleet were confronted by the bulk of the Italian Fleet, was not an accident but the fruit of another Royal Navy decrypted message. The loss of the British destroyers *Lively*, *Kipling* and *Jackal* on 11 May 1942 – the first was sunk by the FIAT BR.20 bombers of the 35° Stormo and the remaining two by German Ju 88 dive bombers – and the night ambush of the British 'Hunt' class flotilla based at Alexandria off El Daba on 29 August 1942 by a pair of MTSM small attack craft belonging to X MAS, which resulted in the torpedoing and total constructive loss of HMS *Eridge*, were likewise courtesy of the efforts of Italian Navy codebreakers.[20]

HMS *Eridge* entering at Alexadria after being torpedoed by an MTSM attack craft belonging to X MAS. Declared a total constructive loss, she later served in Egypt as a depot ship. (Ufficio Storico Marina Militare)

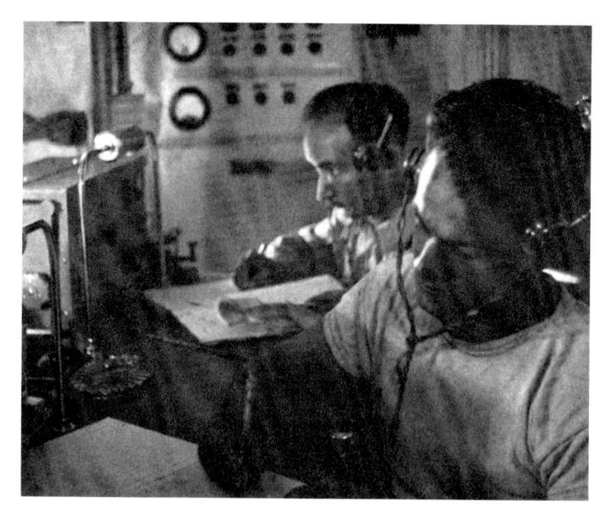

Italian Navy wireless operators at *Supermarina*'s radio centre in 1941. (Ufficio Storico Marina Militare)

The real game, however, was centred around the convoys for Africa. 'Ultra' often facilitated attacks against that traffic by reading the German and Italian Services signals relating to schedules, routes, cargo, and escorts. However, British signals intercepted by the *Regia Marina*'s *Reparto Informazioni*, decrypted regularly within two to five minutes, the SYKO and NYKO messages sent by the RAF and FAA aircraft (and, sometimes, the Royal Navy Naval Cypher messages exchanged between the British submarines and Malta) often allowed *Supermarina* to order immediate countermeasures: redirecting convoys, alerting defences, or allowing the dispatch of reinforcements to the convoy. The claims frequently made by British historians are clearly exaggerated, and Axis losses at sea were much reduced by the work of Italian codebreakers. The Naval Cypher and Naval Code messages were naturally much harder to crack than the SYKO and NYKO traffic, but it was the choice of targets along the enemy chain of command combined with the wise use of the limited assets available, not to mention the efficiency and security of the related communications, that allowed the whole system to achieve these results on a continual basis right up to the Italian armistice of 8 September 1943.[21]

(3) On 10th August, the parent submarine "Scirè" was sunk off Haifa with her human torpedoes on board. Her intentions and her approach had been elaborately followed by Special Intelligence, and she was destroyed according to plan (N.O.I.C. Palestinian Ports 2300C 10.8.42; ZTPI/13518, 13687, 13956, 14018, 14032, 14033, 14068, 14410, 16015; CX/MSS/1258/T19, 1265/T3, 1271/T23, 1275/T7).
(4) I-III/KG 6 from the west, KG 76 from the east.
(5) I and III/KG 26 (torpedo aircraft), I-III/KG 30 (high-level bombers).
(6) I/KG 60 (Ju 88).

The series of decrypted signals which caused the loss of the Italian submarine *Scirè*. (Ufficio Storico Marina Militare)

A poor quality but unique photo of the Sunderland boarded by the Italian torpedo boat *Pilo* on 6 August 1940 off the Libyan coast; its Syko machine and the accompanying documentation were seized. (Ufficio Storico Marina Militare)

Mirror, mirror ...

One of the other pillars of 'Ultra' orthodoxy is that the Axis powers never realised that their traffic was being read. The British went to great lengths to preserve the 'Ultra' secret', but there is also an implication that the Germans were too arrogant and the Italians too stupid to understand what was being done. RA Ratcliff writes:

> [It was] arrogance and complacency about security, as well as the cultural and structural pressures, such as rigid signalling procedures, that limited success of German organisations and their staffs.[22]

The Italians often fare even worse. David Alvarez writes:

> ... the uneven record of Italian signals intelligence in the period 1915–1943 reflected internal problems more than external. The successes of the prewar period were not duplicated during the war, because Rome's services failed

SEGRETO 6

473
1000 N. 27/8/1943
Ore .... del ....

STATO MAGGIORE DELLA R. MARINA
S. I. S. Λ.M.

RAPPORTO SULL' ESAME DEL TRAFFICO INTERCETTATO

1) - Alle 2I15 di ieri un'unità, che risulterebbe uno degli incrociato= ri di base a Malta (tipo DIDO), ha segnalato l'avvistamento di un Smg. in 37° I5' N. - 08° 24' E. - Subito dopo l'unità segnalava che un piroscafo o petroliera era stata silurata e veniva informata del= l'invio di soccorsi da Bona.-

2) - Il convoglio DUNNOCK, che risulta essere il grosso convoglio entra= to recentemente in Mediterraneo, alle I825 di ieri segnalava di esse= re attaccato da aerei in 3708 N. - 08I8 E. mentre procedeva per Est alla velocità di 8 mg.- Alle 2I57 Bona gli inviava scorta aerea.

3) - Movimenti imprecisati risultano in corso fra Malta e Biserta e dal la Sicilia per,ponente, probabilmente verso Orano.-

4) - Unità di un convoglio che si trovavano ieri nel Mediterraneo Cen= trale risultano dirette ad Alessandria.

5) - Movimenti navali imprecisati risultano in zona Tripoli.-

6) - Alle I9I4 di ieri un'unità rilevata in zona Malta ha comunicato co un Comando Divisione Incrociatori, col Com. 2 Squadra da battaglia e Marina Algeri.- Si ritiene trattarsi di una unità sede di comando complesso proveniente o diretta ad Algeri.-

7) - Traffico dia convogli scortati da aerei in zona Siria-Palestina.=

August 27 1943: business as usual for the *Regia Marina*'s codebreakers. (Ufficio Storico Marina Militare)

The scuttling of HMS *Havock* at Cape Bon after her grounding. The photo was taken on 6 April 1942 at 1440 by the MAS boats of Pantellaria which, a few minutes later, boarded the wreck and recovered part of her secret achive. (Enrico Cernuschi)

> to adapt to the new cryptologic world created by the conflict. The Italians maintained the tradition of 'chamber cryptanalysis' although that cryptanaytic model was increasingly anachronistic. In the late 1930s cryptanalysis embarked on an organizational and technological revolution that the war would only accelerate. The Italians, never grasping the significance of this revolution, were left in the dust of those who did.[23]

The truth, of course, is a little different and worth relating at length.

Flying Colours

On 1 May 1941 at the intelligence interrogation centre of Heliopolis (Cairo, Egypt), a British interrogator was interviewing – in French – an Italian officer, the young sub-lieutenant Percivalle Levaro, a survivor of the cruiser *Pola* sunk at Cape Matapan. During that laboured conversation (the French of the Italian prisoner was not good, and he seemed unwilling to talk, possibly because he had a broken arm), a gentleman entered and was warmly welcomed by the interrogator. Speaking at their ease the newcomer said he had been torpedoed in the Atlantic on an auxiliary cruiser, the *Comorin*, finally arriving in Egypt on a cruiser that had forced the Sicilian Narrows. His task was to check and teach someone the techniques developed in Britain to decrypt, with almost no delay, the cyphering machines used by the Germans. Their animated conversation, which resulted from having so many things to tell each other after such a long time, meant that the unproductive interrogation of the prisoner

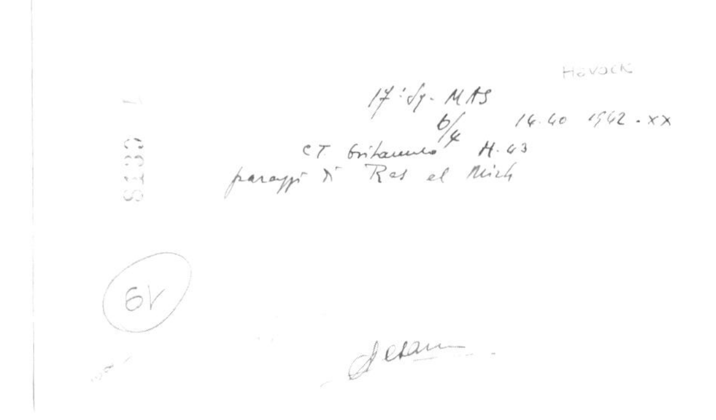

The original caption for the *Havock* photo written by personnel on board the MAS boats. (Enrico Cernuschi)

was abandoned, and Levaro was sent back to the Geneifa Prisoner of War Camp No 306.

The real first name of Levaro was Percy (Percivalle was an Italian version of Percy, or Parsifal, imposed by Mussolini's effort to eliminate foreign names), and he was fluent in English, having had a British nanny when he was a child – as was the case with many Italian naval officers. He had overheard many details about a park, a school and the Germans' foolish habits. Once back in the PoW camp he reported – outside the huts, as they were bugged – to Major (Medical Naval Corps) Pietro Cuscianna, who was likewise from the *Pola*. There was no possibility of escape for Levaro, whose right arm was in a sling, and the final choice fell on Lt Luigi Tomasuolo, third gunnery officer of the *Pola*, a good-humoured type from Rome who had a taste for adventure (Tomasuolo had saved a sailor from drowning the night of Matapan, keeping him afloat until they were rescued by the British).

No complicated plans were made. Once instructed and tested about his recall of every single word overheard by Levaro, Tomasuolo spent two days in his camp-bed with some tobacco leaves under his armpits. The irritation simulated a high fever and he was sent to a military hospital in Cairo. From that moment he had to improvise, armed only with the addresses of the two Catholic churches in the city. The biggest problem for Tomasuolo was that he spoke excellent French but only basic English, although with his stature (tall) and colouring (blond and with blue eyes) he could have easily passed himself off as British. He escaped from the hospital that very night dressed in a British Army officer's uniform freshly delivered from the laundry and of about his size stolen from a nearby room, together with a cap found on a peg in a corridor. If he had been discovered in this disguise he would have been shot, but it worked. When he was outside – luckily, in the camp he had learned how to salute in the British way with the right hand palm facing forwards – he was able to climb unnoticed into a truck after hearing an Egyptian driver say, in French, that he was going to Cairo. Once there, thirsty but still immaculate in his uniform, he was able to read the arrows for the island of Zamalek, where at sunset he entered St Joseph's Roman Catholic Church.

Again he was fortunate. He had been given two possible addresses: St Joseph's and the nuns of the Sacred Heart. The latter location had been raided the previous week by the British Military police, after a Maltese *agent provocateur* had compromised those women by asking for help and caused them to be interned. The Franciscan Friars of Zamalek, two of whom had served as Italian Army chaplains during the Great War, ferried Tomasuolo that same evening, on a cart, to the small villa of a Spanish gentleman, Don Eduardo García, married to an Italian, Clelia Bandini. Here Tomasuolo was later joined by four other Italian officers who had tunnelled out of the Geneifa camp: Midshipman Gennaro Pipitone, Grenadier Captain Umberto Rizzitano and Lieutenants Giorgio Pozzolini and Pasquale Landi, Royal Horse Artillery. Tomasuolo did not tell them his secret, and during the summer the five conceived and rejected numerous escape plans. In September Pipitone was arrested by the Egyptian police during a sortie, but his cover story worked and the other four fugitives remained unmolested.

The course of action finally adopted by Tomasuolo and Rizzitano was the most dangerous: to leave Egypt by sea, embarking on a Turkish steamer with the passports of their generous Spanish hosts, the father and his son; new photographs were substituted for the originals and an embossed stamp improvised for the photo corners.

On 30 October 1941 the couple were at Alexandria and the steamship *Talodi* sailed at noon. The voyage was a nightmare. The rusty tramp steamer embarked on a long and, above all slow crossing, calling in at every possible British-occupied harbour from Gaza to Cyprus. Finally, in late November, they arrived at Iskenderun in Turkey. Their last money (courtesy of Sñr García) was spent on a passage to the mole in a rowing boat; they then made their way to the Italian consulate. Crossing Turkey with new, faked documents took another month. Once they arrived in Istanbul, the Italian naval attaché, who was immediately informed by Tomasuolo about the British ability to read the German 'machine' on a daily basis, preferred not to transmit this information over the airwaves or by telegram, and opted instead to purchase two second-class tickets for his guests on the Simplon Orient Express, while a decoy couple would make the trip on the next train in a very publicly-reserved luxury car.

Tomasuolo and Rizzitano finally crossed the Italian frontier on 10 December 1941 during a snowstorm and took another train to Rome. Here they separated on the platform and the Italian lieutenant, still in plain clothes, caught a cab and arrived at the Navy Ministry, which he attempted to enter as a member of the public by passing through the public entrance between the 'Ancore'.[24] He was permitted only to give reception his name and rank before the *Reali Carabinieri*, the Italian Military police,

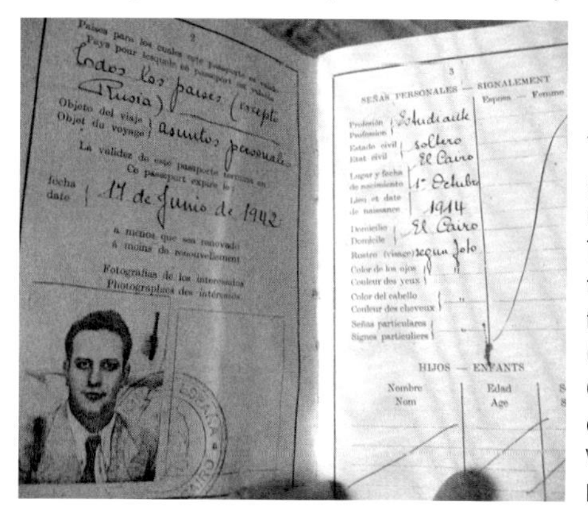

The Spanish passport used by Lt Tomasuolo to escape from Egypt. (Enrico Cernuschi, courtesy of Valerla Isacchini)

Rear Admiral Franco Maugeri in 1943. (Ufficio Storico Marina Militare)

Lt-Cdr Tomasuolo on the day he was decorated by King Victor Emanuel III with the silver medal for gallantry. (Ufficio Storico Marina Militare)

26 June 1942: Mussolini with Admiral Riccardi, Chief of Staff of the Regia Marina, and Admiral Da Zara on board the cruiser *Montecuccoli*. The favourable outcome of the action off Pantellaria (Operation 'Harpoon') of 15 June 1942 was a direct result of the intelligence provided by SIS from 2 June. (Courtesy of Enrico Cernuschi)

seized and handcuffed him. He would spend the next two weeks in Forte Boccea, the military prison.

Tomasuolo was the first Italian Navy officer to escape from a British PoW camp and return home. Before him there had been other episodes, such as the one involving two NCOs who had left Tobruk in January 1941 after the fall of that town, reaching Rhodes in a small sailing boat, but in the case of Tomasuolo the Army counterespionage unit, informed about his arrival at Istambul, suspected that this exploit could be a British attempt to infiltrate a double agent. It was only on 22 December 1941, after his story had been checked and re-checked, that the door of his cell was opened by a smiling Admiral Franco Maugeri, the head of the SIS, and by two sailors, one of whom presented him with a brand-new uniform with the bars and curl of a lieutenant-commander on the cuff and a pin marking his promotion for *meriti di guerra* – an honour which could only be earned in the field.

Following a good lunch and a meeting with the Chief of Staff of the Navy, Admiral Arturo Riccardi, General Ugo Cavallero, Chief of Staff of *Comando Supremo*, was informed that same afternoon that the German Enigma machine was compromised. Cavallero formally told German Field Marshal Kesselring that very afternoon that the British were able to read all (the word 'all' being emphasised) their cyphered messages, including those encrypted by their Enigma machines.[25] However, as usual the Germans refused to believe their ally. The deep mole the British had planted in the *Deutscher Verbindungsstab beim Admiralstab der Königlich Italienischen Marine* (the German Naval Command in Rome) since the summer of 1940 informed his handlers that same night of this new development.

The British response, which had long been planned in readiness for such an eventuality, was swift – possibly too swift according to Maugeri. On 23 December 1941 the German and Italian army counterespionage services were both informed, through different channels, that the security leaks which had hampered the convoys to Libya that autumn had originated not from codebreaking but from HUMINT – the British and Poles had agents in the Taranto, Genoa, and La Spezia areas. In order to persuade their Axis counterparts of the accuracy of this information, some real Allied agents were betrayed, including the old man who had sold, during the Spanish Civil War, the internal and external keys of the Enigma commercial machine. Even a woman, Laura d'Oriano, previously active at the Italian Atlantic submarine base in Bordeaux (France) was sacrificed to protect the 'Ultra' secret. She was arrested at Genoa just after receiving from a well-known British agent in Switzerland an express letter with detailed instructions to confirm information from other agents about the Italian and German traffic with North Africa. D'Oriano was shot by a fascist firing squad on 16 January 1943.

The Germans were thus effectively deceived, as was the Italian Air Force. The *Regia Marina*, however, preferred to believe its own man, particularly as his revelation appeared to confirm the conclusions made by Lt-Cdr Marc'Antonio Bragadin in his July 1941 investigation into the signals broadcast before and during the disastrous Matapan sortie. Bragadin had concluded that the leak of Italian intentions to the British could only have come from the *Luftwaffe* messages. On 30 December 1941 the Italian Navy made the decision to discard the remaining commercial Enigma cypher machines used by the SIS, which dated from December 1936. The final signal stated that they were worn out and were to be destroyed.

Tomasuolo was received by Mussolini on 11 March 1942, together with Rizzitano. When they entered his office at Palazzo Venezia, they were surprised to find present their former colleagues Landi and Pozzolini, who had returned from Egypt via a different route. Tomasuolo asked to return to sea with the fleet, but the dictator replied that fortune must not be challenged too many times. Appointed to the Italian naval base of Pola he became, from 8 September 1943, a hero of the anti-German resistance in Rome and then in northern Italy. Promoted to admiral by the early 1960s, he commanded the NATO Southern Europe Naval Command. He died on 22 December 2010, exactly 69 years after his disclosure of the British penetration of Enigma.[26]

Sources:

Cippico, Aldo, *Dio punisca Anna Tobruk!*, Danesi (Rome 1947).

Cernuschi, Enrico, *'Ultra' la fine di un mito*, Mursia (Milan 2013).

Cernuschi, Enrico, 'Le prede segrete', *STORIA militare*, December 2006/January 2007.

Cernuschi, Enrico, 'Le decrittazioni della Regia Marina', *STORIA militare*, April/May 2007.

Cernuschi, Enrico, 'La guerra dei cavi', *STORIA militare*, October 2007.

Cernuschi, Enrico, 'Marinai e spie', *Rivista Marittima*, May 2012.

Cernuschi, Enrico, 'La guerra dei codici italo-francese', *Rivista Marittima*, January 2016.

Donini, Luigi, 'Sistemi crittografici delle Marine britannica e italiana. Un'analisi comparativa della loro attività nel Secondo conflitto mondiale', *Rivista Marittima*, January 1983.

Donini, Luigi, 'Il servizio crittografico a Punta Stilo', *Rivista Marittima*, October 1987.

Fioravanzo, Giuseppe, 'Il taglio dei cavi telegrafici inglesi e francesi nel Mediterraneo (giugno–agosto 1940)', *Rivista Marittima*, February 1964.

Hinsley, FH, *British Intelligence in the Second World War: Its Influence on Strategy and Operations*, Stationery Office Books (1979).

Hinsley, FH & Strip, Alan, *Codebreakers: The Inside Story of Bletchley Park*, Oxford University Press (1993).

Endnotes:

1 Alexander S Cochran Jr, *Military Affairs*, April 1982. Even harsher is Wesley K Wark, 'Communication in Never-Never land? The British Archives on Intelligence', *The Historical Journal*, 35, 1 (1992).

2 Christopher Andrew, 'Secret intelligence and British foreign policy 1900–1939', in Andrew and Jeremy Noakes (eds), *Intelligence and international relations 1900–1945* (Exeter 1987), 9. The April 1942 C 38 signal later related with that renewed endeavour for force Alexandria had been about a transport by air of naval personnel with top priority; the 3 August 1942 connection was made by decrypting a message about a planned *Luftwaffe* night reconnaissance of Haifa (CX/MSS71275/17) which induced the British to search for a similar C 38 transport message found in the 27 July 1942 records (TNA HW 17795).

3 The Swedish Hagelin C 38 mechanical cypher machine was modified in-house by the *Regia Marina*'s *Reparto Telecomunicazioni* as the 'C 38m'.

4 Typewritten *GC & CS Naval History, vol XX, The Mediterranean 1940–1943*, 216, available in the Bletchley Park Museum. Document quoted by Fabio Ruberti, 'Lo *Sciré*, vittima eccellente di "Ultra" Secret', *Bollettino d'Archivio dell'Ufficio Storico della Marina Militare*, March–August 2010.

5 TNA ADM 223/121. However, the report, dated 1946 and entitled *Notes for Naval Section Output* states that the messages from the 'Italian machine' reached the figure of 400 per month for the first time on January 1942.

6 And tutor of cryptology to the author of this article.

7 *Cryptologia* [A Quarterly Journal Devoted to Cryptology], Volume XIV, Number 2, April 1990.

8 The old commercial (mechanical only) Enigma machine had a theoretical total of 227,304,461,200 periods.

9 The biggest error in the Italian naval wireless organisation was the rigid separation, for security reasons, between the *Stato Maggiore* (Staff) codebreakers and the manufacturers of the *Regia Marina*'s codes and cyphers, who depended instead on the *Servizio Telecomunicazioni*.

10 Marc'Antonio Bragadin, *Il dramma della Marina italiana*, Mondadori (Milan 1982), 200.

11 *Archivio dell'Ufficio Storico della Marina Militare*, Rome (since now USMM), *Fondo Supermarina – Comunicazioni in generale*, Busta 25, Fascicolo 25, *'Servizio delle comunicazioni – Macchina cifrante Hagelin'*.

12 TNA HW 50/15 'Synopsis for F Birch's History on the Italian Naval Section'.

13 FH Hinsley, 'The Influence of Ultra in the Second World War', Security Group Seminar, Babbage Lecture Theatre, Computer Laboratory, 19 October 1993. In his lecture Hinsley does not source these or similar statements. The Italian Navy's actual assessment of enemy strengths are in the *Diario di Supermarina* (the Italian Central Command of the Navy Diary) which reports (6 March 1943), for example, the presence on February 1943 of 12–15 enemy submarines, stating this was a similar figure to January 1943 and the second half of 1942.

14 USMM *Fondo Supermarina, Traffico internazionale, Fondo Intercettazioni Estere* (IE). This total does not include the separate Italian Army codebreaking activity. The British signals decrypted in North Africa between December 1940 and January 1943 number more than 100,000, some of them of naval interest. (*Archivio dell'Ufficio Storico dello Stato Maggiore dell'Esercito, Comando Superiore Forze Armate dell'Africa Settentrionale*)

15 The soft punch card for that electromechanical decyphering machine had been conceived by the Italian codebreakers on 1938, but could not be used before the Italian declaration of war on the United States announced on 11 December 1941, as the management of the Watson SA company in Rome which marketed the IBM machines was 100 per cent American.

16 Franco Fucci, *Le polizie di Mussolini*, Mursia (Milan 1985), 162–182.

[17] This explains why most of the Italian decrypts of the Naval Cypher made on 1940, and available in the *Archivio dell'Ufficio Storico della Marina Militare*, begin with *Decoy* in the address.

[18] For the death of Cdr Giorgio Verità Poeta and the coroner's documentation: *Il comandante Giorgio Verità Poeta. Atti del convegno 18 ottobre 2014*, Inedibus (Vicenza 2016), 91–96.

[19] *British Intelligence*, Vol 1, 210. The exceptions were the Code KKK used by the Italian MTBs found on *MAS 452* off Malta on 27 July 1941 after the death of all her crew, who were killed by a shell fired by a 6pdr coastal battery. After the BBC reported the seizure of the boat, the code was changed on 30 July 1941. The other was the *Cifrario Ridotto* (CR), the tactical cypher for communications between merchant vessels and coastal stations, which was seized at Benghazi on September 1941 and and superseded five months later by a new edition.

[20] Respectively: a) USMM, *Fondo Supermarina, Ufficio Statistica operativa, Danni inflitti al nemico, 28 marzo 1941*, 2; b) USMM, *Fondo Supermarina, Situazione 11 Maggio 1942*; after the codebreaking of the orders modified by Alexandria, *Supermarina* ordered, at 0855 that day, the cruiser *Montecuccoli* and the destroyers *Camicia Nera* and *Aviere* to sortie from Messina to engage the four British destroyers; the destruction of the RN Flotilla by the aircraft prompted the recall of that force; c) USMM, *Fondo Supermarina, Traffico internazionale, Fondo Intercettazioni Estere (I.E.), I.E. 25416, 27 agosto 1942* (110 minutes between the interception of the British signal and the reading by the X MAS command in Egypt of the related cypher message sent from *Supermarina* about that decrypt, *I.E. 25467* and *I.E. 25468, 29 agosto 1942*.

[21] Vincent P O'Hara and Enrico Cernuschi, 'The Other "Ultra"', *Naval War College Review*, Summer 2013, Vol 66, No 3.

[22] RA Ratcliff, *Delusions of Intelligence: Enigma, 'Ultra' and the End of Secure Ciphers*, Cambridge University Press, 2008, 9.

[23] David Alvarez, 'Left in the Dust: Italian Signals Intelligence, 1915–1943', *International Journal of Intelligence and Counterintelligence*, Volume 14, Number 3.

[24] They are the main anchors of the Austro-Hungarian battleships *Viribus Unitis* and *Tegetthoff*.

[25] Ugo Cavallero, *Comando Supremo*, Cappelli (Bologna 1948), 174. See also Antonello Biagini and Fernando Frattolillo, *Diario del Comando Supremo*, Vol I, *Ufficio Storico dello Stato Maggiore dell'Esercito* (Rome 1995), 847. A full description of the Enigma machine was published in Italy by Captain Mario de Monte (head of the *Regia Marina*'s cryptological section between 1939 and 1943 and of the Fascist Republic Navy until 1945) in his *Uomini ombra*, Nemi (Rome 1955).

[26] For Tomasuolo's story see: Valeria Isacchini, *Fughe, Dall'India all'Africa, le rocambolesche evasioni dei prigionieri italiani*, Mursia (Milan 2012), and Paolo Alberini & Franco Prosperini, *Uomini della Marina 1861–1946: Dizionario biografico*, USMM (Rome 2015), 517–18.

THE IJN LIGHT CRUISER *OYODO*

Designed as flagship of a Submarine Flotilla, *Oyodo* served initially as a transport between Japan and its forward bases and subsequently as flagship of the Combined Fleet. **Hans Lengerer** provides an in-depth study of this unusual vessel.

Oyodo was designed as the flagship of a Submarine Flotilla; following her completion, she underwent conversion to serve as the flagship of the Combined Fleet. She was therefore quite different in conception and design from that of other IJN light cruisers.

Before she was built the Imperial Japanese Navy (IJN) used Submarine Tenders (*Sensuibokan*), Special Service Ships (*Tokumukan*) and large outdated warships (*Sentokan*) as flagships of the Submarine Squadrons. They served for supply, repairs and minor modifications, the reconstitution of submarine crews, manoeuvres and training (peacetime), and as command ships. In wartime all operational functions were added. The ships were moored either at the home base (*kichi*) or at an advanced base (*konkyoshi*) and did not take part in front operations under normal circumstances. However, the IJN differed from foreign navies in that there was an elaborate strategy to utilise the flagship of the Submarine Squadron operationally.

Before the decisive battle due to take place when the US main fleet entered Japan's ultimate defensive line, the IJN planned to conduct an extensive attrition operation spanning the Pacific. Submarine Squadrons were to operate as single cohesive units and execute well-planned and coordinated surprise attacks from the moment the US fleet sortied from its bases at Pearl Harbor and on the West Coast until it approached the defined area of the final duel. For this purpose the flagship of a Submarine

This starboard broadside view of *Oyodo* was taken in June 1943 and shows her as completed. Her appearance was unique, with the main guns forward of the bridge, the HA guns on sponsons grouped around the single funnel, and the 45-metre catapult and hangar aft. *Oyodo* was delivered to the IJN by the Kure Navy Yard and commissioned on 28 February 1943. After a short training period she was transferred to Yokosuka NY where the Type 2 No 1 catapult Model 10, produced by the Navy Air Technical Arsenal, was fitted by the manufacturer and the dockyard. She then returned to Kure and entered No 3 dock for final fitting out. This photo was probably taken just after she emerged from the dock. It was discovered in the US National Archives during the 1960s as part of *Oyodo*'s Data Book. (Courtesy of former Captain JMSDF Tamura Toshio).

Squadron could not remain behind in its advanced base but had to move forward to the vicinity of the enemy fleet in order to be able to provide continuous and accurate information about the strength, position and course of the enemy force and to coordinate the successive assaults of the subordinate forces.

There was another important duty for the submarines of the squadrons. They were to advance to the vicinity of the US bases on the West Coast and Hawaii to reconnoitre and observe, to report the movements of the enemy fleet and to attack when the opportunity arose. The escort of these advanced forces by a flagship was considered essential for effective reconnaissance and for the coordination of attacks. A flagship capable of fulfilling these tasks had to have good mobility (*kidosei*) and endurance, command facilities which included extensive and long-range communication systems, and the ability to conduct its own reconnaissance.

The cruiser was considered the ideal platform in which these characteristics could be combined, and initially the IJN used light cruisers of the 5,500-ton type as flagships for the Submarine Squadrons. However, the comparatively low endurance, dated communication systems and limited reconnaissance capabilities of the latter meant that they were poorly suited to the tasks outlined above. This problem was exacerbated by the age of the ships, which were already obsolescent by 1935, by the reinforcement of the US Navy and by the entry into service of large, more powerful Japanese fleet submarines during the 1930s. Despite this, the limitations of the building of cruisers imposed by the London Treaty of 1930, together with domestic fiscal pressures, meant that that the IJN was unable to build new cruiser flagships for the Submarine Squadrons.

In March 1937 the first of two submarine flagships for the Submarine Squadrons entered service. *I-7* and her sister *I-8*, which had been laid down in September and October 1934 respectively, had their own command facilities, and their exceptional endurance enabled them to approach enemy bases and to participate in attacks. However, there remained significant constraints on their operational performance: wireless and reconnaissance capabilities were limited if the flagship approached the enemy on the surface or took part in the attacks. During submerged navigation both command and reconnoitring properties were lost to the submarines in the squadron, and they were therefore unable to fulfil all the flagship missions.

At the end of 1936 Japan withdrew from the arms limitation treaties and financial constraints were relaxed. The construction of a cruiser type with strong air reconnaissance capability as flagship of a Submarine Squadron could now be realised. However, the absence of an existing model led to extensive discussions in the Operations Division of the NGS to determine characteristics. The initial requirements of the NGS, drawn up in mid-1938, are shown in the first column of Table 1. On the basis of these requirements the No 2 design team, headed by then Lt-Cdr Ozono Daisuke[1] and responsible for cruisers, drew up a preliminary proposal under the general direction of Rear Admiral Fukuda Keiji, who headed the Fourth Division of the NTD.

When the proposal was considered by the Technical Conference of the NTD in September 1938, discussion centred around the aviation facilities and the armament. There was concern that mounting the powerful new 45-metre catapult on either side of the ship could result in damage when it was trained outboard, and a single catapult on the centreline was preferred. There was also a preference for stowing all aircraft in the hangar, rather than on the open deck and the catapult(s). The conference also insisted that the proposed high-speed seaplane needed to be ready at the same time as the ship, that a crane be installed for its recovery, and that development of the catapult needed to be accelerated.[2] And although the ship carried aircraft for self-defence, the proposed armament of eight Type 89 12.7cm 40-calibre HA guns was considered inadequate protection against enemy destroyers and was incapable of being used offensively. Without torpedoes the ship would be incapable of opposing an enemy cruiser. It was therefore proposed that the main armament should comprise six 15.5cm guns in triple mountings; these mountings would be available once the cruisers of the *Mogami* class were rearmed. The head of the First Division of the Bureau of Naval Affairs also wanted torpedo tubes mounted aft.

The issues raised concerning defensive capabilities were judged to be of particular importance, and this led to the revision of the design seen in the third column of Table 1. Displacement was increased in order to mount a more powerful anti-surface armament; however, it proved impossible to accommodate torpedo tubes in the narrow space available, and speed and endurance were reduced.

During the discussion of this design in the Technical Conference of the NTD, there were again disagreements regarding the aviation component. The cruiser flagship as currently envisaged depended on an early entry into service of the high-speed seaplane. The latter was designed to reconnoitre an area up to 800nm ahead of the Submarine Squadron and to direct the submarines to an attack position. Time in the air would be about ten hours, and from catapult launch to return the weather might change dramatically, meaning that the seaplane might not be able to land on the water in the vicinity of the ship. Also, communication between plane and ship at these long ranges might be difficult, with the possibility that transmissions might be intercepted or jammed, thereby preventing the execution of the principal task: to direct the submarines of the squadron to the enemy's position. These concerns led to a proposal for the new flagship to be equipped with carrier-based aircraft, possibly in a 'flight deck cruiser' configuration.

This led to a completely new design which, to the best of the author's knowledge, was first revealed by Endo Akira in Issue No 1 of his *Historical Material about Japanese Warships* (*Nippon gunkan shikô*). The NATD's views on the aviation facilities of the flight deck cruiser configuration were expressed as follows:

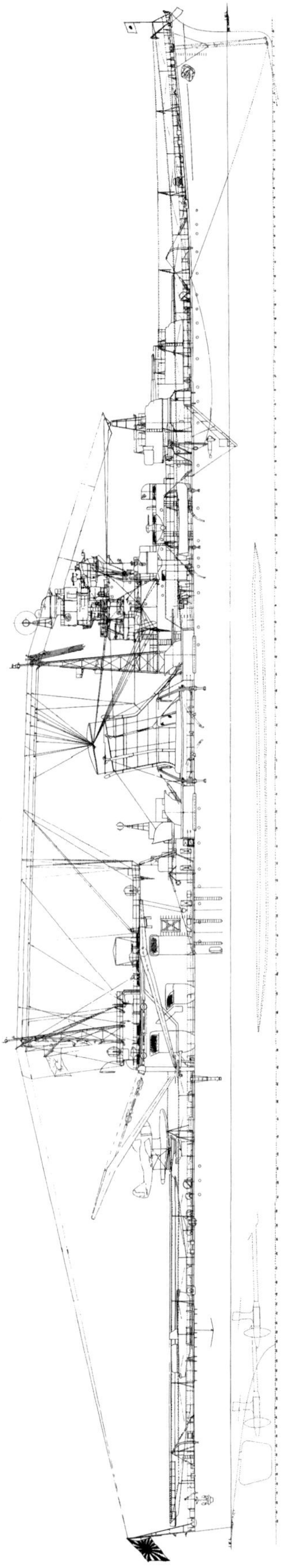

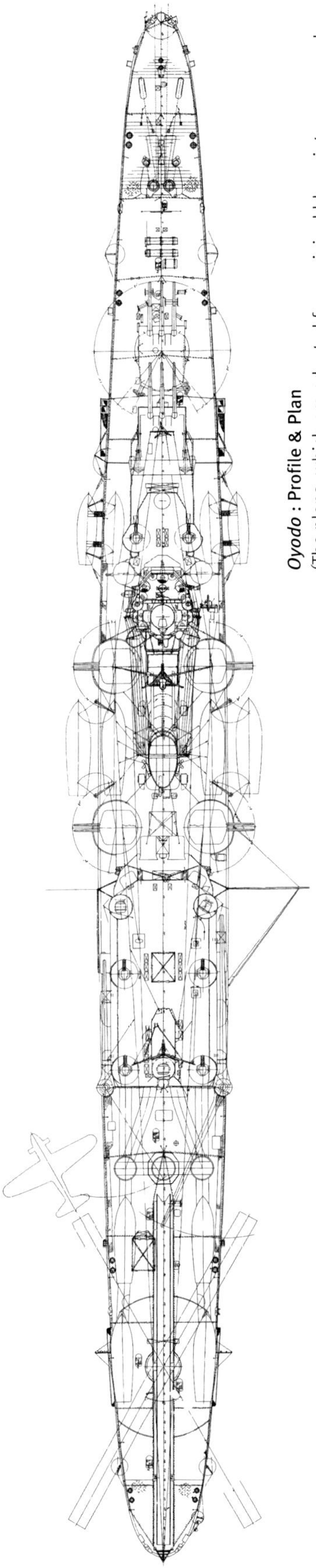

***Oyodo* : Profile & Plan**
(The plans, which are adapted from original blueprints possessed by Mr. Todoka Kazushige, are courtesy of Maeshima Hajime, by kind permission of Hajime Maeshima, Kokubunsha Publishing, Tôkyô)

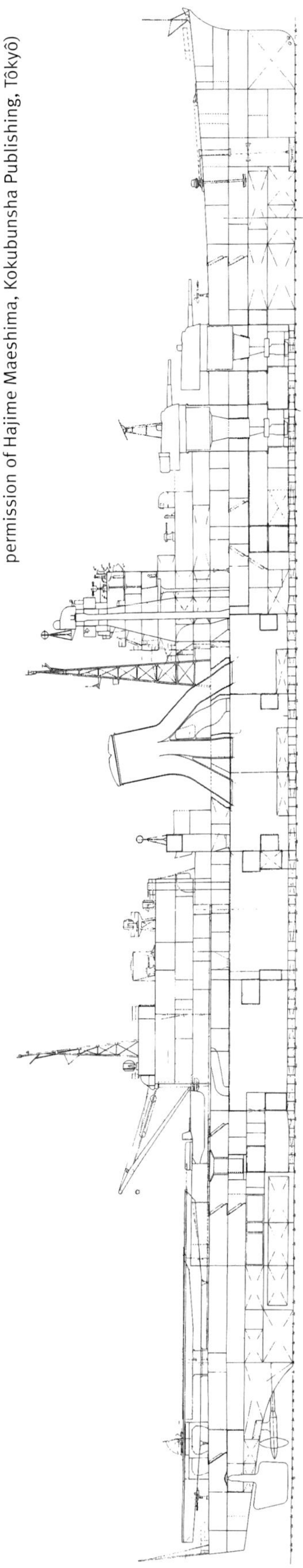

***Oyodo*: Inboard Profile**
(Courtesy of Maeshima Hajime)

- The substitution of carrier planes for the high-speed seaplanes was possible.
- The design should be a flush-deck type with no bridge on the flight deck.
- The flight deck should be about 150m long and 25m wide and its height above the waterline should be 13.8m. The length of the catapult should be 45m.
- Stability should be given special consideration.

The details of the flight deck cruiser design are uncertain, but configuration was probably similar to the US Navy's proposal of the early 1930s, with the two 15.5cm triple turrets at the forward end, and the flight deck extending over the after two thirds of the ship. For the basic characteristics see column 4 in Table 1.

When this design was discussed in the Technical Conference the NATD was still dissatisfied, and proposed the broadening of the flight deck to 25m over the whole length and the lengthening of the hangar aft to increase aircraft capacity. The NTD responded with a proposal to lengthen the flight deck to 170m and to make the width 25m over the whole length. In this configuration either 20 planes of the Type 96 *kansen* fighter type ('Claude') or the Type 97 *kankô* torpedo bomber ('Kate') or high-speed seaplanes[3] could be operated.

The revised flight deck cruiser proposal was submitted to the Higher Technical Conference.[4] After the discussion of the pros and cons the flight deck cruiser was rejected on the grounds of its hybrid character: it was considered to have fundamental flaws both as a carrier and a cruiser.[5] In the end, Design No 2 was accepted and it was decided that the new ship should be built as a cruiser of 8,200 tons standard displacement with the 15.5cm triple turrets forward and the aviation facilities (for high-speed seaplanes) aft.

Preliminary design work on the new cruiser, now designated C-42, was completed on 6 October 1939.

Construction

Two ships were included as 'Cruiser Model C' (*Hei gata*)[6] in the Fourth Fleet Replenishment Program of 1939 with the temporary designations 'warships Nos 136 and 137'. Both were to be built at Kure Navy Yard. Cruiser No 136, later named *Oyodo*, was to be completed in FY 1942; No 137, for which the name *Niyodo* was selected, in FY 1943.

Each ship was to serve as command ship for two of the Submarine Squadrons which were to advance to Hawaii and the West Coast at the outbreak of a war in order to execute the planned attrition of US Navy forces. With the outbreak of war imminent, the laying down of hull No 137 was suspended on 6 November 1941. The war in the Pacific did not progress as planned, and in the wake of the

Table 1: **Design Series for the Future Submarine Squadron Flagship**

	NGS Requirement May/Jun 1938	**No 1 Design 8 Sep 1938**	**No 2 Design Sep/Oct 1938[1]**	**Flight Deck Cruiser (second design)**
Displacement	5,000 tons std	6,600 tons std	8,200 tons std (9,800 tonnes trial)	ca 16,000 tons std (16,600 tonnes trial)
Speed	36 knots	36 knots	35 knots 4-shaft propulsion: 110,000shp	?
Endurance	10,000nm @ 18kts	10,000nm @ 18kts	8,700nm @ 18kts	?
Armament	8 x 12.7cm HAG 18 x 25mm MG ?? x TT	8 x 12.7cm HAG (4xII)[2]	6 x 15.5cm (2xIII) 8 x 10cm HAG (4xII) 16 x 25mm MG (8xII)	6 x 15.5cm (2xIII) 8 x 10cm HAG (4xII) 12 x 25mm MG (6xII)
Flight deck	none	none	none	length 150m (170m) width 19m fwd, 25m aft
Hangar	above-decks hangar	hangar for 4 aircraft	hangar for 4 aircraft	hangar beneath flt/deck 77m (›100m) x 14m x 8m[3] single aircraft lift
Aircraft	6-8 high-speed seaplanes	6 high-speed seaplanes (4 hangar, 2 catapults)	6 high-speed seaplanes (280kts max, 200kts cruise, max range 2,000nm)	9 fighters 9 reconnaissance
Catapults	to enable uninterrupted take-off by all planes	two (both sides of quarterdeck)	one 45m (centreline at stern)	two 45m (depending on aircraft types)
Other	extensive comms, sound detection & u/w signalling gear			

Notes:

1 The decision in favour of the second design was made in Oct 1938. At the same time the flight deck cruiser variant was abandoned.

2 None of the designs featured the torpedo tubes originally specified.

3 The hangar in the first flight deck cruiser design was sized to carry either 16 high-speed seaplanes or 17 Type 97 torpedo bombers ('Kate') or 18 Type 96 fighters ('Claude'). Extending the hangar aft would increase the capacity to 20 aircraft (either high-speed seaplanes, Type 96 fighters or Type 97 torpedo bombers).

Table 2: **Building Data**

Name (hull no):	*Oyodo* (136)
Builder:	Kure NY
Laid down:	14 Feb 1941
Launched:	2 Apr 1942
Completed:	28 Feb 1943
Lost:	28 Jul 1945 (aircraft)
Stricken:	20 Nov 1945

Battle of Midway the construction of No 137 was finally abandoned on 3 August 1942 before her keel had been laid. The order for *Oyodo* had been placed with Kure NY on 6 December 1939; she was to have been laid down in June 1940, but war preparations and delays in the detailed design process and the production of working drawings led to a postponement of eight months.

As with the light cruisers of the *Agano* class, high-quality Ducol steel was employed for the construction of the hull; the armour was exclusively copper alloy non-cemented (CNC). Plate thicknesses were generally increased compared to the smaller *Aganos*; the use of two thicknesses of 25mm on the keel and the adjacent strakes were particular differences.

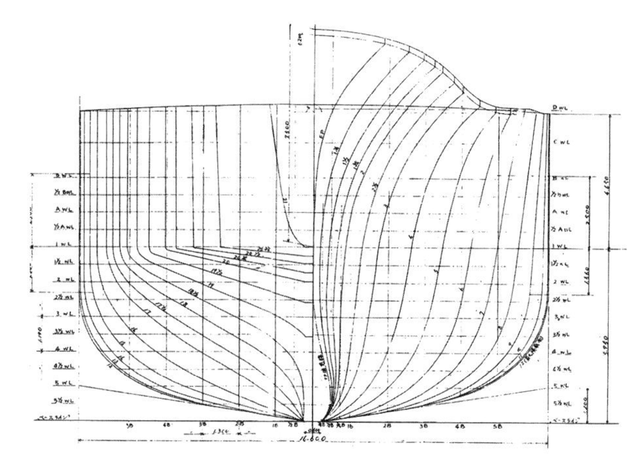

***Oyodo* : Body Plan**
(Courtesy of Maeshima Hajime)

General Arrangement

The hull-form was similar to that of the *Agano* class. The flush deck had the same pronounced sheer to increase the height of freeboard forward. The wave-like structure of the upper deck which was a characteristic feature of the heavy cruisers was again rejected in favour of a simple deck-line with marked sheer at the bow. The body plan (see accompanying illustration) was almost identical to that of the *Agano* class, but with a more pronounced bulb at the fore-foot. As the ship was designed both for great endurance and for high speed, the experiments conducted in the model basin of the NTRI were focused on developing a hull-form which would offer the least resistance at 18 knots and at maximum speed. The hull-form favoured by these experiments was compared with the shapes developed by the Taylor Institute of the US Navy before final decisions were made. The stern was angled aft (the *Agano* class had the classical 'cruiser' stern) and the underside was flattened, as in the *Aganos*, in order to reduce resistance.

The two 15.5cm gun turrets were superimposed forward of the bridge. The configuration of the latter was relatively simple: considerable effort was made to secure functionality with regard to the arrangement of the various command spaces. It closely resembled that of the *Agano* class, and the positioning of the foremast and the single funnel directly abaft the bridge likewise reflected the arrangement in the latter. The six boiler uptakes were combined in a single funnel in order to provide the necessary space for the aviation facilities, which extended from amidships to the stern. The mainmast, which like the foremast was a tripod mast with a light lattice structure, was stepped atop the hangar. The four Type 98 10cm 65-calibre twin HA mountings were on sponsons fore and aft of the funnel at the deck edge on either side, and the 25mm triple MG were distributed in such a way as to provide maximum arcs of fire: there were two mountings forward of the bridge and four on the hangar roof.

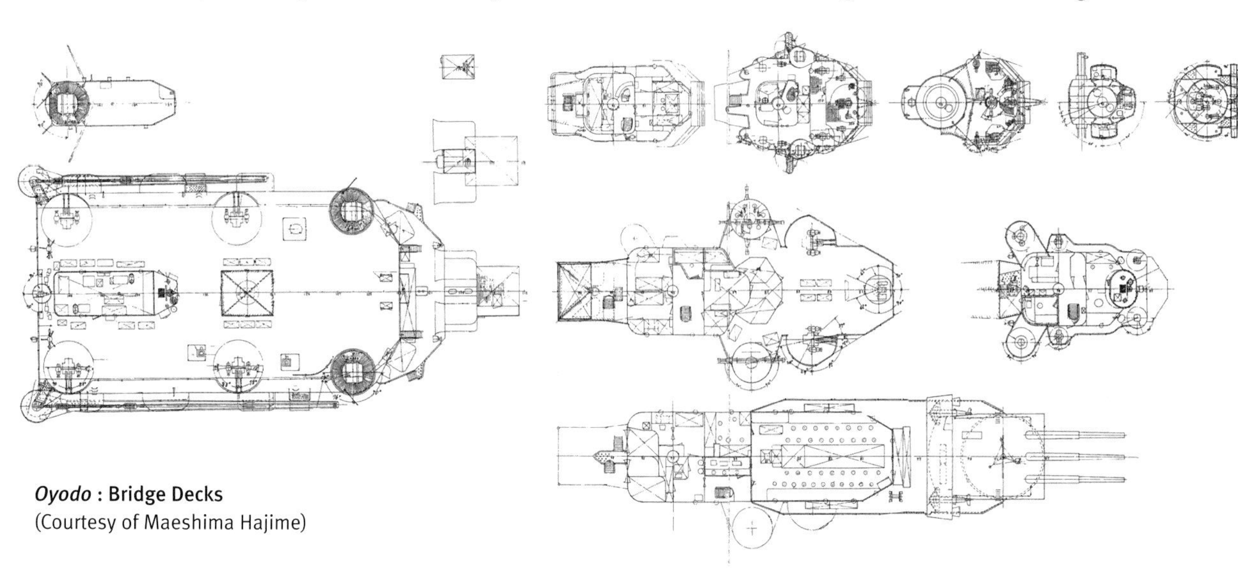

***Oyodo* : Bridge Decks**
(Courtesy of Maeshima Hajime)

The characteristic feature which distinguished *Oyodo* from the *Agano* class was the aviation facilities, which comprised a prominent 45-metre catapult on the centreline aft and a large rectangular hangar, which had a crane at each of the after corners and ventilation ducts for the engine rooms at the sides. The three searchlights were located on the roof of the hangar: one on the centreline abaft the mainmast and one on either side forward of the 25mm triple mountings. The searchlight control consoles were located in the bridge structure.

Protection

The protection system was designed to cope with 15.5cm common projectiles (no range was specified), and 250kg bombs released from horizontal bombers flying at an altitude of 3,000 metres. To attain this goal the side armour was 60mm CNC outboard of the machinery spaces, including the after generator room and the bomb magazine. The middle deck was armoured with 30mm thick CNC plates and rested on the upper edge of the belt armour; it was horizontal throughout, with no slopes at the sides. The citadel was closed fore and aft by armoured bulkheads 35mm thick, reinforced to 50mm over the bomb magazine. The latter magazine, which was located on the centreline abaft the engine rooms, had additional box protection of 35mm on the sides and the forward bulkhead, with a 50mm crown (see drawing).

The forward end of the armoured citadel enclosed the communications tube which connected the main gunnery director with the transmitting station. The armoured deck was then stepped down over the magazines. The vertical armour forward of the citadel was internal: abeam the main 15.5cm magazines (Frames 83-55) it was 75mm, tapering to 40mm at its bottom edge; the box protection was completed by a 50mm crown. Between the main magazines and the forward boiler room (Frames 92-83), in the area of the magazines for the AA guns and the transmitting station, the side wall was 60mm tapering to 30mm, and the thickness of the armoured deck was reduced to 28mm. There was light protection (generally 20-35mm) for the barbettes of the main gun turrets and the ammunition hoists for the HAG and MG.

The armour was manufactured by the Kure NY Steel Division and the protective plating by the Yawata Steel Company. The hull protection was structured so that it contributed to longitudinal strength, as in the light cruisers of the *Agano* class.

The conning tower, which was integrated into the forward edge of the bridge structure, had 40mm on the

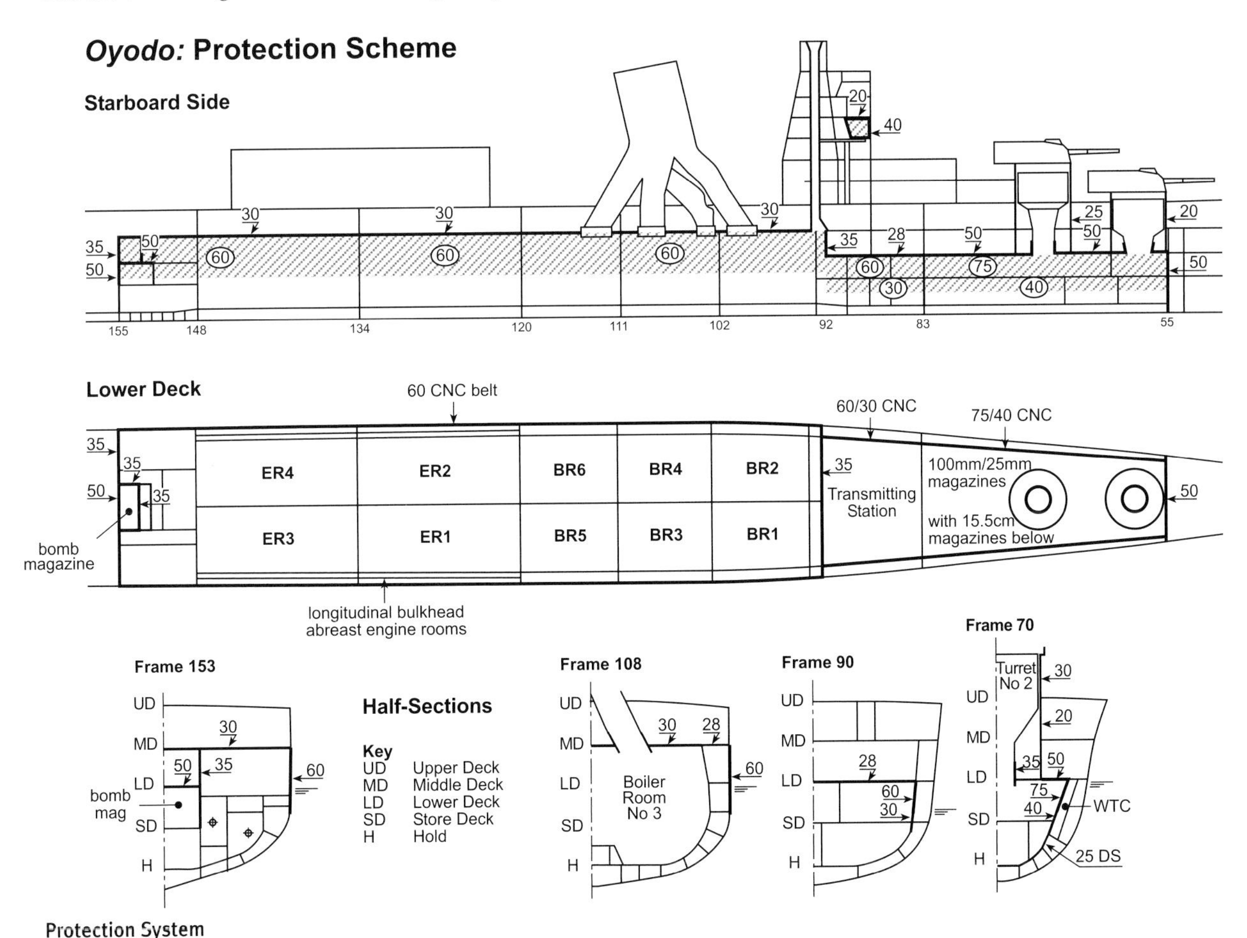

Protection System
(Drawn by John Jordan using information supplied by the author and schematics published in Lacroix & Wells, *Japanese Cruisers of the Pacific War*)

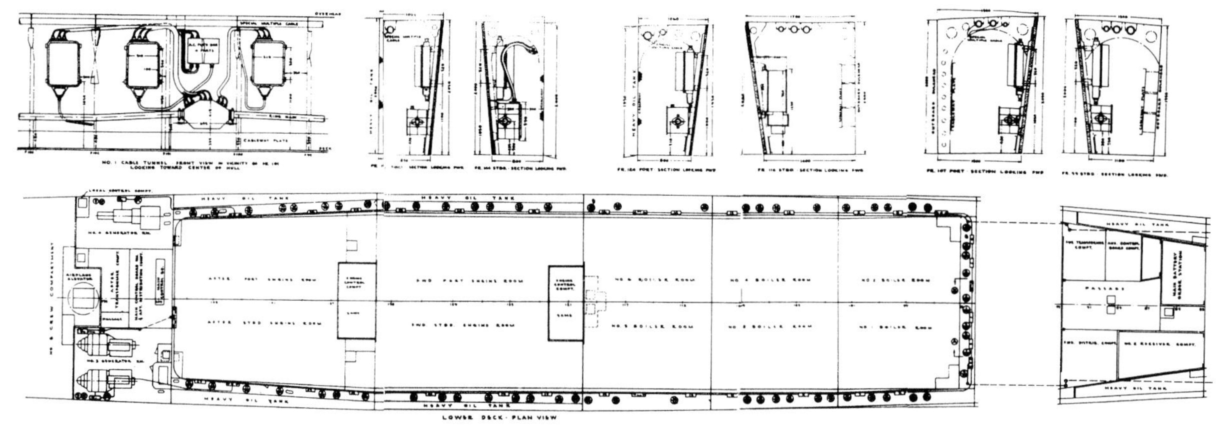

Primary Power Plant and Machinery Spaces
(US Naval Technical Mission to Japan Report S-01-5)

face with a 20mm roof; the sides were protected by 20mm Ducol steel (DS) plates. The steering compartment was in an armoured box with 40mm sides and 20mm (forward) and 25mm (aft) end bulkheads. The ventilation cowls on both sides of the hangar which supplied fresh air to the engine rooms had 16m DS on the sides and 10mm fore and aft; this level of protection was extended to the boiler room ventilation trunking and the uptakes.

No direct underwater protection could be provided beyond the watertight compartments outboard of the magazines. According to *Oyodo*'s Ship Data Book, the designers relied on indirect measures such as close watertight subdivision,[7] the arrangement of heavy oil tanks on both sides of the hull, the fitting of an additional longitudinal bulkhead outboard of the engine rooms with a stand-off to provide a small expansion space, and a damage control system which relied on counter-flooding.

Machinery

In order to attain the designed speed of 35 knots power output was calculated at 110,000shp, to be supplied by four sets of turbines. The arrangement is shown in the accompanying plans. It resembled that of the *Agano* class, except that the centreline bulkhead ran the entire length of the machinery spaces. There were therefore four engine rooms instead of three in the *Agano*, and six boiler rooms *vice* five. The disadvantage of the centreline bulkhead was already known and it is difficult to understand its adoption: the bulkhead would later be largely responsible for the ship capsizing.[8]

Each turbine set comprised HP, IP, and LP ahead turbines; it received the designation No 3 C Model 36 and developed 27,500shp at 340rpm. All three turbines were of the Kampon impulse, single flow (LP double flow) type and ran at 3,632rpm, 3,385rpm, and 2,327rpm respectively. The speed of rotation of the shafts was geared down through a three-pinion single reduction gear. The astern turbines were in the LP turbine casings and developed 6,875shp at 1,471rpm.

Two cruise turbines, designated No 3 A Model 36, were connected to the turbines mounted in the forward engine rooms which drove the wing shafts; the cruise reduction gearing was connected to the main reduction gear. When in cruise mode the steam was sent directly to the cruise turbines, then passed through the HP turbines to facilitate four-shaft operation. At full power the cruise turbines were disconnected from the main gearing and steam was admitted directly to the HP turbines.

Steam was supplied by six Kampon *Ro gô* type oil-burning small watertube boilers with air preheaters producing steam at 30kg/cm^2 (superheater outlet) and 350°C. Steam consumption was an economical 0.36kg/h/shp at high speed.

Fuel stowage amounted to 2,445 tonnes, sufficient for the required endurance of 7,800nm at 18 knots. The three-bladed propellers had a diameter of 3.60m, a pitch of 3.96m and a developed area of 7.56m^2.

Readers interested in an in-depth study of the machinery are referred to Report S-01-4 of the USNTMtJ, 38-42, where detailed data on boilers, condensers, turbines, reduction gearing, and auxiliary machinery can be found. The increase in fire pump capacity noted on page 38 of the report was directly related to the improvement in fire-fighting and damage control capabilities decreed following the Battle of Midway.

The trial results in Table 3 are taken from issue No 1 of *Historical Materials about Japanese Warships*, page 13. Despite the ship being completely slightly overweight, speed was higher than designed and endurance significantly greater than anticipated. Seaworthiness and steering were good; the turning circle at low speed (12 knots) with the rudder at 35° was considerable, but at 8/10 power it was slightly smaller than usual. The turning trials took place on 18 February 1943 and are recorded on page 3 of *Oyodo's* data book.

Steering was by a hydraulically-operated plunger type system; the cylinders were driven by two 60kW electric motors via the two hydraulic pumps, both located in the steering engine room. The customary Janney system was employed. The capstan was actuated by a Ward-Leonard system driven by two 100hp electric motors. The total weight of the machinery including oil and water was

Table 3: **Trials Data**

	Overload	Full Power[1]	12 Knots	Cruising Max	Cruising	Cruising Std
Displacement (tonnes)	10,302	10,381	–	–	–	10,514
Speed (knots)	35.31	35.20	11.76	28.17	21.65	18.28
RPM	346.3	340.3	98.6	192/280[2]	183	153
Horsepower (shp)	115,950	110,430	2,634	41,733	14,840	8,574
Fuel consumption (t/h)	41.0	40.0	3.22	17.0	6.54	4.13
Fuel consumption (kg/shp/h)	0.35	0.36	1.22	0.41	0.44	0.48
Endurance (nm)	2,007	2,051	8,511	3,861	7,714	10,315
Steam pressure (kg/cm^2)	23.2	24.6	15.3	26.2	26.3	24.4
Steam temperature (°C)	339	338	75	323	319	282

Notes:

1 Full Power = 10/10.

2 Inner/wing shafts; trial took place on 31 January 1943.

	8/10 Full Power		12 Knots	
Displacement	10,608 tonnes		10,562 tonnes	
Mean draught	6.17m		6.15m	
Speed (knots)	34.00 knots		12.50 knots	
Ratio rudder area to underwater area	1:52.2		1:52.0	
Metacentric height [GM]	1.37m		1.36m	
	starboard	port	starboard	port
Rudder angle	34.7°	35.0°	33.7°	35.0°
Advance [DA]	713m	735m	497m	516m
Tactical diameter	835m	788m	636m	627m
DA/Lwl	3.77	3.89	2.63	2.73
DT/Lwl	4.42	4.17	3.37	3.32
Max inclination	12.2°	11.4°	2.2°	2.0°

Note:

The rudder was of the balanced type with 6.69m^2 before and 13.75m^2 abaft the axis, making a total area of 20.44m^2. The side area exposed to the wind was 1,731m^2, underwater side area was 1,055m^2 and the ratio of exposed to underwater area was 1.64.

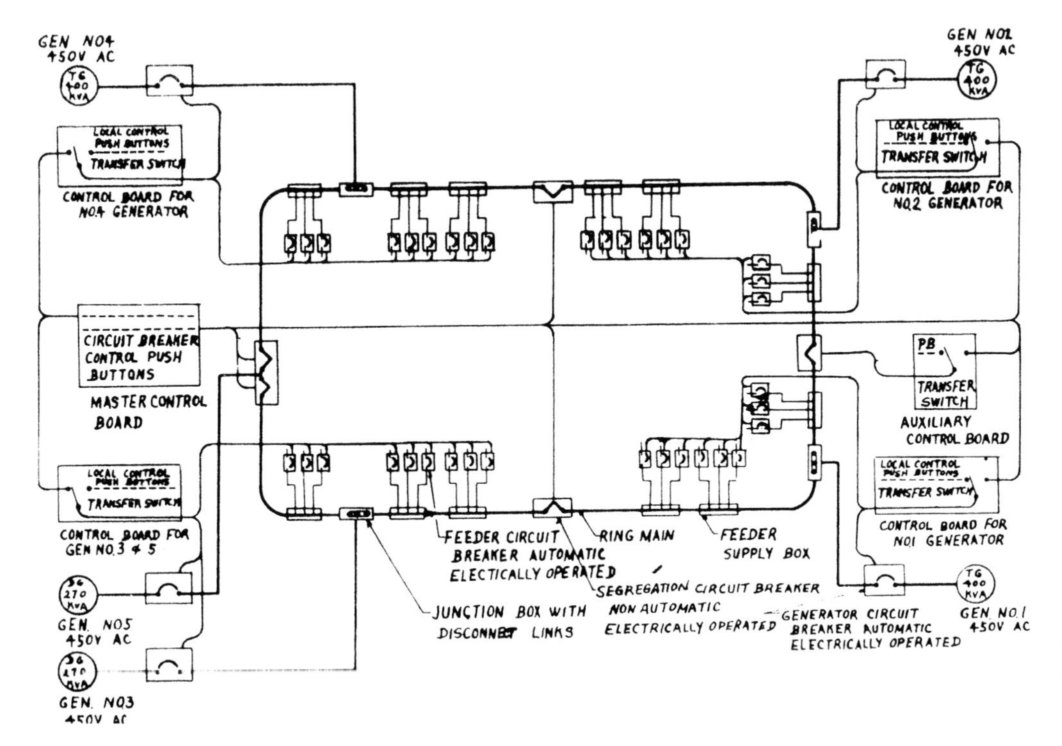

Electrical Main (US Naval Technical Mission to Japan Report S-01-5)

1,961 tonnes (1,765 tonnes in dry condition). In the first condition the weight/output ratio was 17.8kg/shp, in the second 16.1kg/shp.

The electrical generating power (440V as in *Agano* class) was high in comparison to the earlier A-type cruisers: there were three 400kVA turbo-generators and two 270kVA diesel generators, for a total power output of 1,740kVA (= 1,300kW at 75% power factor). The increase in generating power was in part necessary because of the demands of the high-performance W/T equipment outfit.

The reserve power source comprised batteries, as in earlier cruisers. There were two banks each of 53 cells = 320Ah; for steering only there was a third bank of 53 cells, also 320Ah. Two motors generating DC 105V 50kW were fitted. The AC power for the Selsyn motors was supplied by step-down transformers. The motor generators for the searchlights were three 17.6kW units at 88V DC and one 15kW unit at 105V DC. The power for telephone, motor generators and batteries was by two M-G 22V 1kW units and two banks of 11 cells = 320Ah. This equipment was as in the *Agano* class.

Armament

Two 15.5cm triple gun turrets, removed from the *Mogami* class during their conversion to 20cm-gun 'A'-type cruisers in 1939–40, were superimposed forward of the bridge. The 60-calibre 3rd Year type 15.5cm gun had an impressive ballistic performance: maximum elevation was 55 degrees and maximum range 27,400m against surface targets (45°) – for a full account of the development of the triple 15.5cm mounting see Warship Notes on page 195. Fire was controlled by a Type 94 low-angle (LA) system: the director control tower (DCT) was located atop the bridge structure, and the fire control table in the transmitting station below the protective deck between the magazines and the forward boiler rooms. Range data were supplied by a 6-metre rangefinder beneath the DCT and by an 8-metre rangefinder fitted in the upper turret.

The Type 98 10cm 65-calibre HA guns, in four 'A' Model twin mountings, were disposed on either side of the ship abreast the foremast and abaft the funnel. The ballistic data for this gun were a major advance on older models of HA guns in service with the IJN. There were two Type 94 HA fire control systems, the directors for which were mounted on either beam at the base of the bridge. The computers were located in the LA transmitting station.

Close-range anti-aircraft defence was originally to have been provided by six Type 96 25mm twin MG mountings, but this was revised during construction and *Oyodo* was completed with six triple mounts: four atop the seaplane hangar and two forward of the bridge (see line drawings). Their fire was remotely controlled by three Type 95 MG directors located at the forward end of the lower bridge deck and on platforms protruding from either side of the middle bridge deck. As in other IJN warships the 25mm armament would later be considerably reinforced. In March 1944 the number of triple mountings was increased from six to twelve, and eleven single mounts were also fitted. In the course of the AA reinforcement programme after the Battle of the Philippine Sea, additional single 25mm MG were mounted in October 1944, but this weapon proved ineffectual and the Le Prieur sights were unable to cope with the high speeds of the most modern aircraft types of the Allied forces.

The main armament of *Oyodo* was to be her six seaplanes, which were of new, advanced design. The 14 *shi* (trial) *kosoku* (high speed) *suitei* (seaplane) was still at the development stage when the decision was made to embark it in this ship and the *Agano* class; it was designed to reconnoitre areas in which the enemy had air supremacy. In order to attain the high designed speed of 280 knots, trials were conducted with contra-rotating propellers, half- and totally-retractable auxiliary floats, and a detachable main float. However, the Kawanishi-developed plane, first known as the Navy Type 2 High-Speed Reconnaissance Seaplane and then, after its formal adoption, as High-Speed Reconnaissance Plane *Shiun* Model 11 (E15K1), was never embarked in *Oyodo* nor in the *Agano* class. Production was halted in February 1944 after only 15 planes (including the prototype) had been completed because performance did not meet expectations.

The Type 2 No 1 Model 10 catapult was developed specifically for the *Shiun* by the NATA (*Kugishô*). The maximum take-off weight was 4,100kg at 80 knots or 5,000kg at 70 knots; average acceleration was 2.5g. The catapult was designed to launch one seaplane every six minutes. The length of the catapult beam was 44m (effective length 35m), and the width of the rails was 1.5m. The catapult operated by compressed air, and large-capacity pressurised air containers were necessary for the continuous take-off of the six *Shiuns*. The seaplanes were placed on their carriage by a pneumatically-operated lift located at the after end. The volume of compressed air needed for lift and launch was considerable (see *Shôwa Zôsenshi*, Vol 1, 551). *Oyodo's* catapult was used for launch trials of the *Shiun*, but the production version of the aircraft was never embarked.

The large hangar was located abaft the funnel and could accommodate four of the six seaplanes with folded wings. The remaining two planes were to be carried on rails on either side of the catapult. Two 6-tonne cranes were fitted at the after corners of the hangar to handle the aircraft and lift them aboard after landing.

Oyodo was completed without radar or hydrophones. A No 21 air search radar was fitted in April 1943, followed by No 22 surface search and E-27 electronic countermeasures (ECM) in March 1944, and finally No 13 air search in October 1944. At the same time the No 22 surface search radar was modified for fire control, but this modification was unsuccessful: the radar had insufficient definition to provide effective fire control for the guns.

Trials 4: **Armament & Equipment**

Gunnery Equipment

Item	Number
15.5cm guns in two triple mounts (150rpg, total 900, + 8 exercise rounds)	6
Type 98 10cm HA guns in four Model A Mod 1 twin mounts (200rpg, total 1,600 + 2 exercise rounds)	8
Type 96 25mm MG in six twin mounts Model 2 – designed[1] (2,000rpg, total 24,000 + 100 exercise rounds)	12
Type 94 FCS in LA Type 5 director control tower	1
Type 94 HA FCS with director	2
Type 95 MG FCS (one covered, two open)	3
Type 91 smoke generator Model 5 Mod 4	1

Underwater Equipment

Item	Number
Kampon type air compressor pumps Model 3 Mod 1	2
Air condenser flasks Model 2 Mod 1	52
Type 95 depth charges	6
Ignition apparatus storage box (for Type 95 DC)	2
Tool box (for Type 95 DC)	1
DC parachute Mod 1	6
Manual DC thrower Model 1	2
Manual DC thrower (for shipborne boats)	2
Davit for handling DC	1
Small paravanes Model 1	2
Tool box for small paravanes	2
Fathometer Model 2	2
Type 93 sonar Model 3 Mod 1	1

Navigation Equipment

Item	Number
Type 93 No 3 magnetic compass	1
Compass for boats	5
Type 90 magnetic compass Model 2 Mod 1	1
Type Su No 3 gyro compass Model 2 (double system)	1
Ko Type No 2 log meter Type 1 Mod 1	1
Type 98 ship bottom log (together with tube)	1
Type 96 battle tracer Model 1	1
Steering recorder Mod 1	1
Type 90 No 2 fathometer Model 1 Mod 1 (440V AC)	1
Type 91 anemometer Mod 1	1
Type 92 anemometer (rotation at each longitude) Mod 1	2
Type 97 observation balloon for high altitude observation Mod 1	1
Thermometers	1 set
Type 97 identification lamp Model 1	1
Type Ko (A) signal light Mod 1	1
Sextant (meridians)	1

Optical Equipment

Item	Number
Type 93 8-metre duplex RF (main guns)	1
Type 14 6-metre duplex RF (main guns)	1
Type 94 4.5-metre HA RF (HAG)	2
Type 96 1.5-metre RF (navigation)	2
18cm binoculars	2
12cm binoculars (with support)	2
12cm binoculars together with box	2
12cm HA binoculars (including one Model 13)	5
12cm HA binoculars (suspended)	1
6cm HA binoculars	4
Type 13 No 1 direction meter Mod 1	3

Electrical Generating Equipment

Item	Number
Turbo-driven AC generator (dynamo) 400KVA 450V	3
Diesel-driven AC generator 270KVA 450V	2

Projectors

Item	Number
Type 96 110cm searchlight Model 1 (100V)	3
60cm signal light Model 1 AC (100V)	2
20cm signal light Model 1 AC (60V)	2
2kW signal light Model 1 Mod 2 AC (100V)	2

Wireless & Electromic Equipment[2]

Item	Number
Type 95 short-wave No 3 Mod 1 transmitter	1
Type 95 short-wave No 4 Mod 1 transmitter	1
Type 95 short-Wave No 5 Mod 1 transmitter	1
Type 97 short-wave No 6 transmitter	1
Prototype short-wave No 2 transmitter	1
Type 97 short-wave receiver	3
Type 92 long-wave No 4 Mod 1 transmitter	2
Type 91 long-wave Model 1 Mod 1 receiver	3
No 2 radar Model 1	1
Type 91 special No 4 Mod 1 long- and short-wave transmitter	1
Type 92 special receiver Mod 4	18
No 2 medium-wave wireless telephone transmitter Mod 1	2
Type 92 special medium-wave receiver Mod 4	2
Type 93 ultra short-wave transmitter	2
Type 90 ultra short-wave telephone transmitter	1
Type 93 ultra short-wave receiver	2
Type 90 wireless telephone receiver	1
Type 90 ultra short-wave telephone transmitter	1
Type 93 No 1 direction finder	2
Double type underwater signal Model 10	1

Notes:

1 Triple MG mountings were fitted on completion (total 36,000 rounds).

2 Note the unusually comprehensive communications outfit.

Table 4 lists the armament and other equipment of *Oyodo*. It is taken from the ship's 'Log of Specifications' dated 14 March 1943, pages 20–22, with some omissions.

Complement

When completed *Oyodo* was manned by 30 officers, 10 special duty officers, 13 warrant officers, 191 petty officers and 532 ratings, total 776 men. This was very different to the designed complement, which according to page 1 of her official data was 33 officers, 14 special officers and 14 WOs, plus 850 POs and men (total 911); in addition to her basic complement she was designed to embark 13 staff officers to command the Submarine Squadrons. However, *Oyodo* was never used in the configuration for which she had been designed.

Early Wartime Career

When *Oyodo* was commissioned the catapult was not yet fitted. This work was begun on 8 March 1943, and on the 19th work began on plating over the scuttles on the middle deck, together with other works for improving watertightness. While this work was proceeding various official trials took place beginning on 18 March. At the end of April the No 21 air search radar was fitted at Kure NY, the antenna being mounted on the face of the main DCT. On 1 May *Oyodo* was assigned to the Main Body (*Hontai*) of the Mobile Force, but was used mainly for transport duties until February 1944. The first of these operations was carried out in July when she transported troops and materiel from the homeland to Truk and then on to Rabaul, accompanied by the light cruiser *Agano* and the heavy cruisers *Tone* and *Chikuma*. While stationed in Truk 'in readiness' she participated in the much-criticised move of the Combined Fleet to Brown Atoll (Eniwetok) in September and October, following the US air raids on Tarawa and Makin, then Wake.

On 30 December she left Truk in company with the light cruiser *Noshiro* to transport army reinforcements to Kavieng, and experienced her first serious air attack on 1 January 1944. According to her action record she entered Kavieng at 0445 on 1 January on transport mission BO No 3, carrying army units and materiel, and was engaged in unloading when the air raid warning sounded at 0830. Fortunately unloading was completed at 0838 and she made for the open sea. From 0840 enemy planes (recorded as 106 F6Bs, TBMs, and TBFs) approached, and *Oyodo* opened fire first with her 10cm HA guns then with the main guns, which were employed for barrage fire. Between 0857 and 0912 she was attacked by dive-bombers and then by torpedo bombers. Dive-bombers achieved near-misses on the starboard beam and starboard quarter but the four torpedoes launched at her were evaded. She responded to the attacks 'with entire fire power', and ceased fire at 0912 when friendly Zeros engaged the enemy planes. She expended 194 15.5cm and 240 10cm projectiles; her 25mm MGs fired 3,820 rounds, the light machine guns 450. Four Type 95 depth charges were jettisoned by the crew to preclude detonation by strafing. The guncrews claimed to have shot down eight planes (five certain and three in conjunction with other ships), but this assessment is questionable. The near-misses caused 34 holes of about 10 cm diameter in the hull, and damaged part of the ship's armament. Two men were killed and four wounded.

On the way back to Truk *Oyodo* escorted the torpedoed naval transport *Kiyozumi Maru*, arriving there on 4 January. In February she returned to the homeland and embarked on a transport mission from Yokosuka to Saipan (19–26 February). During the return voyage she was assigned to the Combined Fleet.

Conversion to Flagship of the Combined Fleet

Even when she was first completed *Oyodo*'s designed role was being questioned. From a technical point of view, the delayed entry into service of the high-speed seaplane was the biggest problem. The Technical Conference had already stated that the success of the project depended on this, but the *Shiun* was still at the trials stage when *Oyodo* entered service, and there appeared little prospect of the aircraft achieving the required performance criteria. Trial launches of the

A rare aerial view of *Oyodo* taken moored at Truk on 13 September 1943. Note the rails running on either side of the catapult on the quarterdeck. The white painted structure atop the bridge is the Type 94 fire control director for the 15.5cm main guns. The two Type 94 directors for the long 10cm HA guns are located on either side at the base of the bridge structure. (Author's collection)

prototype were conducted using *Oyodo's* special catapult, but the aircraft was never embarked operationally. During the early part of her career she operated only two Zero seaplanes at any one time.

Given that the missions for which *Oyodo* had been designed were no longer technically feasible or appropriate to the strategic situation, it was decided in mid-1943 to convert her to serve as flagship of the Combined Fleet, for which her advanced communication facilities made her eminently well-suited. She would consume much less fuel than the battleship *Musashi*, and her endurance was greater. According to *Senshi sôsho*, Imperial General Headquarters and Combined Fleet (*Daihonei Kaigunbu Rengokantai*) Vol 5, Vice Admiral Ozawa, C-in-C of the Third Fleet, was instrumental in proposing her conversion, and the NGS drew up two alternative proposals. The first required the removal of the main guns and the erection of a structure forward of the bridge housing all operational and administrative spaces and the staff officer accommodation. The hangar was to be removed altogether or reconstructed to serve as living spaces for the ship's complement. The second proposal was for the reconstruction of the command facilities and spaces in the bridge structure, together with the conversion of the hangar to house the operational/administrative spaces and accommodation for the staff personnel.

The first proposal had advantages from the command point of view, but required extensive structural modifications. The second required fewer structural changes and also retained the main armament. This would be the proposal eventually adopted.

The conversion work was undertaken at Yokosuka NY from 6 to 31 March 1944 and was given the highest priority. The hangar was divided into three levels to provide the necessary operational spaces and accommodation, and the long catapult was removed and replaced by the standard Kure Type No 2 Model 5 Mod 1 catapult (length 19.4m). The light AA armament was reinforced at this time (see Armament section). Two of the new triple mountings were located atop the former hangar. They replaced the two searchlights, one of which was moved to a platform on the centreline. Two No 22 surface search radars (horn antennae) were installed on small platforms at the sides of the AA command station. The wireless equipment was reinforced by the installation of transmitters of the latest types.

The primary emphasis during this conversion was on operational and 'defence' capabilities; habitability was accorded a comparatively low priority.

Later Wartime Career

Oyodo served less than six months as flagship of the Combined Fleet, from 1 April to 29 September 1944 (although the official date was 28 November). The day after she completed her reconstruction she was placed under the direct command of the C-in-C Combined Fleet, and the new C-in-C, Admiral Toyoda Soemu, hoisted his flag in her on 4 May. *Oyodo* was moored off Kisarazu in Tôkyô Bay until 29 September, except for the period from 22 May to 28 June when she was at Hashirajima. While she was moored in the latter anchorage Admiral Toyoda's HQ directed the unsuccessful Battle of the Philippine Sea (*A gô sakusen*), which ended with the loss of three valuable carriers and most of the carrier-based air crews.

Taken in May 1944, this photo shows the third C-in-C of the Combined Fleet, Admiral Toyoda Soemu, on the bridge of his flagship, the light cruiser *Oyodo*, which had been modified for the purpose. (Author's collection)

The naval high command then prepared for the next Allied offensive with Operation 'Victory' (*Shô gô sakusen*), for which all fighting ships were needed. *Oyodo* was no exception: her service as flagship was ended, and she returned to the fleet on 29 September, the HQ of the Combined Fleet being transferred ashore. She entered Yokosuka NY on the 30th and remained there until 10 October. In addition to refit and docking she received a further six 25mm single MG to reinforce the close range AA defence; No 13 radar was installed, the antenna being mounted on the mainmast, and the No 22 radar was modified to be used for fire control. During this refit she was assigned to Vice Admiral Ozawa Jisaburo's carrier force, which was to be used as 'bait' for the US carrier task forces, and took part in the Battle off Cape Engaño on 25 October. She sustained minor

Oyodo approaching the sinking *Zuikaku* during the action off Cape Engaño on 25 October 1944. She rescued the C-in-C, Vice Admiral Ozawa Jisaburô, and part of the crew of the last surviving carrier of the Pearl Harbor air attack. Note that the same photograph (but reversed) was published in Vol 2 of Fukui Shizuo's 'albums' as No 2851; the photo was erroneously dated March 1943, and the flight deck in the foreground was wrongly attributed to the aircraft ferry and landing ship *Akitsu Maru* of the IJA. (Author's collection)

damage from two rockets and a bomb near-miss, and took off Vice Admiral Ozawa and part of the crew of his flagship, the carrier *Zuikaku*, before this last survivor of the Pearl Harbor attack slipped under the waves. She had expended 238 rounds of 15.5cm, 964 of 10cm and 32,558 of 25mm; she claimed to have downed more than ten planes, but actual results are uncertain.

On the 27 October she entered Amami-O-Shima (Setsugawa Bay) and sortied on the 29th for Manila, transporting army materials, arriving there on 1 November. On the 8th she anchored at Brunei Bay to replenish and sortied on the 7th for Lingga Roads. On 12 December she left for Cam Ranh Bay. She participated in the San José (Mindoro) Operation (*Rei gô sakusen*) as part of the 2nd Fleet and suffered minor damage from two bombs (one of which failed to detonate) dropped by B-25s on 26 December. The 101st Repair Division (*Kosakubu*) at Singapore repaired her damage from 9 to 29 January 1945. She returned to Japan the following month together with the hybrid battleships *Hyûga* and *Ise* which were used to transport men, gasoline, rubber, tin and other materials; the three escorting destroyers were likewise loaded with materials. Designated 'North Operation' (*Kita gô sakusen*), it was considered to be a great success, but the total load (9,704 tons) was less than a single merchant transport would have brought to

These two photos show the funnel of *Oyodo* damaged by a 500-pound bomb during the attack on Kure of 19 March 1945. On that day Kure and the Hiro Dockyard were subjected to air raids for the first time. The dockyard was damaged and many ships in the naval port suffered damage to varying degrees, among them *Oyodo*. She was approached by tugs and salvage vessels but was judged to be in no danger of sinking. She entered No 3 dock, where these two photos were taken. One bomb struck the forecastle deck to starboard and pierced the middle deck, causing many casualties among her crew. A second bomb detonated in the forward engine room to starboard, resulting in substantial cracks in the outer plating. Damage to other parts of the ship resulted from several near-misses. Following emergency repairs which aimed to restore buoyancy, *Oyodo* sailed for Etajima and was moored at Etauchi. (Author's collection)

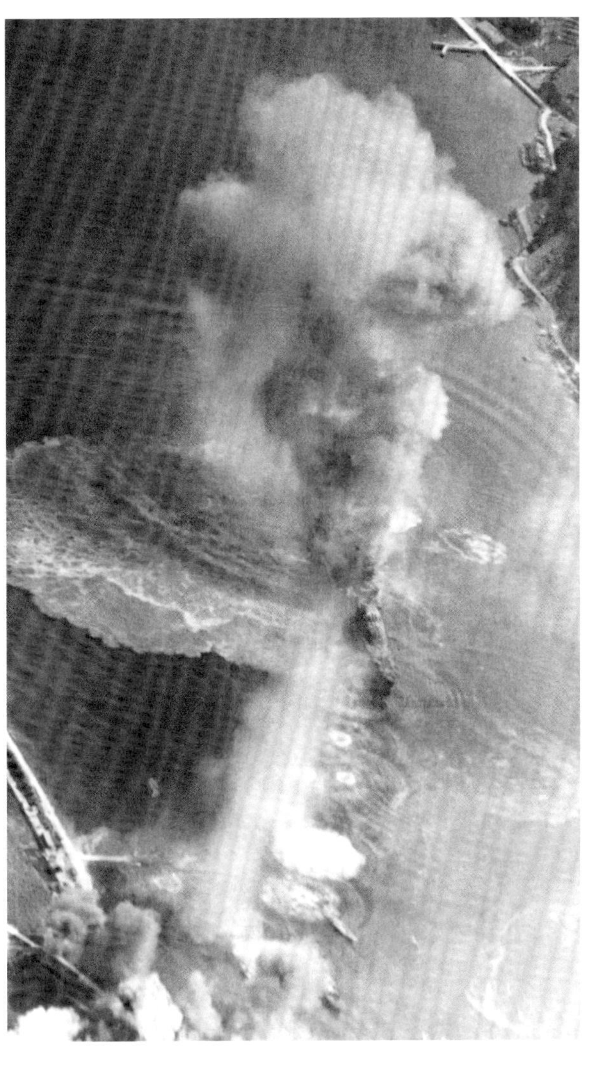

Left: Oyodo under air attack on 28 July 1945 at Etauchi near Kure. She is enveloped in smoke and steam so only her forward half can be seen. Water circles around her and in the foreground show the impact and detonation of many bombs. (US Naval History & Heritage Command, NA 80-G-490225)

Above: A photo of the capsized *Oyodo* taken from the stern. (Author's collection)

Below: This aerial view of the capsized *Oyodo* was taken at Etauchi near Kure. Note the oil fuel flowing out of ruptured tanks at the stern. (NHHC, NA 80-G-490228)

Japan had it been able to break the Allied blockade. One remarkable aspect of this operation is that despite the US Navy being able to decode Japanese radio traffic, only two of the 25 Allied submarines directed to attack the 'Completion Force' found the Japanese ships; they fired a total of twelve torpedoes, all of which missed. The force entered Kure naval port completely intact on 20 February after departing Singapore on the 10th.

When the first large US carrier-based air raid was executed on 19 March *Oyodo* was hit by three bombs, and also suffered a number of near-misses which caused severe flooding and a list to starboard. After emergency repairs she was taken in hand at Kure NY from 23 March to 4 May and then moored at Etauchi (westward shore of Etajima, some 10nm from Kure Naval Station) as a floating AA battery. On 24 July she was hit by five bombs,[9] and capsized four days later after receiving several near-misses to starboard during the third heavy air raid on ships in Kure area. The USNTMtJ investigation team inspected the ship on 29 Nov 1945, when she was

Two more photos showing the capsized *Oyodo* lying on her starboard side at Etauchi. They were taken by the US Navy on 29 July 1945. (NHHC, NA 80-G-351759/60)

lying on the bottom at an angle of 80° with her port side exposed. In report S-06-1, page 48, the damage sustained on the 28th was attributed to 'several close near-misses to starboard'. One near-miss bulged the forward starboard engine room and starboard after boiler room extensively, causing instantaneous flooding of both spaces. Others opened the shell above and adjacent to the waterline. Counter-flooding efforts were unsuccessful, and *Oyodo* capsized to starboard in about 25 minutes. The rapidity with which she capsized was attributed to the presence of the centreline bulkhead in the machinery spaces. This served to confirm the view of Captain Matsumoto Kitaro, who is said to have opposed the installation because he considered *Oyodo* too small to withstand a large degree of off-centre flooding. Theoretically, her maximum list was limited to 15° with one engine room and one boiler room on the same side flooded, but under such conditions there was almost no reserve of stability; any additional flooding would lead to her capsizing.

Oyodo was salvaged in 1947 and broken up at Harima Shipyard, Kure Branch (the former Kure NY) from 6 January to 1 August 1948.

After the end of the war warships which like *Oyodo* were lying on the bottom of the sea in shallow water were salvaged to remove obstacles for navigation, and broken up. This photo was taken during the last stage of the salvage operation.

Conclusion

Oyodo was a highly specialised design conceived for a particular operational sequence planned before the start of the Pacific War. The execution of this operation depended upon the US Navy acting as expected. The forward operation by the IJN submarine squadrons was to have been initiated immediately before the outbreak of the war, but at that stage *Oyodo* had yet to be launched, and by the time she was completed the Japanese forces were already in retreat.

Oyodo's main armament, the *Shiun* high-speed seaplane, never became operational as intended, and as the war in the Pacific developed *Oyodo* lost her *raison d'être*. Her limited fighting power precluded her deployment as a 'fleet fighting ship', and at the start of her career she was used principally for transport operations. The background to the proposal to refit her as Combined Fleet flagship was prompted more by the IJN's lack of

Oyodo was salvaged by Harima Shipyard and then towed to Kure where she entered Dock No 4 for scrapping on 22 December 1947. Note the removal of the outer plating abreast turret 2, revealing twisted frames suffered during the bombing of 19 March 1945. (Fukui Shizuo collection)

Another photo of *Oyodo* in Dock No 4 of the former Kure NY during scrapping. This photo was taken on 27 January 1948 with the work of the breakers already begun. (Author's collection)

fuel and doubts concerning her best tactical employment than by real necessity. With the annihilation of the Combined Fleet in the Mariana and Philippine campaigns she again found herself without a clear mission. After the last transport operation from Singapore to Kure she became an almost stationary target in the Inland Sea due to lack of fuel, and met her fate in the last of the July air raids.

In summary, *Oyodo* owed her existence to the unrealistic strategic and tactical planning of the NGS before the war and to the IJN's adherence to outdated concepts. Doubts surrounding the execution of the planned initial stage of operations halted the conversion of the existing light cruisers into heavy torpedo ships – *Kitakami* was the last – even before the start of the war, and the attack on Pearl Harbor effectively prevented the US Navy from conducting its own anticipated advance into the western Pacific, thereby nullifying the Japanese strategy of forward-based attritional warfare.

Once the development problems with the prototype of the high-speed seaplane became clear *Oyodo*, which was still under construction and had yet to be fitted with her hangar and catapult, might have benefited from redesign to fit her out as an AA cruiser. However, flexibility in thinking was not a strong point of the IJN. As Makino Shigeru has stated in his *Kansen nôto*: 'In the IJN there was a reluctance to abandon long-held ideas even when they had been discredited by events'. This could serve as an epitaph for a cruiser designed to serve as flagship of the Submarine Squadrons.

Endnotes:

1. Technical Captain Ozono was one of the contributors to KZGG. Cruisers are dealt with in Vol 2; it is hard to find a more authoritative source than this series of volumes.
2. The NATD had expressed doubts about readiness dates for the ship, the catapult and the aircraft.
3. It is interesting that the NTD proposed seaplanes as an alternative. The handling of this plane type did not require a flight deck, and it is possible that this reflected considerations regarding later aircraft developments.
4. The timing is uncertain, but most probably in the autumn of 1938.
5. The US Navy had already come to the same conclusion.
6. The heavy cruisers intended to accompany the battle fleet were designated 'Model A' (*Kô gata*), and the light cruisers which served as flagships for the Destroyer Squadrons 'Model B' (*Otsu gata*).
7. WTCs and OTCs were arranged as follows: 94 inside the double bottom; 159 inside the hold and upper bottom: five on the middle deck and 23 between the middle deck and upper deck.
8. USNTMtJ Reports S-01-4 and S-06-1.
9. In Report S-06-1, page 48, only two small bombs (500lb) are mentioned; these struck in the vicinity of the catapult and detonated on impact with the upper deck, causing holes 3–4m in diameter in the deck. 'The upper shell strake on the extreme port quarter was riddled by fragments coming down and outboard ... There was no fire'. In addition, four near-misses off the port bow were reported, but damage (only a few fragment holes well above the waterline and one indentation in the port turn of the bilge amidships (ca 6m long and maximum 0.45m deep) which did not rupture the hull ('and all rivets appeared tight') was judged negligible and without influence upon her capsizing four days later. Lacroix and Wells mention three more hits in *Japanese Cruisers of the Pacific War*, 652, and *Campaigns of the Pacific War*, 347, gives a total of four hits – it also confirms the location of the two hits mentioned in Report S-06-1. The four near-misses are acknowledged by all sources. Japanese sources are generally less precise, with 'many hits and many near-misses'.

COAST DEFENCE AND COAST OFFENCE:

Russian Monitor Designs of the First World War Era

In the years immediately prior to and during the First World War several projects for monitor-type ships were considered by the Imperial Russian Navy. None of these ships were built; however, documentation survives for a number of them, although the available evidence is far from complete. In this article **Stephen McLaughlin** pieces together the history of these projects.

The Baltic: Coast Defence

After the debacle of the Russo-Japanese War one question was repeated over and over again in the Russian press: *Kakoi flot nuzhen Rossii?* – What sort of fleet does Russia need? The debate over this question raged in the pages of the Navy's professional journal, *Morskoi sbornik* (The Naval Digest), as well as in other journals and newspapers. Many different answers were offered: the Emperor and most of the Navy's senior officers wanted to build a new fleet of battleships, but many politicians saw this as nothing more than the desire for a 'prestige' fleet, unsuited to Russia's defensive needs and far too expensive for a country suffering from the economic dislocations brought on by a humiliating defeat and an abortive revolution. Another, cheaper answer favoured by many was a 'mosquito' fleet of destroyers, torpedo boats and submarines intended solely for the defence of the Baltic coasts.

There were also those who offered a third answer: monitor-battleships, heavily armed and armoured but relatively slow. The thinking behind such vessels was that Russia's primary need was for ships to defend the maritime approaches to Saint Petersburg, which meant they would be operating in the narrow waters of the Gulf of Finland. Since the area of operations was so limited, speed and range were not important qualities, and the weight saved in machinery and fuel could be devoted to armament and protection. These considerations led naval constructor GV Svirskii to propose 'a broad-hulled, well-protected, but slow vessel' during the dreadnought design competition in 1906; but this idea was rejected at that time in favour of more conventional dreadnoughts, which eventually materialised in the form of the *Sevastopol* class. Even after the new battleships had been laid down, however, there were proposals for monitor-battleships; one writer noted in 1911 that:

> In the past year our famous and learned naval constructor Guliaev has made calculations showing the economy in weight of machinery and fuel that can be attained if we were satisfied with a speed of 12 knots in the dreadnoughts now under construction. The weight saving in the machinery amounts to 2,300 tons, and that for fuel (for 1,000 miles at full speed) is 900 tons, for a total of 3,200 tons. The weight of artillery [of the current design] is 1,280 tons and the weight of armour is 5,900 tons, so the savings noted above would allow a doubling in the weight of the artillery and an increase in armour thickness of 34%.[1]

Admiral Nikolai Ottovich Essen, commander of the Baltic Fleet, who pushed unsuccessfully for the inclusion of shallow-draft monitors in prewar construction programs.

F.F. Guliaev was a prominent naval constructor, so his views were not to be taken lightly; he had recently argued for the advantages of a ship with a very great beam, which might have provided sufficient stability to carry more and/or larger guns. However, it is unclear how extra guns could have been accommodated in dreadnoughts already under construction, as the writer seems to imply.

In any event, the naval and political authorities were not much interested in slow monitor-battleships. As Prime Minister PA Stolypin pointed out:

> No world power can avoid participating in world politics, or avoid participating in political combinations, rejecting the right to a voice in the resolution of world issues. The fleet is a lever for obtaining this right, this attribute of a great power.[2]

Seen in these terms, Russia needed high-seas battleships that could go wherever the Empire's interests were jeopardised, not slow monitor-battleships limited to the narrow waters of the Baltic. Stolypin, with the Emperor's backing and support from the Foreign Ministry, prevailed in the debate over what sort of fleet Russia would build, and the prospect of slow, powerful monitor-battleships vanished.

However, interest persisted in smaller, more conventional coast defence ships. In May 1909 Rear Admiral NO Essen, commanding the remnants of the once powerful Baltic Fleet, proposed that every year the Navy should lay down three light cruisers, five destroyers, five submarines, one minelayer and two coast defence battleships. The latter were to be shallow-draft vessels intended for use in the skerries, the rock-strewn maze of channels and islets that ran along the Finnish coast. Russia fully expected Sweden to be allied to Germany in a future Baltic war, and a Swedish invasion of Finland, at that time a Russian territory, would have been supported by Sweden's substantial fleet of coast defence ships. These vessels were ideally suited to fighting in the skerries, so Essen wanted similar ships to oppose them. Essen followed up with another recommendation for coast defence ships in a report dated 16/26 December 1909:

> Mastery of the Abo and Åland skerries region, [which is] necessary to oppose a Swedish landing operation on the Finnish coast, can be secured only by creating a special detachment of skerries vessels based permanently at Abo. The core of this detachment should be [composed] of not fewer than six shallow-draft (about 16 feet) coast-defence armourclads with good manoeuvrability, armed solely with heavy artillery (not smaller than 10-inch, and if possible 12-inch [guns]) and 4in anti-torpedo boat guns.[3]

Admiral Essen advocated the construction of a series of shallow-draught coast defence ships to counter the Swedish fleet. Typical of these was *Äran*, name ship of a class of four vessels built 1899–1904. Displacing around 3,700 tons and armed with two 210mm and six 152mm guns, they and the similar ships of the *Oden*, *Dristigheten* and *Oscar II* classes gave Sweden a total of nine ships suitable for operations in the narrow waters surrounding the Åland Islands between Sweden and the Russian territory of Finland. (Leo van Ginderen collection)

Essen again raised the concept in the spring of 1911 when he submitted a lengthy report on the future needs of the Baltic Fleet; Point 11 (of 12) of Essen's report was:

> Starting in 1912, every year [we should] lay down in domestic shipyards two small armourclads able to operate in the Skerries, with an armament of 12-inch guns, stipulating that the first pair shall be ready by the end of 1914, and the second [pair] by the end of 1915.[4]

However, the Naval General Staff was opposed in principle to the idea of 'special purpose' ships – that is, ships that could not form part of a seagoing fleet. So the construction program submitted to the Duma that year did not include Essen's shallow-draft coast defence ships, although in the event they might have proven very useful for defence not only of the skerries but of the Gulf of Riga when the Germans attacked it in 1915 and 1917.

The Black Sea: Coast Offence

While proposals for big-gun warships specially designed for coastal operations in the Baltic centred on the idea of coast defence, conditions in the Black Sea called for coast *offence* vessels. In this theatre Russian planning revolved around the Turkish Straits: the Bosporus, connecting the Black Sea with the Sea of Marmora, and the Dardanelles, connecting the Sea of Marmora with the Mediterranean. Throughout the 19th century Russia worried that the British Mediterranean Fleet might force its way through the Dardanelles as it had in 1878, and then perhaps enter the Black Sea, where Russia's fleet would be too weak to oppose it. Therefore throughout the 1880s and 1890s Russian plans called for an expedition to seize the Bosporus and establish a strong defensive position there, including mobile artillery batteries, mines and shore-based torpedo tubes, in order to keep the British out of the Black Sea.

A strategic game played in 1902 indicated that these plans were in need of review and updating, and so in January 1903 a joint Army-Navy commission was established to look into this matter; among its members was Rear Admiral Grand Duke Aleksandr Mikhailovich, who was a proponent of small battleships. Although the commission focused on the size of the landing force and its transportation requirements, it did not lose sight of the basic purpose of the expedition, which was to create a defensive position in the Bosporus that would deny Britain access to the Black Sea. The existing plans called for the laying of mines in the strait and rapid installation of heavy artillery to protect the minefields. But the commission believed that the guns could not be emplaced before the British Mediterranean Fleet arrived. Therefore, on the suggestion of Aleksandr Mikhailovich, the commission recommended the construction of no fewer than 'twelve small armourclads, of which six should be built in the next five years'.[5] Unfortunately, no further description of these vessels is available.

The disastrous outcome of the Russo-Japanese War (1904–1905) put an end to serious planning for seizing the Bosporus for several years. By the time interest in the project was revived, the rationale had changed: it was no longer a question of keeping hostile powers out of the Black Sea, but of securing the free passage of Russia's export trade though the Straits. By 1910 Russia was emerging from its postwar recession, and the economy began to expand rapidly; but that expansion was dependent on the import of manufactured goods, which were paid for by exports of Russian grain and oil from the Black Sea region. Interest in the Straits was reinforced when the Italo-Turkish War (September 1911 to October 1912) led to two interruptions in Russian grain exports, including a halt of several weeks in April–May 1912 that caused large-scale economic disruption. This was the background to a November 1913 proposal by the Naval General Staff, which recommended that

> ... as the guiding principle of all our military-naval preparations in the near future, the general strategic concept [should be] the preparation of our Baltic and Black Sea fleets for operations not only in defence of our coasts, but also for active joint operations in the Mediterranean and Black Seas, in order to secure in any circumstances Russia's maritime route from the Black [Sea] to the Aegean Sea.[6]

True to its lack of interest in 'special purpose' vessels, the Naval General Staff was thinking in terms of concentrating capital-ship strength in the eastern Mediterranean and the Black Sea, but at least one officer saw a flaw in this scheme. Lieutenant Ivan Anatolevich Kononov, a member of the staff of the commander of the Black Sea Fleet, submitted a report before the war in which he noted that the four dreadnoughts under construction at Nikolaev would give Russia full mastery of the sea, but they could not gain control of the Straits, since 'dreadnoughts were not suitable for battle with forts'. Writing in exile twenty years later, Kononov described how in his report he had argued for the creation of

> ... a special type of armourclad of 12,000–14,000 tons, with a speed of 12 knots, with extremely low freeboard and a very shallow draft, and with a triple, rather than a double bottom, very well armoured sides and upper deck, with 10-inch casemate guns of 30–35 calibres, with a large number of machine guns in the fighting tops and small but extraordinarily rapid-firing artillery. The money allocated for the third and fourth dreadnoughts could have built four such armourclads.
>
> The sole task that these armourclads could have fulfilled was to fight the forts in the Bosporus. All their qualities would have been developed in the direction of their survivability (against mines) and strong armour in order to stand up against the shells of the Turkish batteries...
>
> From the technical point of view the construction of such floating batteries would not have presented any difficulties.[7]

Table 1: **Vickers Monitor Designs for Russia 1915**

	739	740	*Abercrombie*
Characteristics:			
Displacement	5,000 tons	2,750 tons	6,150 tons deep
Length	328ft/100m	275ft/83.81m	320ft/97.53m pp
Beam	85ft/25.91m	65ft/19.91m	90.02ft/27.44m
Draught	8.2ft/2.50m	7.75ft/2.36m	10ft/3.05m
Armament	three 12in/52 (1xIII) four 130mm/60	two 12in howitzers (1xII) two 10in/50	two 14in (1xII) two 12pdr
Protection:			
belt	none	100mm	4in/102mm
barbette	150mm	100mm	8in/203mm
conning tower	150mm sides, 75mm roof	100mm sides, 62mm roof	6in/152mm sides, 2.5in/64mm roof
longitudinal b/hd	75mm	–	1.5in/38mm
transverse b/hds	25mm	50mm	4in/102mm
deck	35mm	25mm	2in/51mm
Machinery	diesels, 3,200bhp	diesels, 2,400bhp	steam, 2,000ihp
Speed	11.5 knots (est)	12 knots (est)	10 knots (designed) 7 knots (actual)
Weight Breakdown:			
Hull & Fittings	2,000 tons	900 tons	2,484 tons
Armour	240 tons	337 tons	682 tons
Protection	650 tons	235 tons	1,170 tons
Armament	1,170 tons	720 tons	845 tons
Machinery	245 tons	200 tons	376 tons
Equipment	150 tons	100 tons	220 tons
Fuel	250 tons	50 tons	200 tons
Margin	345 tons	208 tons	–
Total	5,000 tons	2,750 tons	5,977 tons

Sources: National Maritime Museum, Thurston notebooks, THU/72/07, designs 739, 740; Buxton, *Big Gun Monitors*, 43.

Table 2: **Putilovskii Works Design for a 12in Monitor, Variant 1**[1]

Characteristics:

Displacement	3,995 tons
Length	295ft/89.91m pp, 305ft/92.96m oa
Beam	58ft/17.68m
Draught	12ft/3.66m on an even keel
Armament	two 12in, eight 120mm
Protection:	
main belt	6.5in tapering to 3.5in below the waterline
main battery turret	6.5in sides, 3.5in roof
main battery barbette	6.5in
120mm turrets	4in sides, 2in roof, 1.5in ammunition tubes
conning tower	6.5in sides, 3.25in roof, 2in comms tube
funnel uptakes	2in to a height of 10ft above the upper deck
decks	2.5in + 0.5in
Machinery	four Belleville boilers, two triple-expansion engines, 3,200ihp
Speed	13–14 knots
Endurance	200 tons coal normal; 950nm at 14 knots 400 tons coal maximum, 4,500nm at 9 knots
Complement	275

Weight Breakdown:

Hull	1,259 tons (31.6%)
Artillery	632 tons (16.8%)
Armour	1,467 tons (36.7%)
Machinery	272 tons (6.8%)
Coal	200 tons (5.0%)
Provisions & water	20 tons (0.5%)
Crew & baggage	25 tons (0.6%)
Margin	120 tons (3.0%)
Total	3,995 tons

Note:
[1] Information on other variants has not come to light.

Source: Russian State Archives of the Navy, RGAVMF, f 876, op 173, d 369.

A very simple design for a monitor with a twin 12in turret, drawn up by the Putilovskii Works circa the spring of 1915. The one questionable feature is the location of the 12in turret, which seems too far forward, and would possibly have caused problems with the trim. (© John Jordan, based on a drawing in RGAVMF f 876, op 173, d 369)

Nothing came of Kononov's proposal, but soon after the war began the requirement he had foreseen gained a new urgency. In the autumn of 1914, as the Russian Army's tremendous losses continued to mount, her allies began to worry that Russia might decide to seek a separate peace; as an incentive for staying in the war, on 12 November 1914 Britain agreed to Russian control over Constantinople once victory had been achieved. This was a compelling offer – possession of Constantinople would have fulfilled an ancient Russia goal and provided unhindered access to the Mediterranean. So in February 1915, with the Anglo-French attack on the Dardanelles in prospect, the Emperor ordered that the Black Sea Fleet undertake preparations for an expedition to seize the Bosporus. The prospect of such an operation is presumably what spurred the Russians in the spring of 1915 to ask the British firm of Vickers to run up some sketches of monitors mounting battleship-sized guns.[8] It may be that the Russians had heard something about the British *Abercrombie* class monitors, laid down December 1914. The two monitor designs drawn up by Vickers had the specifications shown in Table 1; the characteristics of the first British big-gun monitor, *Abercrombie*, are provided for comparison. Design 739 was considerably smaller than the first British big-gun monitors, and it is not clear if the hull was to be provided with bulges akin to those of the British ships, since Vickers was not building any monitors at the time and may not have been aware of them. However, if the Vickers proposals were to be fitted with bulges, it is likely that the speed would have been considerably less than estimated – the British monitors were designed for 10 knots, but in fact managed only 6.5–7.6 knots due to their unusual hull form.

At about the same time as the Vickers designs were

delivered, several monitor designs were prepared in Russia. Much uncertainty attaches to these designs, which were prepared by the Putilovskii Works in Petrograd (formerly Saint Petersburg); not even their date is known. It seems likely that they also date from the spring of 1915, before full details of the British monitors were known to the Russians, since they lack anti-torpedo bulges. Although a series of designs was evidently prepared, only a single sketch, with the accompanying characteristics, survives; nor are there any papers to explain the rationale behind them. It seems most likely that these sketches were inspired by the Vickers drawings and were designed to approximately the same requirements. The characteristics of the surviving variant are outlined in Table 2.

The design shows a very simple vessel – low-freeboard hull, no superstructure aside from a conning tower, a single funnel and a pole mast. The machinery would have consisted of two reciprocating engines and four Belleville watertube boilers. The main difficulty in building such vessels in wartime Russia would have been providing the armament and mountings, especially given that the dreadnoughts had triple, not twin turrets. It is possible that it was intended to use a variant of the turrets designed for coastal batteries, although considerable modifications to the mountings would have been required; alternatively, there might have been some hopes that the 12in mountings could be purchased in Britain. The twin 120mm turrets appear to be of the same type as those mounted in the *Shkval*-class monitors of the Amur River Flotilla.

The parallel Vickers and Russian designs do not seem to have received further development, but Russian interest in monitors was by no means dead. In July 1915 the Russian naval attaché in London was instructed to obtain as much information as possible about the British big-gun monitors from the Admiralty. The first of these curious craft had arrived at the Dardanelles on 29 June/ 12 July, and they must have seemed the ideal type of vessel to cope with the shore batteries and support the troops ashore during the projected attack on the Bosporus. The Russians specifically asked for information on the following possibilities:

1 having four to eight monitors built in Britain and shipped, disassembled, to Russia for assembly in Nikolaev.
2 having the same number of ships built in Britain for shipment to Russia, but without guns and ammunition, which would be manufactured in Russia.
3 having only the machinery made in Britain, with the monitors built in Russia to the British design.

In response to these queries, the Admiralty offered only to provide blueprints for the monitors, with notes on desirable changes based on the experience of their operation.

At about the same time that information on the British

The British 14in-gun monitor *Raglan* being towed past Clydebank on 15 June 1915 after leaving Harland & Wolff's Govan shipyard. These vessels aroused considerable interest in Russia, and for a brief time it was hoped that similar vessels could be built in Britain and shipped, disassembled, to Russia for use in the Black Sea. (John Brown, courtesy of Ian Buxton)

Right: Admiral Andrei Avgustovich Ebergard, commander of the Black Sea Fleet, who identified the need for two types of monitor for an assault on the Bosporus – big-gun ships to knock out fortifications, and ships armed with 6in guns to provide fire support to the troops ashore. (Author's collection)

designs was requested, Admiral AA Ebergard, commander-in-chief of the Black Sea Fleet, proposed the construction of two types of monitor for the Bosporus landing: ships with 12in guns for knocking out fortresses, and smaller ships with a broadside of six 6in guns for close support of the troops once they were ashore. In both types the guns needed to be capable of high angles of elevation, and the ships would require lighter guns for use against both aircraft and submarines – against the latter the guns could fire a special type of diving shell.

In the inevitable Russian fashion a commission was set up to study the problem; it included representatives from the Naval General Staff, the Main Administration of Shipbuilding, the Black Sea Fleet staff, and the distin-

Below: The Black Sea Fleet pre-dreadnought *Rostislav*, mounting four 10in/45 and eight 6in/45 guns. She was slated for conversion to a big-gun monitor substitute with the addition of bulges, but she was too busy supporting the Russian army in Anatolia, and so although bulges were made for her, they were never fitted. (US Naval History & Heritage Command, NH 60711)

guished naval constructor AN Krylov, who served as a consultant in constructional matters. The commission set to work in August 1915 and held its final meeting on 1/14 October 1915, by which time it had concluded that Russia lacked the means to build such ships under wartime conditions – the Obukhovskii Works was already working at maximum capacity on field guns for the Army, and no diversion for new naval work was permissible.

Instead of building shore-bombardment monitors, the commission recommended that the older Black Sea Fleet battleships *Sinop* and *Rostislav* be converted into *ersatz* monitors by fitting them with 'side caissons' – that is, bulges. Before the war the *Sinop* had been converted to a gunnery training ship, armed with modern 8in and 6in guns, and could have fulfilled Ebergard's requirement for a close-support ship, while *Rostislav* still retained her 10in/45 guns, making her suitable for the anti-fortress role. The conversion of these old ships was attractive on several levels: it would be quicker than new construction; no new guns or mountings would have to be manufactured; and the recent entry into service of the first Black Sea dreadnoughts meant that these older ships were essentially surplus vessels that could be risked in special operations. Therefore on 8/20 February 1916 the Navy ordered a pair of bulges for *Sinop*. These were not built directly onto the ship's hull, but were manufactured at the shipyard for later installation on the ship. Their fabrication required a couple of months, and on 25 April/8 May *Sinop* arrived at Nikolaev to have them attached in drydock. The bulges covered the ship's side between frames 9 and 75, and had a maximum width amidships of 15 feet (4.6m). In cross-section they had a peculiar five-sided shape – for simplicity of construction the surfaces were all flat, with the top of each bulge extending horizontally from the ship's side below the waterline, while the outboard side was vertical; the bottom angled downward, and then ran up to the turn of the ship's bilge, the fifth side being the ship's hull. These enormous structures would have provided a considerable degree of protection against mines and torpedoes. A similar set of bulges was manufactured for *Rostislav*, but at the time she was busy supporting the Army in Anatolia, so the fitting of her bulges was postponed.

Sinop in drydock in Nikolaev, showing the bulges fitted in April–May 1916. These massive structures would have provided a considerable degree of protection against mines and torpedoes. (Courtesy of Sergei Vinogradov)

The old Black Sea Fleet battleship *Sinop*, as modified in 1912 to serve as a gunnery training ship. Her old 12in/30 guns were replaced by four modern 8in/50 guns (in the large shields abreast the superstructure) and eight 6in/45 guns in the hull embrasures. With the addition of 'side caissons' – bulges – she could have fulfilled the role of a close support ship for the landing force. (Wikipedia commons)

Meanwhile, the Russian high command and the Black Sea Fleet staff pressed ahead with plans for an amphibious assault on the Bosporus. By the spring of 1916, the prospects for a landing seemed favourable – intelligence indicated that the Bosporus defences had been weakened by the transfer of troops to other fronts, leaving only three under-strength divisions in the area, and also that many of the guns of the fortresses had been stripped to reinforce the Gallipoli front. Moreover, because the Turkish railway system was undeveloped and coal was in short supply, it might take as much as two weeks for the Turks and Germans to move substantial reinforcements to the Bosporus. The Russian scheme therefore envisaged a surprise landing by a relatively small force:

> ... after minesweeping the approaches to the Bosporus during the night, the transport flotilla would close in to the coast and before dawn land two divisions with their

Table 3: **Vickers Designs for 'Coast Defence Battle Monitors' December 1914**

	Javary	Design 709	Design 718	Design 719[1]	Design 720
Characteristics:					
Displacement	1,350 tons	1,500 tons	1,800 tons	1,875 tons	1,925 tons
Length, oa	265ft	265ft	275ft	275ft	275ft
Breadth, ext	49ft	46ft	50ft	50ft	50ft
Depth, mld	8ft 6in	9ft 6in	11ft	11ft	11ft
Draft, mean	4ft 8in	5ft 6in	6ft	6ft 3in	6ft 4in
Freeboard	3ft 10in	4ft	5ft	4ft 9in	4ft 8in
Armament	2 x 6in/50	2 x 6in	2 x 8in/50	2 x 8in/50	4 x 8in/50
	2 x 4.7in howitzers	1 x 6in	1 x 8in/50	1 x 8in/50	4 x 6in howitzers
	4 x 47mm	4 x 6in howitzers	4 x 6in howitzers	4 x 6in howitzers	4 x 47mm
	6 x MGs	4 x 3in QFG	4 x 47mm	4 x 47mm	6 x MGs
		6 x MGs	6 x MGs	6 x MGs	
Armour:					
main belt	*all designs:* 3in				
transverse b/hds	*all designs*: 1.5in forward, 1in aft				
turrets	*all designs:* 4in face, 3in sides, 1.75in roof				
barbette	3.5in	3.5in	4.5in	4.5in	4.5in
conning tower	*all designs:* 4in sides, 2.5in roof				
funnel uptakes	??				
deck	*all designs:* 1in				
Machinery	1,200ihp	1,200ihp	1,500ihp	1,500ihp	1,500ihp
Speed	11.5 knots	11 knots	11.5 knots	11.25 knots	11 knots
Weights breakdown:[2]					
Hull & Fittings	580 tons	580 tons	640 tons	660 tons	650 tons
Armour	316 tons	165 tons	211 tons	245 tons	213 tons
Protection	140 tons	158 tons	158 tons	158 tons	–
Armament	138 tons	250 tons	410 tons	410 tons	492 tons
Equipment	49 tons	53 tons	55 tons	55 tons	55 tons
Machinery	117 tons	117 tons	150 tons	150 tons	150 tons
Coal	50 tons	50 tons	50 tons	50 tons	50 tons
RFW	–	10 tons	10 tons	10 tons	10 tons
Margin	–	35 tons	41 tons	37 tons	47 tons
Total	1,250 tons	1,400 tons	1,725 tons	1,775 tons	1,825 tons

Notes:
1 Design 719 was essentially Design 718 but with a forecastle deck added, giving a freeboard of 12ft 3in at the bow.
2 In all designs, the weights as given were 75–100 tons below the quoted displacement. This was not unusual in Vickers' proposals, and was presumably intended to allow for unanticipated weight growth.

Source: National Maritime Museum, Thurston notebooks, THU/72/07, designs 708, 718, 719, 720.

artillery, one on each side of the Bosporus; the landing place would quickly be cordoned off with nets, minefields and patrol craft ... A third division and heavy corps-level artillery would be landed if the situation ashore was proceeding favourably after the landing of the first two divisions; at dawn the shipboard artillery of the entire Black Sea Fleet would energetically support the advance of the landing forces, and take under fire the heavy artillery of the Turkish shore batteries ... [W]hen the troops ashore had seized the batteries, the Fleet would enter into the Bosporus by evening, while the troops would seize the group of batteries in the middle Bosporus by a night attack, which would open the way for the Fleet to advance on Constantinople; after this, part of the transport flotilla would be sent off to embark the second echelon of two divisions at the closest ports of the Black Sea; these would be delivered to the Bosporus on the fourth day [after the landings], that is, before the arrival in Constantinople of any Turkish reinforcements; this second echelon should have been able, along with the first [echelon], to seize Constantinople and the famous Chatalja position [where a Bulgarian offensive had stalled in 1912 during the First Balkan War], which blocked the approach to Constantinople from the direction of the Balkan Peninsula, thereby cutting off the communications between Turkey and her allies.[9]

This was the operation for which *Sinop* and *Rostislav* were being prepared. However, despite its apparent feasibility, it was never put into effect. Although the Emperor supported the Bosporus operation, the Army leadership was generally opposed to any diversion of troops from the fight against Germany and Austria-Hungary, and those forces gathered for the assault had to be used to

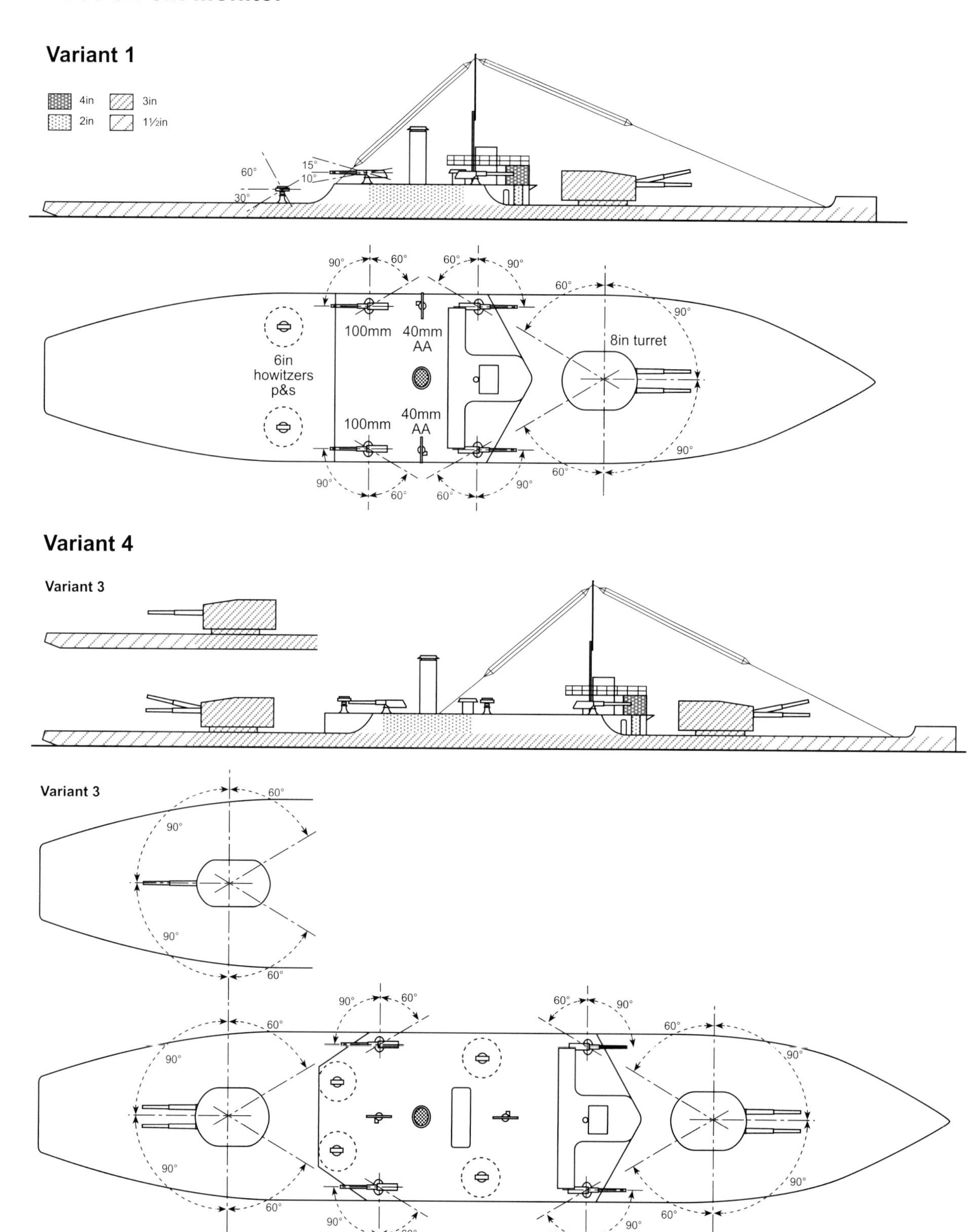

Designs for smaller monitors armed with 8in guns produced by the Navy's Technical Bureau for Ship Design, dated 3/16 March 1915. The vessels show a strong resemblance to the *Javary*-class river gunboats built by Vickers for Brazil, but which were taken into service by the Royal Navy as the *Humber* class. (© John Jordan, based on a drawing in RGAVMF f 876, op 173, d 268)

shore up the Romanians after they joined the Allied cause on 27 August 1916 and quickly collapsed. Any realistic hopes for seizing Constantinople came to an end with the February Revolution of 1917, which soon had a severe impact on the discipline and combat capability of the Russian Army. In the end, *Rostislav*'s bulges were never fitted, and the Russian dream of possessing Constantinople and the Turkish Straits at last came to a bitter end.

Small Monitors for...?

Another series of monitor designs may have been intended to meet the need for inshore close-support vessels, but this is by no means certain as once again we have drawings with no accompanying papers. The story begins in December 1914, when Vickers prepared a number of designs for 'coast defence battle monitors' (as they were dubbed by Vickers' chief designer, TG Owens).[10] All four designs were based on the river monitors of the *Javary* type that Vickers had built for the Brazilians just before the war and which were taken over by the Royal Navy in August 1914, becoming the *Humber* class. The designs prepared for Russia ranged in size from 1,250 to 1,925 tons, the smallest mounting three 6in/50 naval-pattern guns and two 4.7in howitzers, while the largest carried four 8in/50 guns and four 6in howitzers. All the designs had a speed of 11 knots. They were delivered to 'Colonel Tennyson' – almost certainly the Russian naval constructor Lt-Col KA Tennison. Their characteristics are shown in Table 3.

The Royal Navy used the *Javary/Humber* class for inshore operations off the Belgian coast, and it is possible that the designs for Russia were prepared with a similar purpose in mind. In that case the likely role would have been support for the planned amphibious assault on the Bosporus. However, this is by no means certain, since the designs pre-date Admiral Ebergard's call for smaller monitors to serve as artillery support ships, and they do not have the broadside of six 6in guns he specified. Although their shallow draught would have made them suitable for this sort of work, they would probably have experienced difficulty operating in the open waters off the Bosporus. As Ian Buxton notes of their British counterparts, 'in any force of wind they were blown all over the surface of the water'.[11]

The 6in-gun monitor *Javary* passing Clydebank in October 1913 prior to her trials. Ordered from Vickers by Brazil for service on the Amazon River, the three ships of this class were taken over by the Royal Navy as the *Humber* class. The Russians were very interested in obtaining similar ships, and several designs were prepared by Vickers as well as by the Naval Ministry's Technical Bureau for Ship Design in 1915. (John Brown, courtesy of Ian Buxton)

The other possibility is that these monitors were intended for riverine operations. The *Javary* and her sisters were originally designed for operations on the River Amazon, and it is possible that the Russians were interested in similar ships for service on the Danube. In August 1914 the Russians had established the Expedition of Special Purpose (*Ekspeditsiia osobogo naznacheniia*, or EON), to supply arms and equipment to Serbia via the Danube. EON ran convoys of steamers and barges, protected by makeshift escorts, starting in October 1914, but there was always the threat that the powerful Austro-Hungarian river monitors might interfere with these

Table 4: **Designs Prepared by the Navy's Technical Bureau for Ship Design, 3/16 March 1915**

8in/50 Designs

	Variant 1	Variant 2	Variant 3	Variant 4
Characteristics:				
Displacement	1,900 tons	2,150 tons	2,400 tons	2,600 tons
Length	280ft	295ft	315ft	320ft
Beam	49ft	51.5ft	54.5ft	57ft
Draft	6.23ft	6.23ft	6.28ft	6.33ft
Armament	2 x 8in (2xI)	2 x 8in (2xI)	3 x 8in (2xI, 1xI)	4 x 8in (2xII)
	all designs: four 4in/60			
	all designs: six MG (3-line)[1]			
	all designs: two 40mm AA			
Protection:				
main belt	*all designs:* 3in over vitals, 1.5in at ends			
transverse b/hds	*all designs:* 1.5in fore and aft			
turrets	*all designs:* 3in sides, 1in roofs			
barbettes	*all designs:* 3in			
conning tower	*all designs:* 4in sides, 2.5in roof, 2in comms tube			
funnel uptakes	*all designs:* 2in			
decks	*all designs:* 1.5in over vitals, 0.75in fore and aft			
Machinery	1,500ihp	1,600ihp	1,700ihp	1,800ihp
	all designs: two-shaft steam reciprocating engines			
	all designs: Yarrow boilers			
Speed:	*all designs:* 11 knots			

6in/50 Designs

	Variant 1	Variant 2	Variant 3	Variant 4
Characteristics:				
Displacement	1,750 tons	1,950 tons	2,150 tons	2,350 tons
Length	280ft	295ft	315ft	320ft
Beam	49ft	51.5ft	54.5ft	57ft
Draft	5.72ft	5.75ft	5.62ft	5.8ft
Armament	2 x 6in (1xII)	2 x 6in (1xII)	3 x 6in (1xII, 1 xI)	4 x 6in (2xII)
	2 x 6in howitzers	2 x 6in howitzers	4 x 6in howitzers	4 x 6in howitzers
	all designs: four 4in/60			
	all designs: six MG (3-line)[1]			
	all designs: two 40mm AA			
Protection:				
main belt	*all designs:* 3in over vitals, 1.5in at ends			
transverse b/hds	*all designs:* 1.5in fore and aft			
turrets	*all designs:* 3in sides, 1in roofs			
barbettes	*all designs:* 3in			
conning tower	*all designs:* 4in sides, 2.5in roof, 2in comms tube			
funnel uptakes	*all designs:* 2in			
decks	*all designs:* 1.5in over vitals, 0.75in fore and aft			
Machinery	1,450ihp	1,500ihp	1,600ihp	1,700ihp
	all designs: two-shaft steam reciprocating engines			
	all designs: Yarrow boilers			
Speed:	*all designs:* 11 knots			

Notes:
1 A 'line' (*liniia*) was a unit of measurement, equivalent to 0.1in, so a 3-line gun would be .30 calibre.

Source: Russian State Archives of the Navy, RGAVMF f 876, op 173, d 268.

convoys. Vessels along the lines of the Vickers-designed monitors would have been well suited to oppose such interference, although in the larger ships the draught was getting a little deep for operations on the river.[12]

Whatever their intended role, the type was obviously interesting, since in March 1915 the Russian Naval Ministry ordered its Technical Bureau for Ship Design (*Tekhnicheskoe biuro proektironvaniia sudov*) to produce a number of very similar monitor designs. Once again, documentation regarding their purpose is lacking, so all we have to go on are the sketches and characteristics of the ships (see Table 4). There are four variants with 8in guns and four with 6in guns; in general the Russian designs were somewhat larger than their Vickers counterparts and of greater displacement, but with similar armament and armour protection. The configuration of the Russian designs was very similar to that of the *Humber* class, and it seems likely that they were modelled on the Vickers designs prepared earlier. Like those ships they would have been suitable either for operations on the Danube or in support of an amphibious landing. It may be significant that, after Serbia was overrun by German and Austro-Hungarian forces in October 1915, no designs have been found for monitors that could have been used on the Danube.

Black Sea Coast Offence or Baltic Coast Defence?

It was probably in the summer of 1915 that another design process for monitors was set in train, when the state-run Admiralty Works was asked to prepare some studies of monitors armed with a single triple 14in turret. Frustratingly, there is again no known documentation as to the rationale of these ships, just the sketch designs themselves. So the starting point for any discussion of the intended role of these ships must be the characteristics of the designs themselves (see Table 5).

Since these intriguing ships were designed around a 14in turret, a little background on these guns and their mountings is necessary at this point. In 1913 the four battle cruisers of the *Izmail* class were laid down for the Baltic Fleet, and each of these ships was to carry twelve 14in/52 guns in four triple turrets. In addition, the Russian Navy intended to use 14in guns in coastal batteries in the Baltic Sea. The total number of barrels ordered amounted to 82, broken down as follows:

- 48 guns for the *Izmail* class ships
- 12 reserve barrels for these ships
- 18 guns for the Naval Fortress of Revel
- 4 guns for the Naval Proving Ground

Forty of these guns were ordered from the Obukhovskii Works in Saint Petersburg, while 36 were ordered from a new plant being built at Tsaritsyn (later Stalingrad, now Volgograd) with assistance from Vickers. It is not clear who was to manufacture the remaining six barrels.

When war broke out in August 1914 the new plant at

Table 5: **Admiralty Works 14in Monitor Design Variants, 15/28 October 1915**

Characteristics:

Displacement	*Variant I:* 7,400 tons normal *Variants II and III:* 7,550 tons normal
Dimensions	100m wl x 20m x 4.88m
Armament	three 14in/52 (1xIII); 60rpg *Variant I:* six 6in/50; 100rpg *Variants II and III:* eight 6in/50; 100rpg two 63mm AA two 47mm saluting guns two machine guns two submerged torpedo tubes
Protection:	
main belt	275mm, reducing to 200mm fore & aft, 125mm ends
longitudinal b/hds	50mm
main battery barbette[1]	275mm, reduced to 75mm below the upper deck
6in casemate	100mm sides, 50mm roof
conning tower	300mm sides upper level, 200mm sides lower level, 150mm roof, 75mm communications tube
decks	75mm upper deck, 50mm lower deck
Machinery	three 1,000bhp diesels
Speed	12 knots

Notes:

1 No figures are given for the protection of the turret, but it would probably have retained the thicknesses originally intended for the *Izmail* class battlecruisers, ie 300mm (11.8in) face and sides, 150mm (5.9in) roofs.

Weight breakdown:

	Variant I	Variants II and III
Hull	1,470 tons	1,510 tons
Reinforcement 14in turret	100 tons	100 tons
Reinforcement 6in guns	18 tons	25 tons
Reinforcement conning tower	10 tons	10 tons
Wood, cement, insulation	75 tons	75 tons
Internal fittings	85 tons	85 tons
Auxiliary structures	300 tons	300 tons
Boats and fittings	70 tons	70 tons
Mast and Rigging	10 tons	10 tons
Armour	3,070 tons	3,110 tons
Guns and ammunition	1,560 tons	1,623 tons
Torpedo armament	22 tons	22 tons
Machinery	225 tons	225 tons
Fuel	100 tons	100 tons
Shipboard supplies	85 tons	85 tons
Crew & baggage	50 tons	50 tons
Margin	150 tons	150 tons
Total	7,400 tons	7,550 tons

Source: Vinogradov, '14 diuimov dlia pribrezhnoi voiny'.

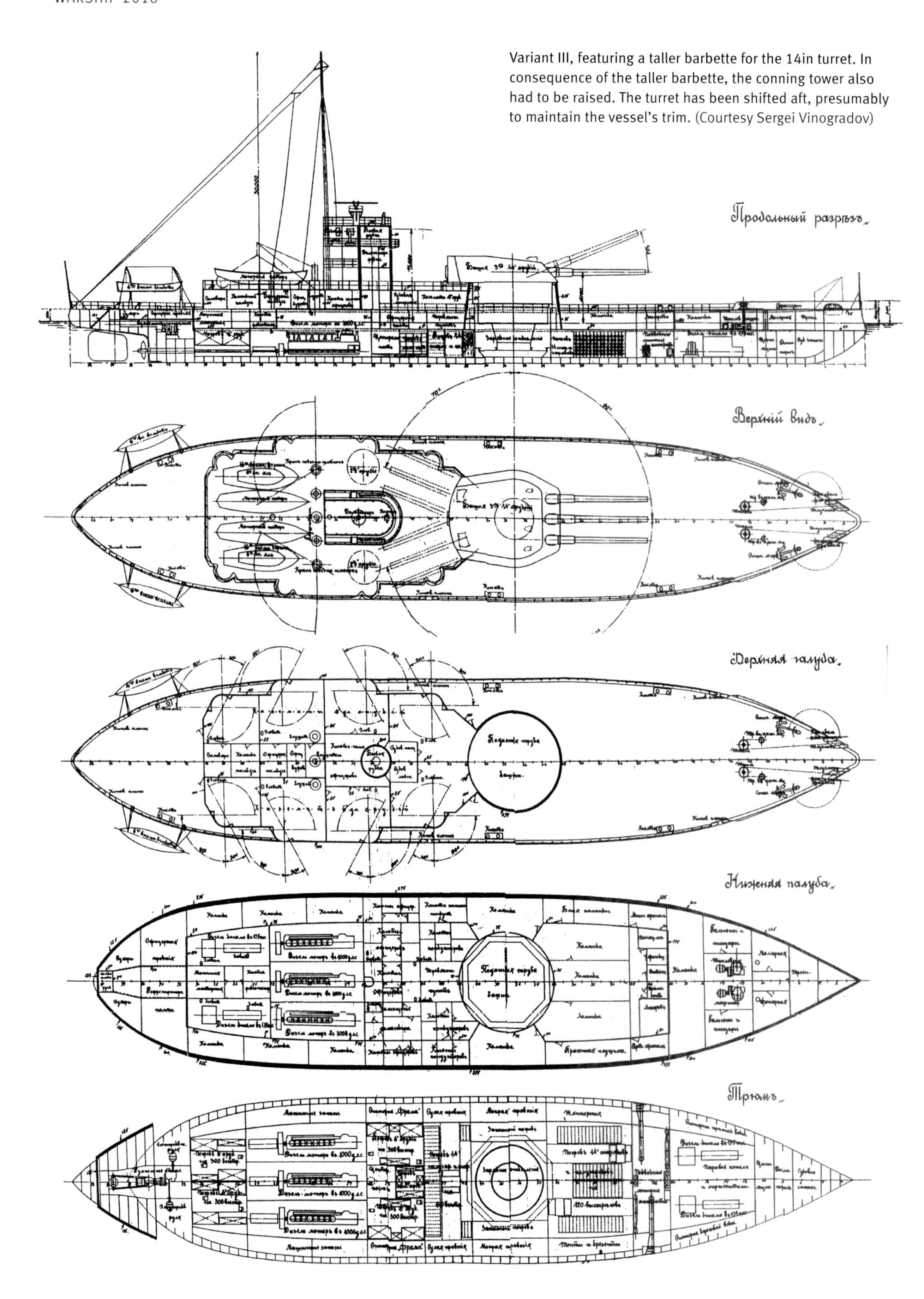

Variant III, featuring a taller barbette for the 14in turret. In consequence of the taller barbette, the conning tower also had to be raised. The turret has been shifted aft, presumably to maintain the vessel's trim. (Courtesy Sergei Vinogradov)

Tsaritsyn was incomplete, so the order for its 36 guns was shifted directly to Vickers in Britain. Both the Russian- and British-made guns were delayed by more urgent war projects, and in the event the Obukhovskii Works never finished a single gun, although ten barrels were in various stages of completion by the time of the February Revolution of 1917. As for Vickers, the firm began manufacture of the Russian guns in 1916, and eventually produced sixteen barrels. Eleven of these were shipped to Russia between September 1916 and May 1917, although one was lost in transit to Arkhangelsk when the ship carrying it was sunk by a U-boat. Thus by the spring of 1917 the Russians had ten 14in barrels on hand.

Obtaining mountings proved even more difficult than getting the guns themselves. The turrets had been designed by the Metallicheskii Works, and orders for the requisite sixteen turrets had been distributed among four Russian factories: the Metallicheskii Works, the Obukhovskii Works, the Admiralty Works, all in Saint Petersburg, and ONZiV in Nikolaev. Unfortunately, the turret design called for a number of foreign-made components – in particular, 8in-diameter spherical bearings to serve as rollers, and these bearings had been ordered in Germany. Naturally, after the war began their delivery was out of the question, and the order for them and the other foreign-made components had to be shifted to domestic factories already overloaded with high-priority war work. As a result, in 1915 the Naval Ministry declared that the armament of the *Izmail* class ships 'cannot be manufactured quickly enough for their participation to be expected in the current war'.[13] The battle cruisers were therefore downgraded to a second-priority project, and work on them slowed.

However, if there was no possibility of obtaining a complete suite of guns and turrets for a battle cruiser, by August 1915 it seemed likely that at least a few complete mountings might soon be available. In that month a Russian company delivered eighty 8in bearings for the turrets (although in the end only eleven of these passed inspection); the Metallicheskii Works had almost completed one turret, and most of the components for a few more were in an advanced stage of work. So by the summer of 1915 the possibility existed of obtaining one or perhaps two mountings from the Metallicheskii Works and a number of barrels from Vickers. This must have raised the question of how best to make use of these powerful guns, and this in turn presumably led to the idea of mounting them in a monitor or two. By 15/28 October 1915 three variants for ships with a single triple 14in turret had been prepared, all with the same dimensions, speed and protection.

The timing of these designs suggests that they were an offshoot of the commission established to investigate the possibility of building monitors to support an amphibious assault on the Bosporus. That body had begun work in August 1915, and the Admiralty Works' design office must have started drawing up these monitor sketches around the same time; they were completed just two weeks after the commission held its last session in October. But aside from the low freeboard of 3 feet (0.91m) and shallow draft, these designs seem unsuited to inshore work off an enemy's coast – despite access to the British monitor design, bulges were not fitted, and in fact the underwater protection was very weak, with a single unarmoured bulkhead set about 12 feet (3.6m) inboard of the outer hull, making their operation in

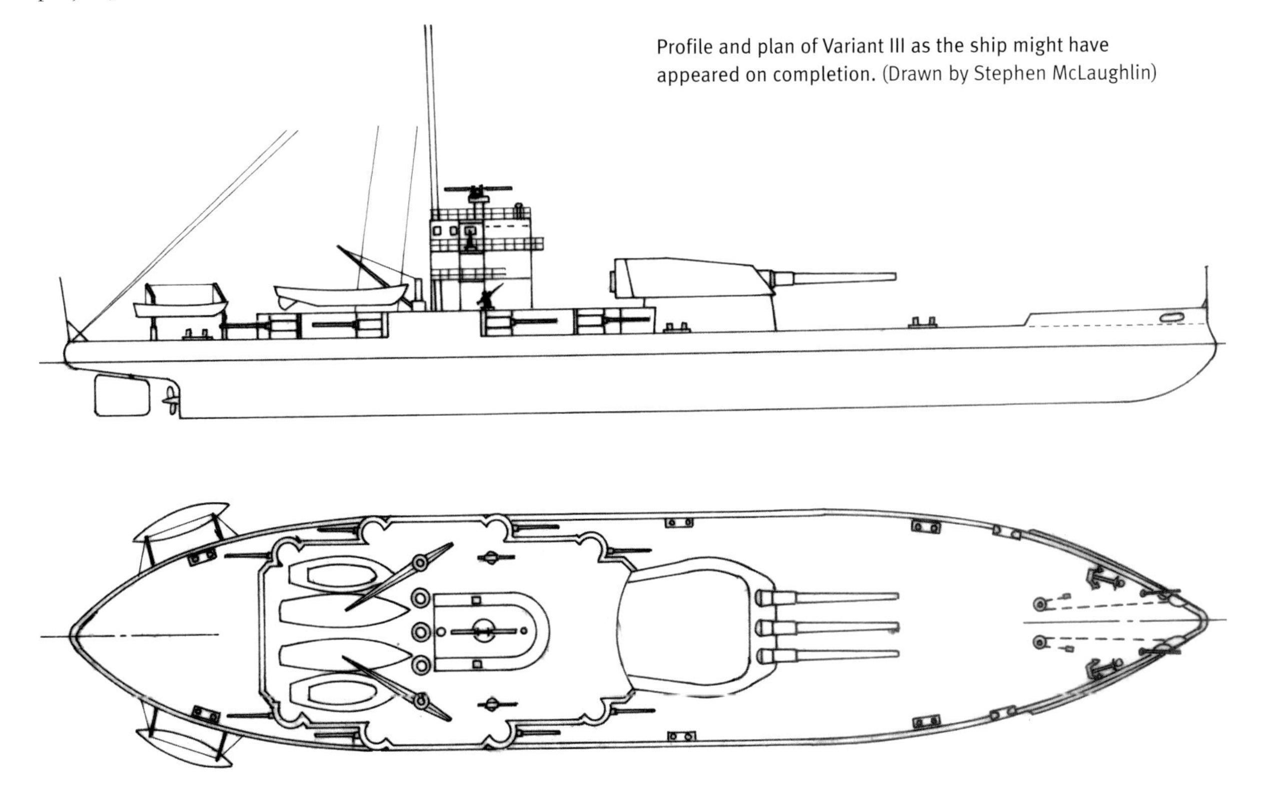

Profile and plan of Variant III as the ship might have appeared on completion. (Drawn by Stephen McLaughlin)

mine-strewn coastal waters extremely hazardous. Moreover, they were complex designs, with casemate-mounted 6in guns, a longitudinal splinter bulkhead analogous to that fitted in dreadnoughts, and even underwater torpedo tubes. Particularly noteworthy is the battleship scale of the armour protection of these designs – indeed, at 275mm the belt was heavier than that of *any* Russian dreadnought (the *Sevastopol* class had a 229mm belt, the *Imperatritsa Mariia* class, 262.5mm, the *Izmail* class, 237.5mm, and the never-completed *Imperator Nikolai I* had a 270mm belt). There was even a complete battleship-style conning tower – surely an excessive (and heavy) luxury in such small vessels. These were not at all the sort of 'quick-and-dirty' vessel that a navy could risk off an enemy coast, such as the British monitors or the Italian *Faà di Bruno*.

In sum, the designers at the Admiralty Works seem to have been aiming for a sort of mini-battleship rather than a monitor for coastal operations. The unusual technical features might be explained by a simple misunderstanding between the commission and the Admiralty Works regarding what was required, perhaps further confused by 'an inertia in the tactical thinking' of the Naval General Staff.[14] Certainly this is the view of Sergei Vinogradov, who has studied these designs more closely than any other historian. There is, however, a second possibility: that these ships were intended not to support the Bosporus landing, but were to be used 'in the mine-artillery positions of the Baltic Sea', as two other writers, Gribovskii and Chernikov, maintain.[15] The concept of the 'mine-artillery position' formed the basis for Russia's prewar strategy for the Baltic theatre: a minefield laid at the entrance to the Gulf of Finland, flanked by powerful shore batteries on the Finnish and Estonian coasts, with battleships providing coverage for the central portion of the minefields. To break into the Gulf of Finland, an enemy – assumed to be Germany in all Russian planning prior to the First World War – would have had to sweep lanes through the minefields, but the minesweepers could have been brought under fire by the flanking coast batteries or the battleships positioned behind the minefields. Meanwhile, the German battleships would have been unable to protect their minesweepers because the minefields would have prevented them from closing within effective range of the Russian battleships. With this combination of defences, it was believed that even the powerful German High Seas Fleet could be kept at bay for a considerable time.

At the outbreak of war, the coastal batteries intended to protect the mine-artillery position were far from complete; nevertheless, the basic concept continued to serve as the cornerstone of Russia's Baltic strategy. With their heavy artillery, these monitors could certainly have helped defend mine barriers against German incursions, while their own exposure to enemy mines and torpedoes would have been minimal, so the weak underwater protection would not have been an issue – although the underwater torpedo tubes still defy rational explanation.

Their shallow draft would also have made such monitors well suited to operations in the Finnish skerries, a need Essen had foreseen before the war. But if the Baltic was indeed their intended field of operations, it seems

Table 6: **Chernikov's 1916 Monitor 'Reconstructions'**

	Coast Defence Monitor	Armoured Gun Vessel
Characteristics:		
Displacement	7,550 tons	11,000 tons
Dimensions	100m x 20m x 4.88m	112.9m wl x 28.6m max x 5.26m
Armament	three 14in/52 (1xIII)	nine 14in/52 (3xIII)
	eight 6in/50 (8xI)	twelve 4in[1]
	two 76.2mm AA	two 76.2mm AA
	two 47mm saluting	
	two MGs	
	two submerged TT	
Protection:		
main belt	275mm, 125mm fore & aft	256mm, 25mm, 15mm
longitudinal b/hds	75mm, 100mm	?
transverse b/hds	150mm, 125mm, 100mm, 75mm	?
main battery barbette(s)	275mm upper, 75mm lower	?
casemates	100mm, 50mm 'cover'	100mm
conning tower	300mm sides, 150mm roof, 75mm tube	?
decks	75mm upper, 50mm lower	20mm, 12mm, 15mm
Machinery	three-shaft diesel	two-shaft diesel
	3,000bhp	?
Speed	12 knots	16 knots

Notes:
1 These would presumably have been the 4in/60 gun, production of which began in 1911.

Source: http://otvaga2004.mybb.ru/viewtopic.php?id=145.

more probable that they were destined for the Gulf of Riga. This large bay had not been part of the original Russian prewar defensive strategy, but it became a vital area for supporting the Army's seaward flank during the war. In August 1915 the German Navy mounted two attacks on the western entrance to the gulf, the Irben Strait, successfully breaking into it on the second attempt. Among the defenders was the predreadnought battleship *Slava*, but her old 40-calibre 12in guns lacked the range to counter German dreadnoughts. Fortunately for the Russians, although the German Navy was able to break through into the Gulf of Riga, the operation was soon abandoned due to lack of support from the Army.

A monitor along the lines of the Admiralty Works designs would have been ideal for this theatre of operations, with guns heavy and long-ranged enough to deter even the latest German dreadnoughts from attempting to force the Irben Straits. The timing of the designs is again suggestive: they were completed in October 1915, and may well have been a response to the August campaign in the Gulf of Riga. It is therefore possible that the designs were an attempt to bolster the defences of the Irben Straits. Indeed, one recent Russian article claims that the 'experience of combat actions in the Gulf of Riga' provided the impetus for these designs, although no archival source is given for this assertion.[16]

Given the lack of documentation, all of this is speculation. For these intriguing monitor designs we have only

Russian 14in Monitors

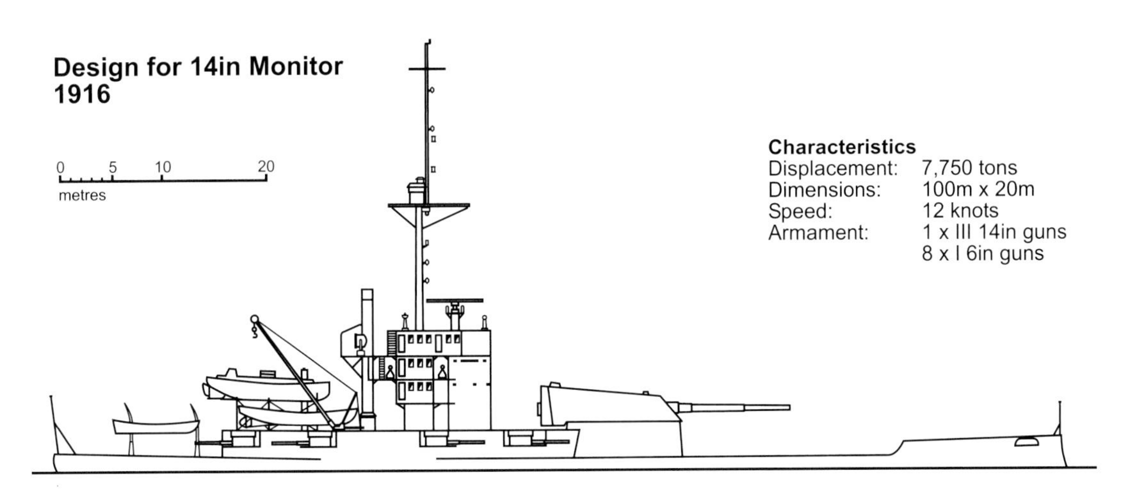

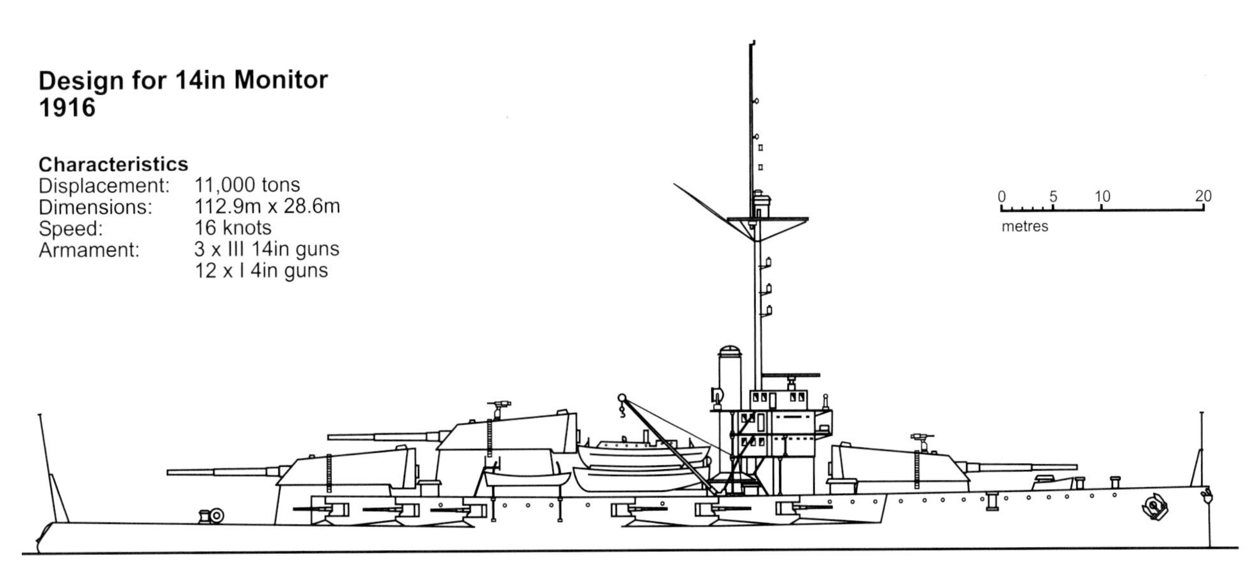

A pair of 'reconstructions' of 1916 designs for monitors, signed by the Russian naval historian Il Chernikov. The upper drawing shows what appears to be a development of the 1915 monitor designs, with a single 14in turret. The lower drawing, however, shows a much more ambitious vessel, with three triple 14in turrets; it is labelled an 'armoured gun vessel for the Baltic Fleet', and amounts to a very powerful type of coastal battleship. Unfortunately the source of the drawings is unknown, and they may be apocryphal. (© John Jordan, based on sketches published on the internet)

Table 7: Guns of the Monitor Designs

	14in/52	12in/52	8in/50	6in/50	130mm/55
Gun characteristics:					
Calibre	14in/355.6mm	12in/304.8mm	8in/203.2mm	6in/152.4mm	5.12in/130mm
Dates	1917 (trials)	1907 (ordered)	1905	1908 (trials)	1913 (production)
Weight	82–83 tons	50,600–50,700kg	14,397kg	6,850kg	5,290kg
Barrel Length	18,490mm/52 cal	15,850mm/52 cal	10,160mm/50 cal	7,620mm/50 cal	7,150mm/55
Bore length	17,927.3mm/50.41 cal	15,419.5mm/50.59 cal	9,959mm/49 cal	7,467mm/49 cal	6,939.8mm/53.38 cal
Rifled length	14,954.25mm/42 cal	12,852mm/42.2 cal	8,167mm/40.2 cal	6,121mm/40.16 cal	5,862mm/45.1 cal
Rate of fire	3rpm (designed)	1.66–1.33rpm	3rpm	4rpm	5–6rpm
Projectiles & performance:					
Weight	747.8 kg	470.9kg	112.2kg	47.3kg	36.86kg
Charge	203kg	132kg	39kg	17kg	11kg
MV	731.5m/sec	762m/sec	807.7m/sec	823m/sec	823m/sec
Range	23,200m @ 25°	23,228m @ 25°	16,827m @ 19.72°	17,385m @ 30°	15,364m @ 20.18°

	120mm	4in/60	2.5in AA[1]	40mm AA Vickers
Gun characteristics:				
Calibre	4.72in/120mm	4in/102mm	2.5in/63.5mm	1.575in/40mm
Dates	1908	1911 (trials)	?	1914 (ordered)
Weight	3,150kg	2,850kg	1,300kg (incl mounting)	229kg
Barrel Length	6,000mm/50 cal	6,284mm/61.6 cal	2,413mm/38 cal	1,576mm/39.25 cal
Bore length	5,803mm/48.36 cal	?	2,266mm/35.7 cal	?
Rifled length	5,042mm/42.02 cal	5,284.5mm/51.81 cal	?	1,393mm/34.82 cal
Rate of fire	?	12–15rpm	?	200rpm
Projectiles & performance:				
Weight	20.48kg	17.5kg	3.73kg	0.91kg
Charge	7–8kg	2.4kg	0.8kg	0.093kg
MV	823m/sec	823m/sec	686m/sec	622m/sec
Range	13,359m @ 20°	15,364m @ 30°	6,804m @ 20°	ca 1,200yds[2]

	12in Howitzer Vickers Mk. II	6in Howitzer Model 1910[3]	4.7in Howitzer Vickers
Gun characteristics:			
Calibre	12in/304.8mm	152.4mm	4.72in/120mm
Dates	1916	1911 (trials)	?
Weight	?	831.3kg	0.57 tons
Barrel Length	160in/13.33 cal	1,829mm/12 cal	89.9in
Bore length	?	?	18 cal
Rifled length	?	1,417mm/9.3 cal	?
Rate of fire	?	6	10
Projectiles & performance:			
Weight	750lb/340.3kg	41kg	45lb
Charge	?	1.83kg	1.78lb
MV	?	335.3m/sec	1,200fps
Range	11,340yds	7,767m	ca 9,000yds @ 45°

Notes:

1 Shirokorad gives the calibre of the 2.5in AA gun as 63.3mm, but 63.5mm seems to fit the bore and barrel lengths better.

2 Range for Vickers 40mm is effective AA range.

3 The Model 1910 152mm howitzer was the most modern Russian weapon of this type, but it is possible that older weapons would have been mounted in the monitors.

4 The 47mm saluting guns specified for the 1915 monitor designs would presumably have been the 47mm adopted in the 1880s; as of 1 January 1901 the Naval Ministry had 963 of these guns in stock, and although a large number must have been lost during the Russo-Japanese War, a considerable number were still available in 1914–1917.

5 The 10in/30 or /35 suggested by Kononov for his monitor proposal does not correspond to any gun in Russian service; the only gun of this calibre in naval or land service was the 10in/45 dating from the 1890s. Possibly Kononov's memory was faulty on this point, or perhaps he envisaged cutting down existing 10in/45 guns to convert them to howitzers.

Sources: Shirokorad, *Entsiklopediia otechestvennoi artillerii*; Titushkin, 'Artilleriia russkogo flota v 1877–1904 gg'.

the bare facts of their characteristics and timing to guide us in interpreting their intended purpose. It may well be that the Naval General Staff had some other employment in mind for these 'pocket dreadnoughts'. In any case, they are fascinating vessels.

Postscript

It is possible that Russian interest in monitors persisted into 1916, but the evidence is equivocal. Two sketches have been posted on the Internet, apparently the work of the noted Russian naval historian II Chernikov. They are labelled 'reconstructions', which suggests that Chernikov found some description of the designs in the archives, but not the actual drawings; their characteristics are given in Table 6. One sketch, described as a 'coast defence armourclad of 1916', appears to be a simple development of the 1915 designs; the second sketch, however, is far more unusual, not to say bizarre. It is labelled an 'armoured gun vessel for the Baltic Fleet of 1916', and shows an 11,000-ton ship carrying three triple 14in turrets. This seems an overly ambitious design for wartime conditions, and is massively over-gunned for its size. Unfortunately, the original source of these drawings has not been traced; Chernikov himself died a few years ago, and his personal papers have yet to be sorted and made public, so for the present these two designs must remain no more than tantalising possibilities.[17]

Acknowledgements:

As so often in the past, I am indebted to Sergei Vinogradov for much of the material used in this article, and to John Jordan for turning a rather confused batch of material into fine drawings of the monitor designs. Ian Buxton generously provided photographs of British monitors. My wife Jan Torbet has once again provided invaluable service as an editor *par excellence*.

Sources:

Russian State Archives of the Navy (RGAVMF), f 876, op 173, d 268; f 876, op 173, d 369.

Thurston notebooks, THU/72/07, Caird Library, National Maritime Museum, Greenwich.

Vinogradov, SE '14 diuimov dlia pribrezhnoi voiny', *Morskaia kampaniia*, 2009, no 7, 12-25.

Endnotes:

1 N Lushkov, 'Kakoi nam nuzhen bronenosnyi flot?', *Morskoi sbornik*, vol 314, no 5 [May 1911], 80.

2 EF Podsobliaev, 'Kakoi flot nuzhen Rossii? Po materialam diskussii, sostoiavsheisia nakanune Pervoi mirovoi voiny', *Novyi Chasovoi: Russkii voenno-istoricheskii zhurnal*, no 34, 1996, 61.

3 RGAVMF f r-1529, op 2, d 5, l 40; reference courtesy of Sergei Vinogradov, e-mail to author, November 2013.

4 AIu Emelin and K L Koziurenok, 'N O fon Essen, A V Kolchak i razrabotka programmy usilennogo sudostroeniia Baltisskogo flota', *Gangut*, no 24, 2000, 30–31, 43.

5 MA Petrov, 'Podgotovka Rossii k mirovoi voine na more', *Gosudarstvennoe voennoe izdatel'stvo*, Leningrad 1926, 85–86.

6 IaM Zakher, 'Konstantinopol' i prolivy (ocherki iz istorii diplomatii nakanune mirovoi voiny)', *Krasnyi arkhiv*, vol 6, 1924, 68.

7 IA Kononov. Letter to the Editor, *Morskoi zhurnal* [Prague], whole no 82/83 [10/11], October–November 1934, 19–23, 159–163.

8 Although these designs are not dated, other designs in this number sequence date from the spring of 1915.

9 Aleksandr Bubnov, *V tsarskoi stavke: Vospominaniia admirala Bubnova*, Izdatel'stvo imeni Chekhova, New York 1955, 282–283.

10 National Maritime Museum, Thurston notebooks: designs 708, 718, 719, 720.

11 Ian Buxton, *Big Gun Monitors: Design, Construction and Operation 1914-1945*, Naval Institute Press, Annapolis 2008, 94.

12 In February 1915 the Royal Navy ordered the *Fly* class gunboats for use on the Danube in the event that the Allies succeeded in breaking through the Dardanelles and into the Black Sea; these vessels, described as 'China gunboats' to conceal their actual purpose, eventually formed the Tigris Flotilla.

13 SE Vinogradov, *'Izmail' – sverkhdrednout Rossiiskoi imperii*, Modelist-konstruktor, Moscow 2001, 19.

14 Vinogradov, SE, '14 diuimov dlia prebrezhnoi voiny', 17.

15 VIu Gribovskii and II Chernikov, *Bronenosets 'Admiral Ushakov'*, Sudostroenie, St Petersburg 1996, 121.

16 Emelin and Koziurenok, 'N O fon Essen, A V Kolchak', 44–45 n 16.

17 Information provided by Sergei Vinogradov, e-mail to author, received 9 June 2016.

Note about dates: All the dates in this article are given in the form Russian/Western calendar.

MODERN NAVAL REPLENISHMENT VESSELS

Conrad Waters conducts a survey of the latest developments in replenishment at sea.

The availability of an effective replenishment at sea (RAS) capability is an essential requirement for the conduct of current 'blue water' naval operations.[1] Today's warships are often capable of unsupported deployment in peacetime conditions. Indeed, a number of recent classes – for example, Germany's F125 *Baden-Württemberg* 'stabilisation frigates'– have been specifically designed with extended constabulary taskings in mind. However, in the absence of specialised replenishment shipping, such missions are only possible if friendly harbours are available for refuelling and re-provisioning. Even in peace, considerations of geography and/or politics may mean that this is not the case. Clearly, the availability of such facilities is even less likely to be assured in time of war. In their absence, a modern navy's freedom of action is essentially determined by its ability to replenish its fighting vessels at sea.

The extent and nature of a navy's RAS requirement will inevitably reflect a number of variables. The more important include a navy's likely area of operations; the availability of supporting bases in these localities; and the extent of the warfighting capability it wishes to support. For example, it can be argued that the British Royal Navy was slower than the US Navy to develop effective RAS capabilities during the days of empire because of its access to a global network of dockyards and refuelling stations that was unmatched by any other naval power.[2] The most intensive RAS requirements are inevitably associated with aircraft carrier deployments. The volumes of ordnance and aviation fuel that are likely to be expended in such operations need a very high RAS capacity. Moreover, it is desirable for the vessels providing this capacity to have sufficiently powerful and sophisticated propulsion machinery to remain on station with a potentially fast-moving carrier group.

In spite of the clear importance of RAS and a greater focus on expeditionary activities, procurement of replenishment vessels was limited in the immediate post-Cold War era. This probably reflected the impact of restricted funding for new construction and the reduction in fleet size experienced by many of the established blue water navies. A consequence of this has been a significant increase in the average age of logistic support fleets across the world to an extent that replacement can often be deferred no longer. As a result, many acquisition programmes for replenishment ships are now underway. The present article aims to consider some of the factors impacting these new designs and to assess the resulting ships. Table 1 summarises the characteristics of the principal designs considered.

Replenishment at Sea: Historical Background

It is helpful, as a starting point, to place current replenishment ship designs in the context of the historical development of replenishment at sea. The steady transition of navies from sail to steam during the 19th century resulted in a number of new logistical challenges. Among the most significant was the new generation of ships' seemingly insatiable appetite for coal, a bulky fuel that was also inherently difficult to transfer. Essentially, this meant that either a network of coaling stations or a large fleet of colliers was needed to provide a fleet with strategic mobility. The latter also required a safe location where coaling could occur while at anchor. Although successful trials of coaling at sea were conducted by both the Royal Navy and US Navy, this evolution required calm weather conditions and never achieved a sufficiently rapid rate of transfer.

The increasing use of oil fuel during the early years of the 20th century made refuelling at sea a much more practical proposition. The Royal Navy carried out experiments with underway oiling as early as 1905. However, other fleets – particularly the US Navy – were subsequently to take the lead. It seems that the Americans were the first to use a refuelling at sea operationally when, in May 1917, the US Navy's fleet oiler *Maumee* (AO-2) successfully refuelled the six destroyers of the Fifth Destroyer Squadron on transit to the British Isles in a single day.[3] The US Navy remained in the forefront of conducting refuelling at sea in the inter-war era. This reflected the realisation that Japan was now the most likely enemy. As a result, the fleet would have to operate across the vast expanse of the Pacific to come to the relief of the then-American territory of the Philippines in the event of hostilities.

The far-sightedness of the US Navy's approach was demonstrated during the deployment of fast carrier task forces against Japanese forces for extended periods in the later years of the Second World War. However, these operations required more than just a refuelling capacity; the intensity of operations also required frequent replenishment of munitions and other stores. This resulted in

Two photos of the now-retired Royal Navy aircraft carrier *Ark Royal* replenishing from the AOR-type replenishment oiler *Wave Knight* – designated a 'fast fleet tanker' in British parlance – in 2010. The images depict simultaneous refuelling from a replenishment at sea (RAS) station and vertical replenishment (VERTREP) of solid stores by means of a helicopter. Replenishment at sea is a prerequisite for the effective support of modern 'blue water' naval operations, particularly when carrier-based operations are being conducted. (Crown Copyright 2010)

the acquisition of a whole new fleet of replenishment vessels that utilised a rudimentary rig to transfer these supplies at sea. The ships that formed this expanded supply 'train' were largely basic conversions of existing tanker and freighter designs. A consequence was a number of operational limitations. Notably, the resulting ships tended to be restricted to providing just one category of supply, significantly extending the duration of hazardous and vulnerable replenishment evolutions. In addition, the converted ships had insufficient speed to keep pace with the carrier groups. The standard process was for a replenishment group to operate a shuttle service from a fixed resupply base to a (comparatively) safe rendezvous point some way back from the theatre of operations. The combatant forces would periodically retire from the front line to allow the replenishment rendezvous to take place.

The US Navy was not the only fleet to pay particular attention to replenishment capabilities in the inter-war period. The German *Kriegsmarine,* which had lost its overseas bases as a result of the First World War, also acquired a flotilla of supply ships and tankers to support raiding activities in the North Atlantic and beyond. In the late 1930s it laid down the *Dithmarschen* class of logistic support vessels. Considerably in advance of ships being acquired by other navies, these could supply fuel, dry stores and munitions, as well as providing repair and hospital facilities.[4] They also had a much faster turn of

Table 1: **MODERN REPLENISHMENT SHIP DESIGNS**

Class	*Cantabria*	*Aotearoa*	*Tidespring*	*Maud*	*Berlin* (Type 702)	*Karel Doorman*
Type:	AOR	AO/AOR	AO (Fleet Tanker)	AOR (LSV)[1]	AOR	JSS
Date:[2]	2010	[2020]	2017	[2018]	2001	2015
Navy:	Spain & Australia	New Zealand	United Kingdom	Norway	Germany & Canada	Netherlands
Builders:	Navantia, Spain	HHI, S Korea	DSME, S Korea[3]	DSME, S Korea	ARGE EGV, Germany[4]	Damen, Romania[5]
Designers:	Navantia, Spain	Rolls-Royce, UK	BMT, UK	BMT, UK	ARGE EGV, Germany	Damen, Netherlands
Number:	1 (SP) + 2 (AUS)	1	4	1	3 (GMY) + 2 (CAD)	1
Unit Cost:[6]	US$300m	US$350m	US$200m	US$170m	US$420m[7]	US$480m
Displacement:	19,500 tonnes	24,000 tonnes	39,000 tonnes	26,000 tonnes	21,000 tonnes	28,000 tonnes
Dimensions:	174m x 23m x 8m	166m x 25m x 9m	201m x 29m x 10m	181m x 26m x 9m	174m x 24m x 7m	205m x 30m x 8m
Propulsion:	Diesel; 1 shaft	CODLAD; 2 shafts	CODLOD; 2 shafts	CODLOD; 2 shafts	Diesel; 2 shafts	IEP; 2 shafts
Speed:	21 knots	16 knots	15+ knots	18 knots	20 knots	18 knots
Range:	6,000nm	6,400nm	8,000nm	Not known	Not Known	10,000nm
Capacity (liquid):	7,300 tonnes diesel	8,000 tonnes diesel	17,000 tonnes fuel	7,000 tonnes diesel	7,900 tonnes diesel	6,800 tonnes diesel
	1,400 tonnes aviation	1,500 tonnes aviation	(included above)	300 tonnes aviation	500 tonnes aviation	900 tonnes aviation
	200 tonnes water	250 tonnes water	1,300 tonnes water	Not known	1,300 tonnes water	400 tonnes water
Capacity (solid):	750 tonnes cargo	No stated capacity	No stated capacity	200 tonnes cargo	600 tonnes cargo	See text
	20 containers	12 containers	8 containers	40 containers	78 containers	See text
Facilities:	4 x beam RAS	2 x RAS beam	3 x RAS beam	2 x RAS beam	2 x RAS beam	2 x RAS beam
	1 x heavy crane	1 x heavy crane	2 x cranes	1 x heavy crane	2 x heavy cranes	1 x heavy crane
	Stern refuelling	Stern refuelling	Stern refuelling	Stern refuelling		
Aviation:	2 x helicopters	1 x helicopter	1 x helicopter	2 x helicopters	2 x helicopters	6 x helicopters
Accommodation:	164 personnel	98 personnel	ca 110 personnel	ca 100 personnel	ca 240 personnel	300 personnel

Notes:

There is considerable variation in published data on the ships covered, and the above table should be considered as indicative only. Specific notes are as follows:

1 *Maud's* official designation is a logistic support vessel, reflecting her multi-role support capabilities that extend beyond replenishment.
2 Dates refer to commissioning of the first unit of the class.
3 Final fitting out was carried out by A&P Group in Falmouth, UK.
4 The two Canadian Ships will be built by Seaspan in Canada.
5 Final fitting out was carried out at Damen's yard in Vlissingen in the Netherlands
6 Unit costs are based on converting local currencies to US$ amounts at mid-2017 exchange rates. Given differences in cost calculations from country to country, they should be treated with particular caution.
7 Cost and subsequent data refers to the third ship of the class, *Bonn*, which was commissioned in 2013.

The replacement of coal by oil in naval propulsion facilitated the development of refuelling at sea techniques. The US Navy was quick to assume leadership in this area, using the fuel ship *Maumee* to facilitate the deployment of destroyers to the British Isles in 1917. (US Naval History & Heritage Command, NH 93098)

The German *Kriegsmarine*'s *Dithmarschen* class of logistic support vessels – which could meet a broad range of supply requirements – had a significant impact on post-Second World War US and Royal Navy thinking. *Dithmarschen* herself was taken into US Navy service as *Conecuh* (AOR-110) and used to test the feasibility of the replenishment oiler concept. (US Naval History & Heritage Command, 80-G-678091)

The fast combat support ship *Sacramento* (AOE-1) replenishing the battleships *Wisconsin* (BB-64) and *Missouri* (BB-63) in the course of Operation 'Desert Shield' in 1991. The fast combat support ship, developed to operate with carrier strike groups as a 'one stop shop' for all replenishment needs, is arguably the ultimate development of the replenishment ship concept but is an expensive investment. *Sacramento* used half the propulsion plant intended for *Missouri* and *Wisconsin*'s uncompleted sister *Kentucky* (BB-66) to achieve the speed necessary for carrier operations. (US Navy)

speed – significantly in excess of 20 knots – than other contemporary oilers, although replenishment was usually undertaken when stationary rather than underway. Two of the class survived the war to be taken into US Navy and Royal Navy service. Here – in modified form – they were used to develop 'one stop' replenishment concepts under which a single ship could transfer a wide range of liquid and/or solid stores while underway. As with the *Dithmarschen* class, many of the resulting new generation of logistic support vessels were specially designed as resupply ships rather than being simpler merchant ship conversions. Another advance during this time was the introduction of improved RAS rigs that facilitated the replenishment process.

The next stage of replenishment ship development in the US Navy was the construction of fast, multi-product combat support ships. These could operate as part of a carrier task force and supply all its logistical needs in a relatively quick and efficient fashion. The first of these, *Sacramento* (AOE-1), was laid down in 1961 as one of a class of four ships. Displacing in excess of 53,000 tons in full load condition, she could accommodate 177,000 barrels of fuel, 2,150 tons of ammunition and 750 tons of other stores when delivered in 1964. Maximum speed was 26 knots.[5] The design was also innovative in incorporating both a flight deck and hangar for up to three medium helicopters, thereby helping support the developing concept of vertical replenishment (VERTREP). A class of four slightly smaller *Supply* (AOE-6) fast combat support ships was commissioned between 1994 and 1998.

In spite of the undoubted capability provided by these ships, the AOE concept had limited application beyond the US Navy. Given the absence of significant carrier forces, few other navies had a need for replenishment with the volume of solid stores provided by a large combat support ship. Moreover, few could justify the expense of both acquisition and operation associated with the high speed of an AOE. As a result, the most numerous type of modern replenishment vessel is the multi-product replenishment oiler or AOR. These typically carry a smaller volume of dry stores and ammunition than a full-scale AOE and operate at slower speeds.

Even the US Navy has found it hard to afford AOE operation in the post-Cold War era, taking all but two of these ships out of service by 2017. However, its solid replenishment ammunition cargo ships (AKEs) are equipped with a limited refuelling capability. Similarly, its oilers (AOs) have a modest ability to perform underway transfers of solid stores. One of the main weaknesses inherent in this approach is the slower speed of these vessels. They were initially intended to operate primarily as 'shuttle ships' transporting fuel and stores from a resupply port to the carrier group in similar fashion to their Second World War predecessors. Now they also have to operate as substitute 'station ships' for the AOEs with the carrier groups. This is presumably justified by the reduced risk of operations being challenged by a 'near peer' opponent following the collapse of the Soviet Union.[6]

Although modern replenishment fleets have come to be dominated by the ubiquitous AOR-type replenishment oiler, inevitable variations in the operating requirements of different navies are reflected in similar variations between the ships currently in service. In addition, many of the new generation of vessels now entering service after the post-Cold War hiatus in construction exhibit significant design and technological changes from their predecessors. To a large extent this reflects the evolution in mercantile operating conditions and practices during a period when few new naval logistic support ships were constructed.

Environmental Considerations

Among the strongest of these influences on the mercantile sector – particularly with respect to commercial product tankers – has been the implementation of more stringent environmental regulations. In particular, the US Oil Pollution Act of 1990 and revisions to the International Maritime Organization's MARPOL 73/78 regulations from the early 1990s onwards have resulted in the phasing-out of single-hulled tankers and their replacement by double-hulled vessels.[7] Naval ships have typically been able to obtain exemption from these regulations. Indeed, many single-hulled replenishment tankers remain in service today. However, there are significant practical and ethical hazards associated with such an approach. A number of existing ships have even been expensively-modified to a double-hulled standard. More significantly, the replacement AORs being now brought into service are invariably double-hulled.

A good example of an evolutionary transition to the new double-hulled requirement is provided by the Spanish Navy's *Cantabria*. Commissioned in 2010 at a cost of around €250m (ca US$300m), she is effectively a development of the 1990s-era *Patiño* and *Amsterdam*. The two earlier ships were acquired as a result of a joint programme between the Netherlands and Spain and, in turn, owed something to the influence of the earlier Dutch *Zuiderkruis*. The new ship's full load displacement of ca 19,500 tonnes is some 2,000 tonnes greater than her predecessor and she has slightly larger dimensions. However, total liquid cargo capacity is modestly reduced from that provided in *Patiño*, reflecting the design trade-offs associated with the double-hulled format.[8] Typical liquid load includes a relatively higher proportion of aviation fuel than that embarked in the

The Spanish *Patiño* – pictured here in June 2017 – is broadly typical of the AOR–type replenishment oilers put into service by many mid-sized 'blue water' fleets in the later part of the Cold War era. She was built as part of a joint procurement programme with the Netherlands and has a near sister, *Tacna* (the former *Amsterdam*), in Peruvian service. A single-hulled ship, she no longer meets modern environmental requirements. (Leo van Ginderen collection)

replenishment ships of mid-ranking European fleets, reflecting the need to support the Navy's AV-8B Harrier II component. Specific provision has been made to ship containerised cargoes.

Other enhancements over the original design include improvements to the sensor suite and installation of the indigenous SCOMBA combat management system, currently being rolled out across the Spanish fleet. *Cantabria's* design is otherwise quite similar to that seen in *Patiño*, with two combined RAS stations for both liquids and solids on each side of the ship and a single stern refuelling station. Two medium-sized helicopters can be embarked and supported for VERTREP. The propulsion system links two diesel engines to a single shaft line in direct-drive configuration. Maximum sustained speed is in the order of 20 knots. Improved medical facilities encompass a small but comprehensively-equipped hospital facility. It incorporates both operating and intensive-care facilities. This reflects a trend – enlarged upon below – towards utilising the space inherent in large logistic ships to provide a range of supplementary capabilities.

As well as the single ship commissioned into the Spanish Navy, the *Cantabria* design has also achieved some success in export markets. In May 2016, Spanish builders Navantia signed an A$646m (ca US$520m) contract to deliver two slightly modified variants of the class to the Royal Australian Navy (RAN). The order was placed to meet the requirements of Australia's Project SEA 1654 Phase 3 and involves the replacement of two existing single-hulled oilers. The RAN has also recently taken delivery of new amphibious assault ships and air defence destroyers designed by Navantia. This meant that selection of the Spanish AOR provided a level of commonality with respect to systems and equipment, providing associated support efficiencies. However, the Spanish ship has fared less well in a number of other competitions. This may reflect the view that the class's lengthy antecedents arguably mean that it no longer reflects the latest thinking in replenishment ship design.

One potentially obsolescent feature in the *Cantabria* class design is its continued use of a single-shaft propulsion system. This approach is by no means unique to Navantia's ships; for example the two fleet tankers of the *Deepak* class delivered by Italy's Fincantieri to the Indian Navy in 2011 also adopt this configuration. However, commercial product tankers are increasingly adopting a twin screw configuration. This trend is, again, largely a reflection of environmental considerations. In particular, there have been increasing concerns over pollution related to the much higher statistical likelihood of total propulsion loss arising in a single screw ship. Moreover, the enhanced low-speed manoeuvrability provided by a twin-shaft design reduces the likelihood of collision. The latter

The Spanish replenishment oiler *Cantabria*, pictured here in Australian waters in 2013, is a double-hulled evolution of the earlier *Patiño* and *Amsterdam*. She retains single-shaft propulsion; many new 'clean sheet' designs feature twin screws. (Royal Australian Navy)

factor clearly has direct relevance to replenishment vessels, given the inherent difficulties involved in RAS operations.

Other arguments supporting twin-screw propulsion include the operating benefits arising from the ability to carry out maintenance on one of the propulsion lines without taking a ship out of service. Gains in propulsive efficiency can also be achieved by dividing installed power between two plants. Of course, twin-screw propulsion is not a new feature in naval logistic support vessels. It has frequently been adopted in 'high end' replenishment vessels such as AOEs and sophisticated AORs. However, for all of the reasons highlighted above, it seems likely that the use of twin shafts will become a more universal feature of replenishment oiler designs in the future.

One of the most recent and interesting orders for a new twin-shaft AO/AOR design is represented by the Royal New Zealand Navy's fleet tanker *Aotearoa*.[9] The contract for her construction was placed with South Korea's Hyundai Heavy Industries in July 2016 at a cost of NZ$493m (ca US$350m). The 24,000-tonne ship is a replacement for the existing oiler *Endeavour*, a converted single-shaft mercantile design. The relatively high cost of the new vessel is explained by a requirement to operate in support of New Zealand's bases in Antarctica. This has involved the specification of thicker, higher-grade steel to allow summer and autumn operation in first year ice in line with the Polar Class 6 classification society notation. A significant amount of 'winterisation' includes side ballast and flight deck heating, adaptations to the main crane and an enhanced power system. The Antarctic resupply role has also mandated a substantial overall liquid cargo capacity that is supplemented by a desalination plant that can produce an additional 100 tonnes of water each day. The overall design is a modification of the Rolls-Royce commercial 'Environship' concept. It incorporates an Environship Leadge (Leading Edge or axe) bow and a combined diesel electric and diesel (CODLAD) propulsion system. The latter improves overall fuel efficiency and range by utilising electric motors drawing power from the ship's generation system for lower speed propulsion.

Evolved Mercantile Designs

Aotearoa is also representative of another important trend impacting current replenishment vessel acquisition in that she is a purpose-built naval evolution of a mercantile concept. In this, she differs from the majority of previous replenishment ships. These have tended to be either (i) more basic conversions of merchant ship designs or (ii) bespoke auxiliary vessels developed by a naval design team, albeit incorporating some commercial features.

Both traditional practices have their limitations. Mercantile conversions are undoubtedly relatively quick and cheap to procure. They have been the route by which many navies first developed a RAS capability and remain popular with second-tier fleets. However, most commercial vessels are likely to be too slow to operate comfortably with front-line naval vessels and may lack the accommodation necessary for an augmented naval crew. The incorporation of helicopter facilities – essential for VERTREP – can be difficult due to the limited space available abaft of the fuel stowage compartments.[10] By contrast, purpose-built auxiliaries will be able to avoid these pitfalls but are typically more expensive to produce. This will particularly be the case if there is widespread adoption of naval practices. This is an obvious danger if the vessels are designed by a warship design team, who – moreover – are unlikely to have benefited from exposure to the latest commercial developments.

One of the strongest proponents of the evolved commercial tanker approach has been the British BMT

The Royal New Zealand Navy's AO/AOR-type fleet tanker *Aotearoa* is based on Rolls-Royce's 'Environship' concept and optimised for service in Antarctic conditions. Her distinctive 'Leadge' displaces water more efficiently, creating less drag and aiding RAS evolutions. (Rolls-Royce Group Plc)

Defence Services naval design consultancy. It was quick to recognise that relatively limited numbers of orders for naval auxiliaries had resulted in the emergence of a design lag compared with commercial practices. Accordingly, it started work during the mid-2000s on a series of concepts to bridge the gap. The resulting AEGIR portfolio is based on a series of Baltic-type commercial product tanker designs developed by Norway's Skipskonsulent. Four AEGIR concepts were initially produced. Three of these – the AEGIR 10, AEGIR 18 and AEGIR 26 – were scalable variants of a fleet tanker ranging from 18,000 tonnes to 26,000 tonnes deadweight. The fourth – the AEGIR 18R – was a multi-product replenishment oiler with much greater dry cargo carrying capacity. Common features to all the concepts included double hulls, a two-shaft, twin-skeg hull form and combined diesel electric or diesel (CODLOD) propulsion.[11] In AEGIR, the last-mentioned involves the supply of propulsive power to each shaft line either by a main diesel engine or by a hybrid electrical machine, each coupled to a reduction gearbox.

The first order for an AEGIR-based design was achieved in 2012 when the British Ministry of Defence placed a contract with South Korea's Daewoo Shipbuilding & Marine Engineering (DSME) for four fleet tankers. The resulting 39,000-tonne 'Tide' class is a modification of the AEGIR 26 concept and is optimised for liquid replenishment. In this sense the ships are more oilers (AOs) than multi-product AORs. The design is equipped with two starboard and one port RAS stations, together with a stern RAS rig. A flight deck and hangar for a single helicopter are also provided to allow VERTREP. Work on the first vessel – named *Tidespring* (see drawing) – commenced in 2014. However, problems at the outfitting stage meant that it was April 2017 before she arrived in the UK. Here she was equipped with

***Tidespring* (UK)**

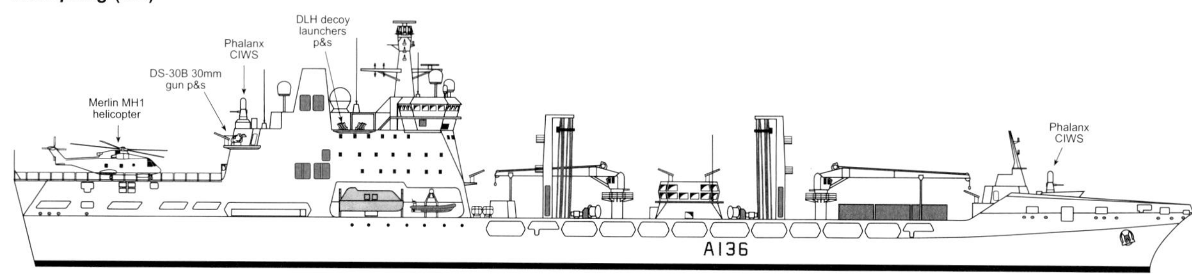

0 10m 20m 30m 40m 50m

© John Jordan 2017

***Tidespring* (UK)**

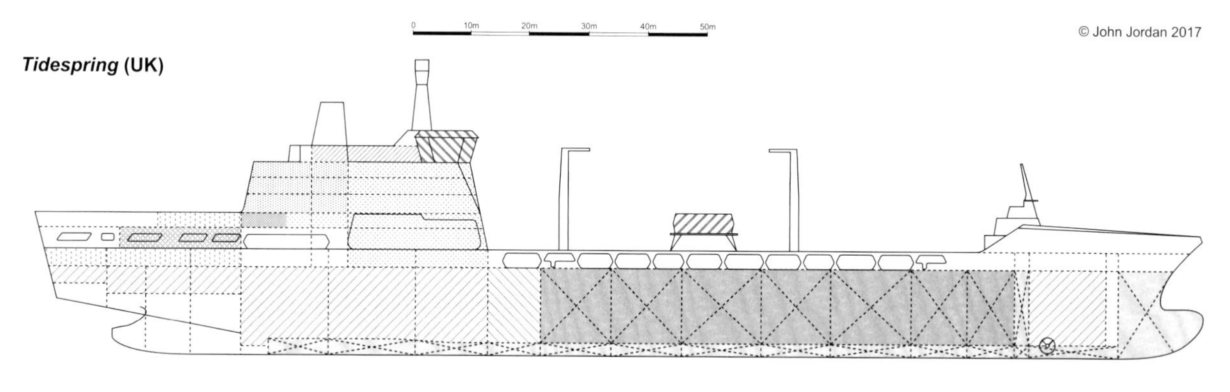

***Maud* (Norway)**

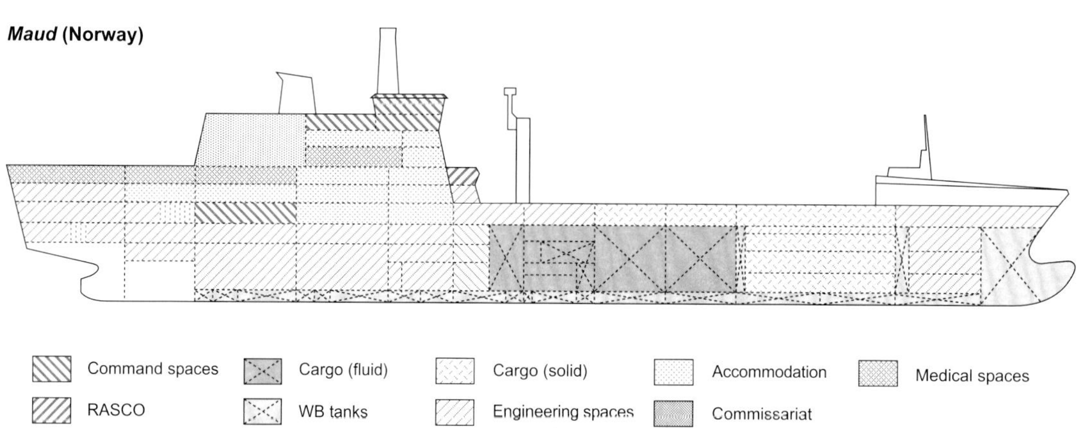

© John Jordan 2017

The British BMT design consultancy has developed a range of AEGIR trademarked replenishment vessels, all evolved from a Baltic tanker design. The first to be completed is the Royal Fleet Auxiliary AO-type oiler *Tidespring*, pictured here at Falmouth in April 2017 after delivery from South Korea. She is referred to as a (fleet) tanker in British service. (Crown Copyright 2017)

military and other sensitive equipment prior to commencement of final trials and work-up. She entered service in November 2017. The other three ships will follow before the end of 2018.

An indication of the efficiencies involved in adopting an evolved mercantile design is provided by the costs associated with the four-ship contract. These amounted to around £600m (ca US$800m). Of this total, just over £450m was spent with DSME and the remainder in the UK for customisation, design services and other support services. These figures suggest that a 'Tide'-class vessel can be acquired for around two thirds the cost of Spain's *Cantabria* in spite of being a much larger ship with approaching double the liquid stores capacity. Against this, *Tidespring's* solid stores capacity is limited to just eight TEU (twenty foot equivalent unit) containers and she also lacks the Spanish ship's sophisticated combat management and hospital facilities. Her quoted maximum speed of 'in excess of 15 knots' suggests she might experience difficulties maintaining station with a fast-moving carrier group.

Multirole Capabilities

Although the range of capabilities of the 'Tide' class is therefore somewhat restricted, the evolved mercantile approach can also be applied to a true multirole ship. This is evident from the design of the only other AEGIR-based ship ordered to date, the Royal Norwegian Navy's logistic support vessel *Maud*. A significantly modified derivative of the AEGIR 18R concept, she was ordered from DSME at a cost of 1.3bn Norwegian Kroner (ca US$170m) in July 2013.

While the 26,000-tonne *Maud* (see drawing) retains a significant liquid transportation capacity, she is able to accommodate substantial amounts of dry cargo as well. This includes up to 40 TEU containers and 200 tonnes of ammunition. Alternatively, a mix of vehicles and boats can be shipped. Replenishment facilities include two RAS positions, a stern reel and a high-capacity 25-tonne crane. However, *Maud's* role extends beyond that of a multi-product AOR insofar as she is equipped to carry out additional roles as a depot ship for smaller warships, a command and control facility, or a hospital ship. For example, she is equipped with mooring positions, hose and power connections and a side ramp to allow other ships to berth alongside, as well as accommodation, catering and leisure facilities for their crews. Alternatively, this surplus accommodation can be expanded into a 48-bed hospital ward adjacent to the ship's medical centre when *Maud* is operating in a humanitarian role.

Maud's multi-function capabilities reflect a trend that has already been touched upon previously towards util-

The Norwegian logistic support vessel *Maud* is another derivative of the AEGIR portfolio. She is essentially an AOR-type replenishment oiler modified to undertake a wider range of logistical tasks, including acting as a mobile base ship for warships and submarines. (BMT Defence Services)

ising the space inherent in most replenishment ships to allow the performance of additional roles. This practice is by no means a new one; the French Navy's later *Marne*-class replenishment oilers that were built in the 1980s were redesigned as *bâtiments de commandement et de ravitaillement (BCR)* – command and replenishment ships – by dint of being equipped with facilities and accommodation for an admiral and his staff. However, the widespread reduction in the size of many established fleets during the post-Cold War era has tended to make such a practice more of an imperative. This appears to be particularly the case with medium-sized fleets investing in bespoke naval designs.

One of the more impressive recent multirole auxiliary designs is represented by the German Navy's Type 702 *Berlin* class. Three of these 21,000-tonne ships were commissioned between 2001 and 2013 as part of a plan to increase the fleet's capability for extended deployment in the aftermath of the Cold War. The last unit of the class, *Bonn* (see drawing), incorporates a number of improvements with respect to accommodation and propulsion in comparison with the first two vessels. These were based partly on the lessons learned from operating the earlier ships.

The official designation of the class is EGV or *Einsatzgruppenversorger* (task force supply ship), reflecting its primary AOR type role. The ships are able to accommodate significant volumes of both liquid and dry stores, with a capacity for up to 78 containers. Replenishment facilities include two RAS positions, two heavy cranes primarily used for cargo-handling, a stern refuelling position and two embarked helicopters for VERTREP. Propulsion is provided by two main diesel engines driving twin shafts.

The large container-carrying capacity is particularly important in facilitating deployment in secondary roles, most significantly that of a hospital and humanitarian support ship. In this configuration, a modular MERZ marine emergency response centre based on German Army Medical Service practice is embarked. Comprising 26 containers, it is installed in the form of a two-storey deckhouse that is linked up to the ship's electrical distribution network. The facility contains examination, operating and laboratory areas and can be used in conjunction with other accommodation in the ship to house more than forty patients. *Berlin*-class vessels also have appropriate accommodation and communications facilities for use as task group flagships, for example in command of NATO Standing Maritime Groups. They have carried out trials with personnel landing craft embarked to assess their suitability for amphibious support.

The capabilities provided by the *Berlin* class come at a price. *Bonn* is understood to have cost around €350m (ca

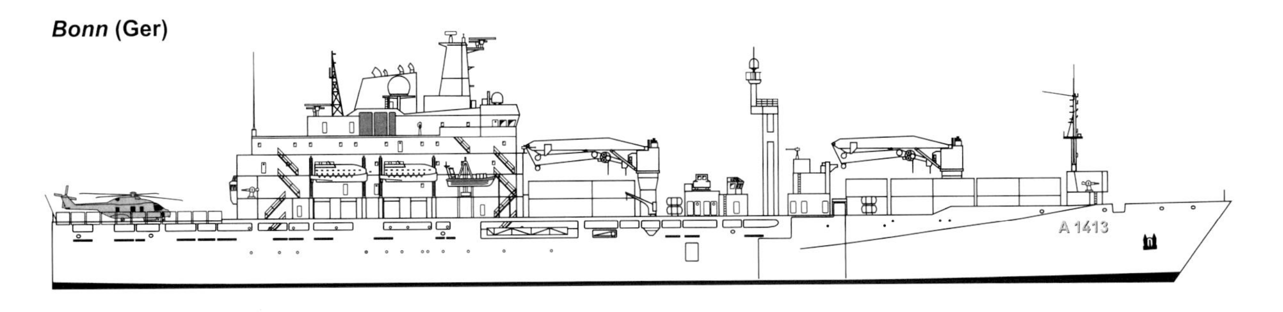

0 10m 20m 30m 40m 50m

A good example of a modern AOR design with enhanced multi-role capabilities is provided by Germany's Type 702 *Berlin* class; the slightly enhanced third unit, *Bonn*, is shown here. The class can support a containerised hospital facility and also has good command & control and some amphibious support capabilities. (Leo van Ginderen collection)

US$420m) to complete. Expenditure on two modified variants to be built by Seaspan in Canada for the Royal Canadian Navy could amount to twice as much. These ships will replace the recently-retired, 1960s-era AORs of the *Protecteur* class. They will leverage the multirole flexibility inherent in the *Berlin* design to undertake additional roles, notably that of strategic sealift. Canada originally hoped to acquire even greater flexibility under a joint support ship (JSS) concept that was intended to support a broad range of replenishment, logistical support and amphibious requirements. Although the joint support ship title has been retained for the programme, the vessels now being acquired will provide a somewhat lower level of capability.

The Joint Support Ship

One fleet that has successfully implemented the JSS concept is the Royal Netherlands Navy. Also known as the Joint Logistic Support Ship, the programme originates from the examination of the fleet's structure in the post-Cold War era under the Naval Study of 2005. This mandated the retirement of a number of existing front-line ships and their replacement by vessels better suited to extended deployment on lower-intensity stabilisation missions.[12] The resulting JSS was approved in 2009. Construction was split between Damen's yards at Galati in Romania and Vlissingen in the Netherlands, the latter being responsible for final outfitting and systems integration. This approach helped to hold down construction

A schematic illustrating some of the key features of the Dutch joint support ship *Karel Doorman*. She combines replenishment, sea lift logistics support and amphibious sea basing capabilities in a single hull. (Royal Netherlands Navy)

costs to a little over 400m (ca US$480m). Named *Karel Doorman*, the ship was deployed operationally to support medical activities aimed at countering the Ebola epidemic in West Africa in late 2014. She was not formally commissioned until April of the following year.

Karel Doorman (see drawing) essentially incorporates fleet replenishment, logistical transportation and amphibious sea-basing capabilities in a single, 28,000-tonne hull. In overall arrangement, she bears a close resemblance to an amphibious transport dock (LPD), combining a substantial forward superstructure with an extensive helicopter deck. The bulky superstructure provides space for accommodation and for a hangar which can house up to six medium helicopters. Two heavy helicopters are able to operate from the flight deck simultaneously. The similarity with an amphibious ship extends to the incorporation beneath the flight deck of a vehicle deck that can house a wide range of vehicles or solid stores. However, the need to find space for liquid stores means that there is no well deck; instead, there is a berthing point for landing craft transfers at the stern and provision for smaller personnel landing craft to be

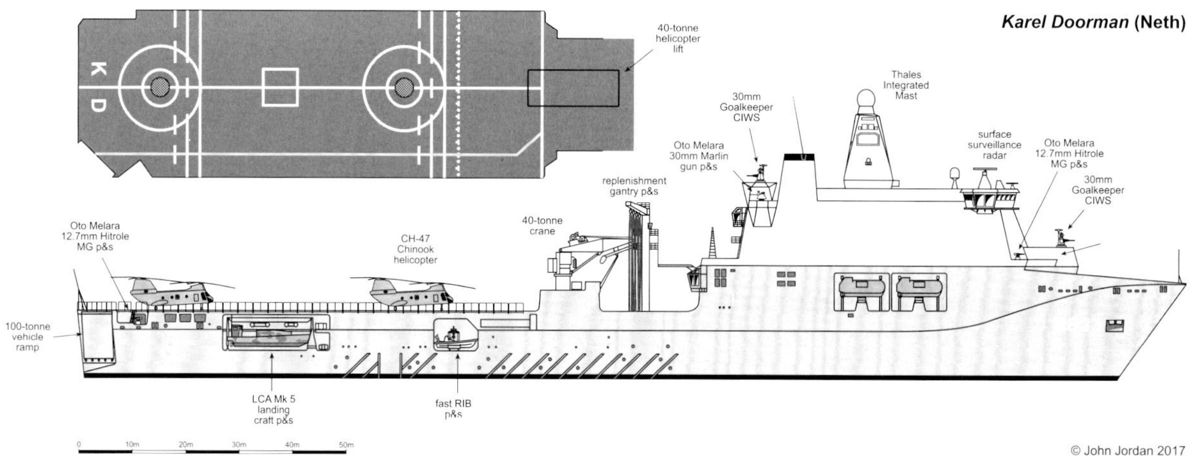

carried on davits. Liquid cargo capacity is similar to previous Dutch replenishment vessels, and there are two RAS positions and one heavy crane.

Karel Doorman utilises an integrated diesel-electric propulsion system. This provides a flexible and efficient power arrangement. However some reliability issues have been experienced.[13] Sensors and communications facilities are derived from those installed in the oceanic patrol vessels of the *Holland* class. Like the latter ships, they are fitted with a Thales Nederland integrated mast atop the superstructure. The overall level of surveillance capability provided by the phased arrays and electro-optical surveillance equipment installed exceeds that of many surface combatants.

The overall combination of functions represented by the JSS is logical considering the synergies inherent in the replenishment, sealift and amphibious support missions. The type has particular attractions for a medium-sized blue water navy that wishes to undertake the full range of these activities but which is unable to afford separate ships for the various roles. The inherent drawback is the increased capital and operating cost associated with a ship that can provide this broad range of capabilities.

Conclusion

It can be seen from this brief review that the current generation of replenishment ships have evolved significantly from their Cold War predecessors. While these changes are not always readily apparent from an external perspective, environmental and technological developments mean that features such as double hulls and twin-shaft propulsion are becoming increasingly prevalent. The use of evolved commercial designs – offering much of the capability of a bespoke naval auxiliary at a considerably lower price point – also seems to be gaining some traction. Another factor worth noting is that these changes in overall ship design are being accompanied by similar enhancements to ship equipment. Reference to various diesel-electric propulsion combinations has been made in the text; all have potential benefits in terms of economy, endurance and safety. Another area of progress is in the area of RAS handling equipment. Here various initiatives are in train to allow the safer, speedier transfer of heavier cargo loads.[14]

Turning to different replenishment ship configurations, it seems unlikely that any significant challenge will emerge to the current dominance of the AOR-type multi-product replenishment oiler. This provides an appropriate balance of replenishment functions for most fleets wishing to conduct blue-water operations. It also offers a degree of flexibility to provide additional capabilities, for example with respect to the provision of medical facilities, extended logistical support or a basic command infrastructure. The extent to which this optionality will be taken up will be driven by specific fleet requirements; it is likely to be more popular with those navies that lack alternative platforms to carry out these roles. The larger

The new Dutch joint support ship *Karel Doorman* pictured undergoing replenishment trials with the air defence frigate *De Zeven Provinciën* in June 2014. The JSS concept supports a broad range of capabilities but is a relatively expensive way of providing a replenishment capability. (Royal Netherlands Navy)

The fast combat support ship USNS *Supply* (T-AOE-6) replenishing the carrier *George H W Bush* (CVN-77) in June 2017. The high expense involved in constructing and operating AOEs mean that few have been built, but their inherent suitability for supporting carrier operations may result in greater popularity as more nations attempt to develop this capability. (US Navy)

fleets will probably continue to build ships optimised for one or more primary replenishment functions; this is evident in the specification of the new RFA 'Tide' class and also the US Navy's forthcoming *John Lewis* (T-AO-205) class of fleet oilers.

One open question is whether the increasing importance attached to aircraft carrier operation among emergent blue-water naval powers will result in the AOE-type fast combat support ship concept returning to favour. China's recent commissioning of the first vessel of this type outside the United States Navy certainly suggests that it has recognised the value of the concept for this type of deployment. Other countries could well follow suit.

Sources:

This chapter has been researched from contemporary industry and government literature, as well as press releases and news reports. The following sources provide more enduring reading material, particularly from a historical perspective:

Friedman, Norman, 'The Fleet Train', *Conway's History of the Ship: The Eclipse of the Big Gun*, Conway Maritime Press (London, 1992), 165-171.

Kimber, Andy and Vik, Arne Magne, 'Future naval tankers – bridging the environmental gap – the cost effective solution', paper presented at the World Maritime Technology Conference 2006, London, BMT Defence Services Ltd (Bath, 2006).

Miller, Marvin O, Hammett, John W and Murphy, Terrence P, 'The Development of the US Navy Underway Replenishment Fleet', SNAME Transactions, Vol 95 1987, Society of Naval Architects and Marine Engineers (Alexandria VA, 1987), 123–158.

Office of the Chief of Naval Operations, Underway Replenishment NTTP 4-01.4, Department of the US Navy (Arlington VA, various editions).

Steigman, David, 'Naval Auxiliaries', *Conway's History of the Ship: Navies in the Nuclear Age*, Conway Maritime Press (London, 1993), 121–132.

Waters, Conrad, 'AEGIR Type Support Vessels', European Security & Defence 5/2016, Mittler Report (Bonn, 2016), 63–66.

Wildenberg, Thomas, *Gray Steel and Black Oil: Fast Tankers and Replenishment at Sea in the US Navy, 1912–1992*, US Naval Institute Press (Annapolis MD, 1996).

Endnotes:

1 The term Replenishment at Sea (RAS) is used by the Royal Navy, NATO and many Commonwealth nations. The corresponding US Navy terminology is Underway Replenishment (UNREP). There is also considerable variation in the terminology used to describe different types of replenishment ships; this results in a degree of inconsistency. Principal types in service today can be summarised by reference to the following US Navy nomenclature:

AOE – Fast Combat Support Ship: These vessels carry a comprehensive range of fuel, dry stores and ammunition.

They were developed to support carrier task groups and have sufficient speed to keep station with the carrier group they are tasked with replenishing. Very few vessels of this expensive type have been built. They are currently in service with the United States and China.

AOR – Replenishment Oiler: This category covers the majority of replenishment vessels in service today. They are multi-product replenishment ships that typically have greater liquid than solid replenishment capabilities, although the balance varies from class to class. They also tend to be slower than an AOE.

AO – Oiler *or* Fleet Replenishment Oiler: These vessels primarily undertake replenishment of fuel and other liquids but often have a limited dry stores replenishment capacity. The broadly equivalent British term is a (fleet) tanker. Note that tankers in US Navy service refer to ships that are intended to transport fuel but do not have a primary replenishment role.

AKE – Ammunition Cargo Ships: These are focused on replenishment of solid stores and ammunition, but the ships currently in service with the US Navy have a limited liquid refuelling capability as well. They are similar in concept to the older AFS combat stores ships, which have now been retired. The British Royal Fleet Auxiliary (RFA) service's current and future solid support ships fall into the AFS category.

JSS – Joint Support Ship (Non-US term): These vessels are multirole auxiliaries that typically combine replenishment, logistical transportation and amphibious functions. The term has been used to describe AORs with enhanced logistic functions, but the Dutch *Karel Doorman* is arguably the only true JSS in service today.

2 Equally, it can be argued that today's Royal Navy maintains a much greater RAS capacity than the broadly similarly structured and tasked *Marine Nationale* because the naval facilities in France's still extensive network of overseas *départements* are able to fulfil much of its navy's lower-intensity support needs.

3 *Maumee* was initially ordered and operated as Fuel Ship No 14. She was designated as a fleet oiler (AO) in 1920. The refuelling operation was necessary because the smaller US Navy destroyers of the era did not have the necessary endurance to cross the Atlantic without this support.

4 Five *Dithmarschen* class supply ships were completed between 1939 and 1943. One was the *Altmark*, from which nearly 300 prisoners of war captured during the South Atlantic cruise of the 'pocket battleship' *Admiral Graf Spee* were released by the Royal Navy destroyer *Cossack* in February 1940.

5 *Sacramento* and her sister *Camden* (AOE-2) each employed half the propulsion plant of the uncompleted *Iowa* class battleship *Kentucky* (BB-65).

6 Interestingly, China's People's Liberation Army Navy (PLAN) has adopted the AOE concept to support its own, expanding aircraft carrier fleet. Its latest Type 901 replenishment vessel is similar in size to the US Navy's *Supply* class. The adoption of a gas turbine propulsion system also means that it is expected to have a broadly equivalent maximum speed of around 25 knots. The lead ship of a class of at least two units was commissioned at Guangzhou on 1 September 2017.

7 The full name for MARPOL 73/78 is the International Convention for the Prevention of Pollution from Ships 1973, as modified by the Protocol of 1978. It was developed by the International Maritime Organization to reduce maritime pollution. It contains a number of annexes, each relating to the regulation of various categories of pollutants. More than 150 countries, representing around 99 per cent of total global mercantile tonnage, have signed the convention to date.

8 The impact of utilising double-hulled construction is quite significant. The US Navy has stated that modifying the design of its last three *Henry J Kaiser* (T-AO-203) oilers to double-hulled standard reduced cargo capacity by 17 per cent. It also increased construction time from 32 to 42 months.

9 *Aotearoa* is the Maori term for the country of New Zealand.

10 An alternative approach to mercantile conversions – possibly reflecting the difficulties inherent in adapting a commercial tanker – is to use a container ship hull as the basis of a multi-role replenishment ship. This solution has been adopted by the Canadian Project Resolve consortium. It acquired the container ship *Asterix* for rebuilding as an interim multi-product AOR for the Royal Canadian Navy pending the delivery of purpose-built vessels. The ship is due to enter service around the end of 2017.

11 Skegs are sternward extensions of the hull structure. In the AEGIR design they are used to enclose the shaft lines. The author understands the first large warships built to this configuration were the two US Navy battleships of the *North Carolina* (BB-55) class, which were commissioned in 1941. Early skeg designs were frequently associated with severe vibration but this issue appears to have been overcome. Recent studies suggest that current twin-skeg hull forms require around 6 per cent less propulsive power at normal operating speeds compared with a single-screw ship of the same cargo capacity.

12 The Naval Study also resulted in acquisition of the sophisticated oceanic patrol vessels of the *Holland* class. These were considered to be more suitable for many of the Royal Netherlands Navy's evolving constabulary missions than the more heavily-armed Cold War-era frigates.

13 More specifically, design faults with the ship's electric motors took the ship out of service for approximately a year from March 2016 while the problem was rectified.

14 For example, Rolls-Royce has been working with the UK Ministry of Defence to trial a new Heavy Replenishment At Sea (HRAS) transfer rig that could allow the weight of solid stores transferred in a single load from the current British limit of two tonnes to five tonnes. This would permit transfer of complete F-35 jet engines from the proposed new generation of RFA solid support ships to the aircraft carriers of the *Queen Elizabeth* class.

LOST IN THE FOG OF WAR:

Royal Navy Cruiser Designs for Trade Protection 1905–1920

The increasing focus by the Royal Navy on small, fast cruisers to scout for the battle fleet in the North Sea during the early 1900s did not preclude the continuation of design work on cruisers for trade protection, some of which were built but many of which remained on the drawing board. Using the notebooks of some of the most prominent British naval constructors of the day, **David Murfin** looks at the history of these designs and presents a more coherent and detailed account than has been possible hitherto.

In 1900 the Royal Navy battle fleet matched the two next largest fleets combined. Adding the numerous cruisers defending trade, policing the Empire sea routes, and scouting for the fleet made that preponderance even greater. Between 1906 and 1916 the battle fleet was renewed with 'dreadnought' battleships. Smaller and faster cruisers were designed to fend off destroyers carrying torpedoes of increased range and destructive power. No large cruisers were built for commerce protection, although proposals ranged from limited modifications of existing designs to completely new ones. Their details, sometimes sparse, are examined here, filling some gaps by deduction – and a little speculation.

Strategic Factors

The French and Russian fleets, the main threats in the 1890s, favoured a *guerre de course* against trade over a fleet action. The numerous British cruisers were spread across a worldwide network of naval bases, with battleships deployed in several separate fleets. The Mediterranean Fleet was largest, countering both the French fleet stationed there and the Russian Black Sea fleet. It was also available to reinforce Atlantic operations or to defend British interests in the Far East, where the China Station also included large cruisers and some battleships. At Tsushima in 1905 the Russian battle fleet was effectively eliminated, but a direct German challenge to Royal Navy supremacy was growing.

That threat was met by formal agreements with other powers: the 1904 *Entente Cordiale* with France and the 1902 Anglo-Japanese Alliance, strengthened in 1905 and cemented by Japan's battleships being of British design and construction. These agreements made it possible for the Royal Navy to concentrate its forces in home waters; a battle cruiser squadron was retained at Malta. By 1914 the Grand Fleet of modern battleships was concentrated at Scapa Flow to avoid the threat from torpedo craft, especially submarines, in the North Sea. The change was not simply geographical but driven by technical advances.

Admiral Fisher, appointed First Sea Lord in 1904, was largely responsible for these measures, scrapping some 150 ships (particularly small, old cruisers and corvettes) and reorganising the geographical fleets into three fleets with different levels of availability: ships in full commission; ships 50% manned with experienced crews; and ships in reserve under care and maintenance parties. This lowered maintenance and crew costs, releasing money for new vessels. Large-scale fleet actions in the North Sea would require new types of ship with modern machinery and weaponry, leaving trade protection to the still-numerous older cruisers.

Technical Factors

The advent of wireless telegraphy (radio) affected the cruisers' mode of employment. From about 1850 cable telegraphy transmitted messages around the world under sea and over land. Fleet despatch vessels which previously needed to voyage home contacted their bases via the nearest cable station. W/T enabled communication ship-to-shore and between ships.

Fleet exercises in 1899 with seven trial sets proved its value. Without radio, a scouting line with ships spaced at nearly twice the limit of visibility maximised the chance of finding an enemy ship or fleet. A successful scout must either shadow the enemy or turn to inform the main fleet, while the other scouts continued unaware. Radio allowed a scout to inform both while maintaining contact. By 1905 technical developments made W/T standard equipment on all RN ships larger than destroyers, with a chain of seven shore stations around the British coast. The presence of enemy ships could be reported, powerful reinforcements summoned, broad tactical instructions sent ship-to-ship, and messages sent between ship and shore at ranges up to about 100 miles, then farther via the cable telegraph network. The main long-range W/T station at Poldhu in Cornwall could send messages to ships 1,000 miles away. Coding for secure communications and code-breaking to gather intelligence to guide deployment became priorities.

The first three photographs illustrate the ships available to the Australian 'Fleet Unit' at the outbreak of the First World War. This is the flagship, HMAS *Australia*, one of three 'battle cruisers' of the *Indefatigable* class which followed the revolutionary *Invincible* class. Powered by turbines, they could hunt down and destroy any enemy cruiser of the day. (Allan C Green collection, State Library of Victoria)

HMAS *Sydney*, a typical 'Town' class light cruiser and a prominent member of the Australian 'Fleet Unit'. (Leo van Ginderen collection)

In 1906 HMS *Dreadnought* revolutionised naval strategy and tactics. The all-big-gun armament, which was later allied to director-controlled salvo fire, soon quadrupled battle ranges from 3,000 yards. Turbines meant higher speeds reliably sustained. *Dreadnought*'s 21 knots, 3 knots more than earlier battleships and close to the 23 knots of the latest armoured cruisers, rendered the latter obsolescent. Most of the old, small cruisers which Fisher scrapped were never intended to serve with the fleet and were far too slow for modern conditions; they were also no longer needed with the advent of W/T. 'Scouts' leading destroyer flotillas after 1905 made 25 knots, with turbines in the later ships, but destroyers were soon even faster. The latest and largest armoured cruisers were to fight for information and thwart enemy scouts, but also to provide a fast supplement to the battle line, to concentrate fire on the enemy van or to deal with damaged stragglers, as at Tsushima.

Large Armoured Cruiser Designs

In line with this thinking, the 1906 construction programme featured three improved dreadnought battleships and three 'dreadnought armoured cruisers' of the *Invincible* class (17,250 tons, eight 12in guns, speed in excess of 25 knots), intended to outclass all existing armoured cruisers, clear the seas of commerce raiders, and perhaps even replace battleships in the longer term. The new Liberal government, in the face of the 1907 financial recession, ordered three more battleships, but schemes for improved 'dreadnought' armoured cruisers and an even more expensive 'fusion design' with the armour and armament of *Dreadnought* and the speed of *Invincible* were abandoned. The response to *Von der Tann*, Germany's powerful reply to the *Invincibles*, was just one more armoured cruiser (*Indefatigable*) in 1908. She was slightly cheaper than *Invincible* – £1,430,000 plus guns £98,500[1] – due to lower steel prices, but the cost of sufficient *Invincibles* for general cruiser tasks, particularly hunting down lone ocean raiders, would have been excessive. Germany built better-armoured, faster ships for fleet use, and Britain replied with the even more expensive 'battle cruisers' of the *Lion* class (26,000 tons and 28 knots, with eight 13.5in guns on the centreline).

New large cruisers to protect trade had to cost less. The most modern 6,000-ton second class protected cruisers carried eleven 6in QF guns, the older (and larger) ones sixteen, with fourteen in the belted armoured cruisers of the 10,000-ton 'County' class. Later cruisers

HMS *Encounter*, one of the last ships of the Victorian second class protected cruiser type; on a displacement of 5,880 tons she carried eleven 6in QF guns, but maximum speed was only 20.75 knots. Loaned to Australia to bridge the delay in completing HMAS *Brisbane*, she was one of a group of cruisers which would need to be replaced by more modern units of the 'Town' type, or possibly by Atlantic Design A. (Allan C Green collection, State Library of Victoria)

of the type (*Devonshire* class) had some of the 6in QF guns replaced by the heavier 7.5in, while the latest armoured cruisers had between four and six 9.2in to defeat thicker armour. Their coal-fired boilers and reciprocating engines gave 21-23 knots, which was insufficient to scout for turbine-powered dreadnoughts or catch 25-knot German light cruisers.

In 1907 Jellicoe proposed 25-knot ships with turbines and an *Invincible* layout with 9.2in guns, to deal with all armoured cruisers except a few with larger guns: the German *Blücher*, built with 8.2in guns in ignorance of the *Invincibles*' 12in; *Riurik*, built by Vickers for Russia, with 10in; and the four Japanese *Ibukis* and *Tsukubas* with four 12in. Jellicoe claimed *Invincibles* would sink these ships and his 9.2in-gun cruisers would outrun them – although the RN aimed to outnumber stronger ships, not flee.

Sir Phillip Watts produced two designs: 'E' with four twin 9.2in and wing turrets as *Invincible*; the other 13,000 tons, 510ft pp with three twin 9.2in for £1.1million, its only midships turret on the centreline. No drawings are available, with further details for 'E' only (see Table 4, page 162). A 1907 three-funnel design drawn by S Payne[2] had four 9.2in twin turrets *en echelon* as in *Indefatigable*, with six 6in QF guns in casemates, a 5in belt and a 1in deck. These schemes were not pursued.

Fleet Units

Invincibles as general purpose cruisers were too expensive. To help protect Empire trade routes a 1909 Imperial Conference asked each Dominion navy to provide a 'fleet unit', whose major vessels would be an armoured cruiser of the *Indefatigable* class and four light cruisers. A set of ⅛in = 1ft blueprints held at the National Archives, entitled 'New 2nd class Colonial Cruiser 1910–11 Programme' depicts a variant of the new *Bristol* class, longer ranged than current scouts. Such 'units' might combine to form a Pacific Fleet. Although New Zealand, Canada, and Malaya offered various other contributions, only Australia came near to fulfilling this dream. The battle cruiser *Australia*, ordered in 1909, commissioned in 1913 as the flagship of the Royal Australian Navy, then joined the Grand Fleet. *Sydney* and *Melbourne*, built in Britain to the *Chatham* version of the 'Town' design, commissioned in 1913. *Brisbane*, Australia's third *Chatham*, built at Sydney's Cockatoo Island dockyard, commissioned in 1916 with machinery and armaments from Britain delayed by the war. The fourth ship, *Adelaide*, with design changes compounding similar delays, was eventually completed in 1922 as a modified *Birmingham*.

Prelude to War

The Fleet Unit scheme proved largely abortive. Germany's latest small cruisers could scout for the fleet or act alone on the high seas as raiders, but the 1912 German Naval Law specified ten cruisers for foreign service by 1920. RN cruiser needs were reviewed. The battle fleet needed light cruisers as scouts, and to work with and against destroyers in fleet actions. Older second class and third class protected cruisers were too slow and insufficiently protected to meet modern German cruisers. Destroyers were larger, tougher and faster, posing a greater torpedo threat, with speeds around 30 knots. The small scouts built since 1905 for destroyer work made only 25 knots, with protection now considered inadequate and gunpower weak. Of existing ships, only the new 'Towns' were suitable. Their 6in guns could deal with German light cruisers and destroyers. At 5,400 tons they could sustain their 25 knots better than the scouts or destroyers in North Sea waves, but as the only RN ships apart from the battle cruisers with the necessary speed and endurance to find and catch enemy ocean raiders, they were needed elsewhere, and too costly (£350,000)[3] to build in sufficient numbers for both roles. The Admiralty Chief of Staff, Admiral FT Troubridge, addressed the problems in 'Future War requirements in respect of cruisers, March 1912' here summarised:

> The RN had in April 1912 (excluding ships more than 20 years old) 98 cruisers compared to Germany's 49 – exactly the required 100% preponderance. But Germany was laying down two more each year and to maintain that 100% the RN needed to build four each year, and another fourteen ships over the next eight years to replace ships becoming over-age, as all the German ships were newer. That meant at least eight light cruisers laid down in 1912.

Troubridge had 'Towns' in mind, but eight new 3,500-ton light cruisers of the *Arethusa* type entered service from 1914. These were large enough to steam close to their designed 30 knots in a seaway, and carried two 6in and six 4in guns, with a 3in armour belt able to resist 10.5cm (4.1in) shells from German light cruisers. They and the numerous 'C's derived from them proved admirable fleet cruisers, but even the late war 'D's of 4,750 tons needed more range to protect trade. Concentration on fleet cruisers, then war in 1914, meant that Victorian and early Edwardian armoured and protected cruisers were still needed for ocean service, despite further design work in 1912 and 1913.

Cruiser Designs, Small, Intermediate and Large

Peacetime ceremonial, patrol and policing functions on foreign and colonial stations still engaged many older small vessels, some carrying sail, despite Fisher's 1904–05 cull. The 25-knot scouts were too short-ranged for such work and for the trade routes. A 1910 experimental 'New *Amphion*' with a third central shaft driven by a 6000bhp diesel engine to increase range, allied to 12,000shp turbines on the outer shafts, was cancelled when Sir Arthur Wilson succeeded Fisher as First Sea Lord. (Appearance would have been as *Amphion* except that the space between third and fourth funnels was doubled. *Amphion* had two shafts, the other 'Scouts' four – see Cover 263)

In 1912 a Committee on the 'Design of Cruisers for Foreign and Colonial Stations' reviewed 'peace cruiser' functions, seeking a suitable design (if expedient, for both peace and war) to replace ships on colonial stations over the next eight years, at minimal cost.[4] The Foreign and Colonial Offices wanted 'suitably impressive' warships with effective gun power, able to carry landing parties of armed sailors or marines. In August Rear Admiral Slade, the senior naval Committee member, agreed an outline specification with the Third Sea Lord and DNC for a short, manoeuvrable ship with a straight stem, a protective deck, coal- and oil- fired steam turbines plus two masts with sails (fore & aft rig) to extend range. Characteristics were to be 2,000–2,500 tons, draught up to 14ft, six 4in guns, 5,000shp for 20 knots on trials, sea speed 17–18 knots, cost ca £100,000.

Design was entrusted in September to SV Goodall (DNC 1936–44). This, his first design, provided valuable experience for his next, HMS *Arethusa*. The adoption of belt armour instead of an armoured deck contributed to hull strength, keeping displacement to 1,750 tons. His 4in guns, which had higher command, had better ahead fire than did the Committee's outline, and he suggested geared turbines for fuel economy, but a Committee review on 4 October demanded additions, taking displacement to 2,000 tons. (see Table 1 and Figure 1).

The Committee reported on 1 February 1913. A minority report by Captain Ballard, the committee's representative on the War Staff, argued that the low sea speed of 17–18 knots made the design very vulnerable to faster small modern cruisers with greater gun power, and unable to catch armed merchant cruisers raiding commerce as the Committee proposed; established Admiralty policy was already meeting that threat with guns for British merchant ships. Ships with a modern light cruiser's speed and fighting power exceeded the 'minimum cost' in the Committee's terms of reference. Churchill's minute dated 5 February expressed a First Lord's view: new British cruisers must be designed solely to fight modern warships in order to execute strategic plans. Two superior British light cruisers (more if of lesser power) must be built for each German light cruiser. Military considerations alone governed their design, making them unnecessarily large and costly for Colonial duties; older, time-expired warships could fulfil those, although these were increasingly expensive to maintain. 'A steam hammer will, in addition to its ordinary work, crack a nut. But nut-crackers continue to be made.' Following the pattern of unarmed surveying vessels, a very few weaker ships might be built specially for 'peace

Table 1: **Goodall's Colonial Cruiser Designs**

	Goodall	Goodall	*Espiegle*
Date	Sep 1912	Oct 1912	1903
Displacement	1,750 tons	2,000 tons	1,070 tons
Length, pp	300ft	320ft	185ft
Beam	35ft	38ft	33ft
Draught, mean	11ft 9in	11ft 9in	11ft 3in
Power	5,000shp	5,000shp	1,400ihp
Speed trial	20 knots	20 knots	13 knots
Fuel	400 tons coal + 100 tons oil	400 tons coal + 100 tons oil	
Range	5,000nm at 10kts	5,000nm at 10kts	3,000nm
Armament:	6 – 4in (300rpg)	2 – 6in (200rpg) 4 – 4in (300rpg)	6 – 4in
Protection:			
Belt	2.5in	2.5in	
Deck	1in mags/steering 0.5in other	1.5in mags/steering/ machinery	
Complement	173	190	150
Cost	£115,000	£125,000	ca £80,000
Weights:			
Hull/protection	1,050	1,210	
Armament	90	143	
Machinery	235	245	
General	150	162	
Board margin	30	40	

Notes:

1 Unless otherwise stated, Displacement for unbuilt designs in this and the following tables is Legend, and Draught and Speed calculated for Legend displacement; in the Weights section, the figures represent tons.

2 For comparison, the 'most recent sloop' *Espiegle* (*Cadmus* class) is used; sails (except *Espiegle*, none) were unused after 1904, and yards removed 1914; served on Colonial stations during the First World War.

cruiser' duties, to be laid up in time of war with their crews transferred to fighting cruisers.

The Naval Staff rejected the Committee design as expensive for its peacetime duties, yet too slow and weak to fight effectively in wartime, and too costly to be exposed to loss. New construction should exclude inferior fighting ships. The Chief of Staff thought shorter and handier older sloops, gunboats and small cruisers matched the required duties better than the Committee's 320ft design; improved *Espiegle*-type sloops might be

1,750-ton Colonial Cruiser: Profile & Plan

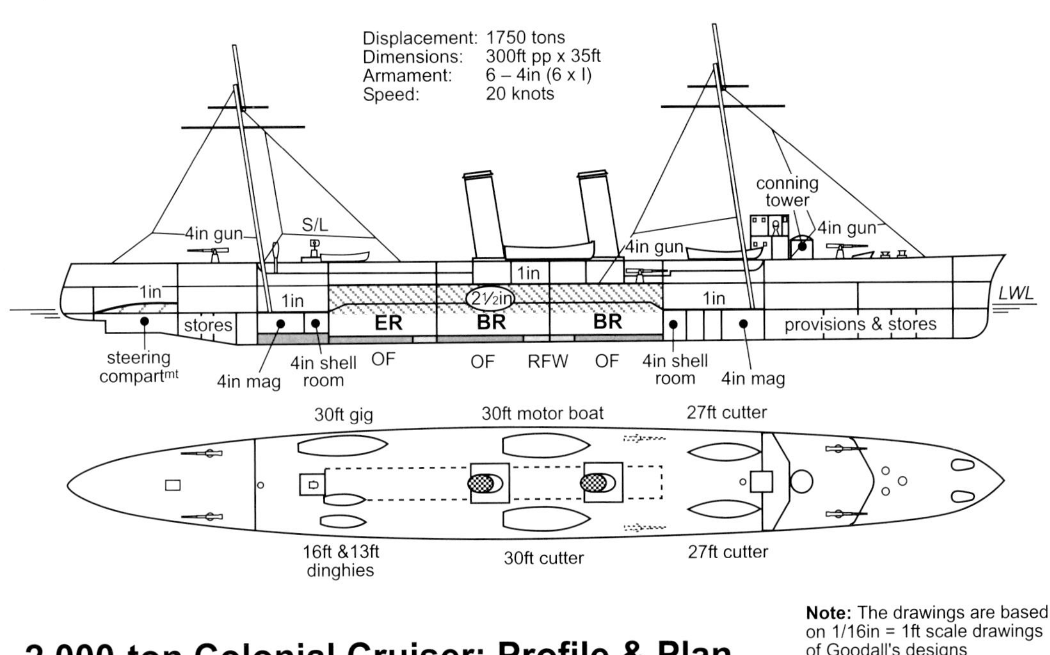

2,000-ton Colonial Cruiser: Profile & Plan

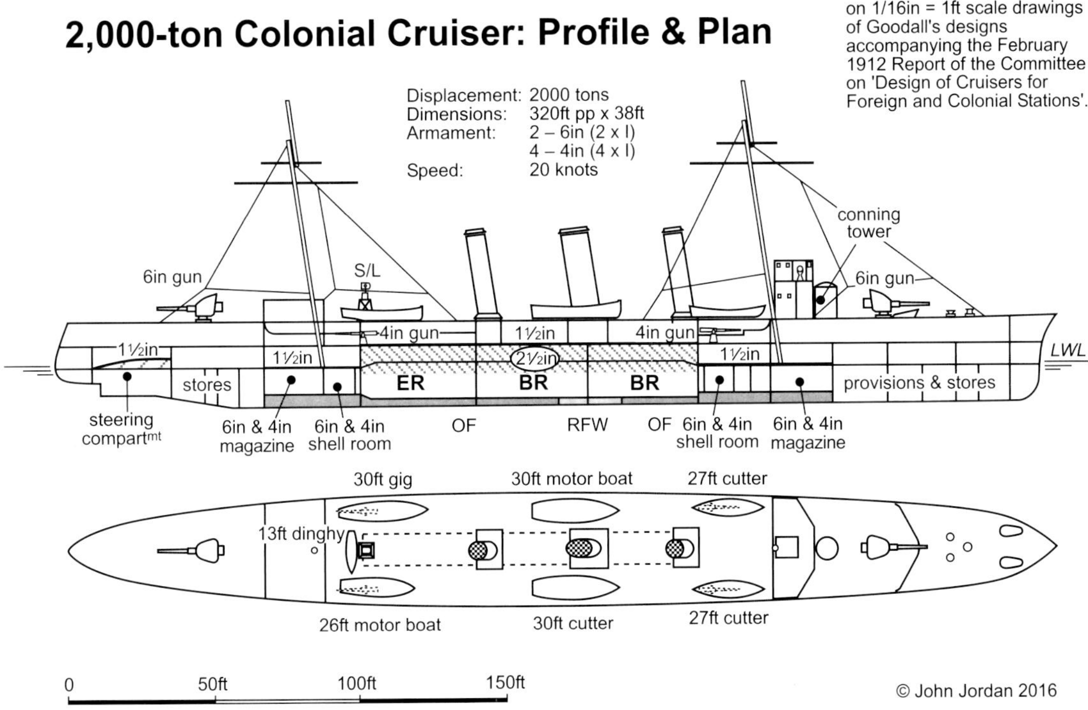

Fig 1: 3rd Class Cruisers 1912.
That was the official designation for the small protected cruiser to police distant stations and 'show the flag'. Goodall's initial design was a compact 1,750 tons. A review insisted on two 6in guns, landing guns and other additions, raising displacement by 250 tons and cost by £10,000. The ship was castigated by the Chief of Naval Staff as 'neither fish, nor fowl, nor good red herring'. Both designs carried sails, not in general RN use since 1904, to extend range; their fore and aft rig is indicated, but neither set of original plans makes precise details clear.

suitable. DNC would prepare a new design, with the Foreign and Colonial Offices informed.

In August 1913 'Sloops for the Pacific' again considered small ships with long range, perhaps using auxiliary sails. Sail required short, beamy ships; economic steam power long narrow ones. DNC reviewed past compromises. The main yards of the 960-ton *Torch* class of 1894 had been removed to simplify handling the sails. By 1913 seamen knew even less of canvas; it would be simpler to increase the bunkerage of the most recent design of sloop. The issue was put aside, and with it the formerly ubiquitous Victorian 3rd class cruiser.

'Armoured Cruisers for Atlantic Waters'

First described relatively recently, these designs merit more detailed study. Brown[5] drew somewhat inaccurately upon the notebook recording Tennyson d'Eyncourt's designs, built and unbuilt. Like Admiralty Form A, the left-hand page listed dimensions, speed, armament, armour, etc, the right-hand page the estimated weights of hull, armament, protection and machinery, with (in the notebook) a simple hand-drawn profile of the design. Friedman[6] began with the 1913 designs, quoting the more detailed explanatory memos with scale drawings supporting the Form A 'legend' – the specification for designs submitted for Board of Admiralty approval, DNC guaranteeing calculations that such ships would be stable and float level.

Five notebook double pages illustrate four distinct though closely related designs for 'Armoured cruisers for Atlantic Waters'. The three entries dated October 1912 (see Table 2) are among the earliest in d'Eyncourt's Notebook and were ascribed by Brown to William Berry, then head of the cruiser section. Design A was of 6,150 tons. Sketches of B1 and B2, best regarded as two approaches to a design of 8,000 tons, were drawn on longer slips of paper stuck and folded into the relevant notebook pages. These are the only source of information on A, B1 and B2, although later comments in memos on B3 and B4 may also apply to them.

Designs B3 and B4 were dated July and August 1913. Under d'Eyncourt's supervision, they differed markedly from the 1912 ships and from one other. In terms of their size and chronology these designs form 'a little known bridge' between the 'Towns' and the 'Improved *Birminghams*' of the *Hawkins* class, but they were more heavily armoured.

The German Naval Law of 1912 required ten cruisers for foreign service by 1920. Sixteen ships were laid down by 1916 (all around 5,500 tons with eight 5.9in guns, developments of earlier German designs and roughly equivalent to RN 'Towns'). Eight were completed during the war. As noted above, to maintain 100% preponderance the RN needed to build 34 ships by 1920, replacing the protected cruisers of the *Diadem* class and the smaller 'County' class armoured cruisers built around the turn of the century to protect trade. The

Table 2: 'Atlantic' Cruisers 1912

(B1 and B2 columns merged to emphasise similarities – see Fig 2 for differences)

Class	A	B1	B2	*Devonshire*
Displacement	6,150 tons	8,150 tons	8,000 tons	10,850 tons
Length, pp	500ft	540ft	540ft	450ft
Beam	49ft 6in	54ft 6in	54ft	68ft 6in
L/B ratio	10.1	9.9	10	6.6
Draught, mean	16ft	18ft		25ft
Power	40,000shp	40,000shp		22,000ihp
Speed	28 knots (load)	28 knots (load)		22.75 knots
Fuel	1,000 tons oil	1,100 tons oil		1,750 tons coal
Armament:	2 – 7.5in (2 x I)	4 – 7.5in (4 x I)		4 – 7.5in (4 x I)
	8 – 6in sided	8 – 6in sided		6 – 6in sided
	8 – 12pdr	8 – 12pdr		
	2 – 21in TT sub	2 – 21in TT sub		2 – 18in TT sub
Protection:				
Side	4in	6–4in + 2in ends		6in + 2in
Deck	1.5in lower aft	1.5in lower aft		2in
	+1in upper	+1in upper		
Cost (excl guns)	£550,000	£750,000		£740,000
Weights:				
Hull	3,100	4,230	4,180	2,300
Protection	680	1,100	1,000	660
Armament	730	960		2,086
Machinery (+Eng's stores)	800+40	800+40		585
General	350	450		69
Board margin	50	70		n/a

HMS *Carnarvon*, a 10,850-ton armoured cruiser of the *Devonshire* class, which Atlantic Design B1 or B2 may have been designed to replace. (Allan C Green collection, State Library of Victoria)

25 ships of these three classes of medium-sized cruisers were a substantial part of RN resources for trade protection, but with heavy reciprocating machinery and old-fashioned boilers they could not catch modern cruisers. The 18 smaller 6in-gun protected cruisers of the *Eclipse* and *Highflyer* classes also continued to serve, although slow and obsolescent. Table 2 compares data for Designs A and B1 or B2 with *Devonshire*, suggesting that they were intended simply to replace existing RN cruisers rather than to match any specific (but in mid-1912 unknown) design features of German ships. The *Devonshires* replaced the unreliable forecastle and quarterdeck twin 6in turrets of the *Monmouths* and the two forward pairs of casemate 6in guns (unusable in moderate seas) by four single 7.5in turrets during building. Modern machinery allowed B1 or B2, though 2,000 tons lighter, to carry a very similar armament to *Devonshire* at higher speed and with similar protection. The parallel was even closer in retrospect. In 1915 *Devonshire*'s remaining 6in casemates were plated over and the guns moved to upper deck single mounts. Design A, smaller and cheaper, was more like a stronger 'Town' and a natural replacement for the *Eclipse* and *Highflyer* claases, but could also replace the larger ships if Germany, with few medium-sized armoured cruisers, simply continued building light cruisers, which 'A' with two 7.5in could catch and outgun.

A handwritten note appended to the August 1913 entry for B4 in d'Eyncourt's notebook read:

> War Staff notes on designs A, B1, B2, B3 and B4 17th October 1913 (Report Book II). For protection of ocean trade when *Edgar*, *Hyacinth*, *Diadem* and 'County' classes become worn out. Provided radius of action at least 5,500 miles with fuel stowage of which not less than 50% is coal, designs B4 and A are preferred to larger ships B1, B2, and B3. Radius of action named is a sine qua non and if B4 and A are deficient in this respect, either displacement should be increased as necessary, or armour reduced. If B4 and A have this radius of action as they stand, they are considered extremely well suited in general to our prospective requirements. Vertical masts and funnels preferred. Good flare in bow and overhanging stem also considered an important feature. Bearings on which TTs are fixed should be considered tentative.

At least one of the 'Atlantic' designs could well have been built in some numbers had not the war intervened. The mixed-fuel version of B4 was the clear favourite to replace older cruisers, with 1,600 tons of fuel for her heavier engines (de-rated somewhat for greater reliability), although the B4 burning oil only was lighter and faster, and might have been preferred as trust grew in oil fuel. Either could easily have had vertical masts and funnels like 'A'. However, a decision on their adoption was deferred for 12 months; by then Britain was at war, forced to rely on existing cruisers to protect trade.

The 1912 'Atlantic' designs were expensive, with B1 and B2 costing twice as much as the £350,000 'Towns'.

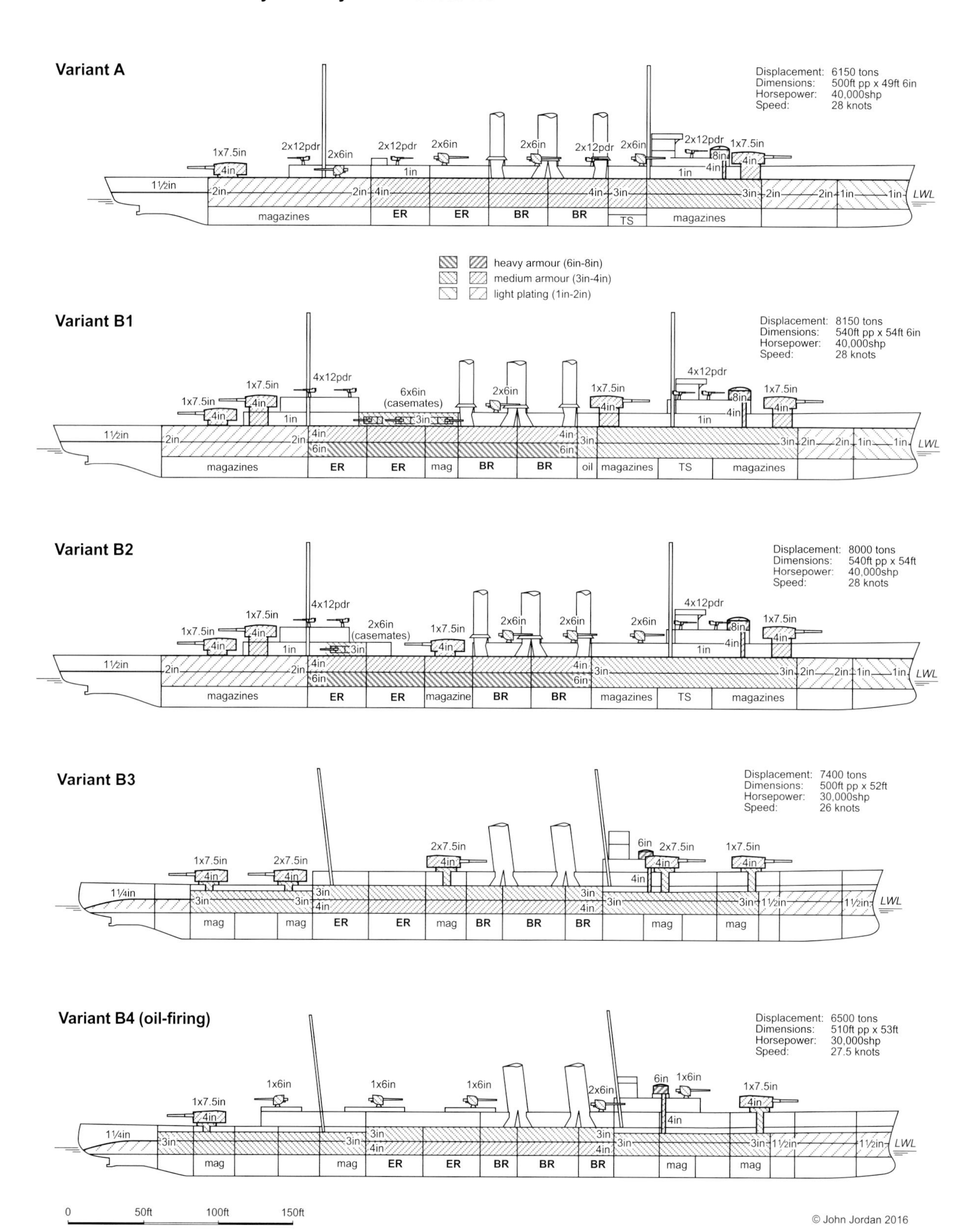

Fig 2: Cruisers for Atlantic Service.
Profiles of five designs are shown. The 1912 designs rely on the sketches in d'Eyncourt's Notebook. NMM Cover 319 contains $\frac{1}{16}$in = 1ft detailed plans for the 1913 designs, but neither has drawings for the variant of B4 with mixed coal and oil firing.

Higher power from oil-fired turbines and a length/beam ratio close to 10 (rather than the 8.6 of the 'Towns' and the later *Hawkins*) gave 28 knots compared to 25 knots for the smaller 'Towns', which were their immediate predecessors, and 20–23 knots for older ships with similar armaments and/or purpose (*Edgar*, *Eclipse*, *Highflyer*, *Devonshire*). B1 and B2 had a narrow 6in belt on the waterline, with 4in above; 'A' had a uniform 4in belt. All had a mixed armament of 6in and 7.5in guns. 'A' had two 7.5in, B1 and B2 four. Six of B1's 6in guns and two of B2's were in casemates.

In d'Eyncourt's notebook, the sketch of Design A has three vertical funnels, but those of B1 and B2 omit funnels, masts and bridgework. It is reasonable to suppose the same installations, as all three had three boiler rooms and the same weight of machinery (800 tons) for 40,000shp and 28 knots, although A (lighter with less draught but the same length/beam ratio) might be expected to be faster. The supporting memo for B3 (1913) stated that its machinery (900 tons, 30,000shp) would be 'of the same type as the light cruisers now being constructed [ie *Arethusas*] but rather heavier and with less full power to be more reliable on extended trips'. *Arethusa*'s machinery was of the lightweight destroyer type, 765 tons; rated at 30,000shp, it could deliver 40,000shp on a 4-hour overload designed to give 30 knots for short periods. (*Invincible*'s 40,000shp machinery weighed 3,300 tons.) Figure 3 shows the probable arrangements of 'Atlantic' boiler rooms, B1 and B2 having the standard three-room layout like A (and *Arethusa*), so drawn here with three funnels. The 1912 'Atlantics' are assumed to have machinery somewhat more robust than *Arethusa*'s, but of the same type to provide 40,000shp on the same basis. *Arethusa*'s range was 5,000nm on about 800 tons of oil.

The 1913 designs B3 and B4, about the same length as 'A' with displacements between 7,400 and 6,500 tons, were the subject of more detailed submissions to the Board. Specifications and weight breakdowns on Form A in Cover 319 were supplemented by explanatory memos, $\frac{1}{16}$in = 1ft profile and deck plans and armour profiles; the designs were described simply as 'Cruisers for Atlantic Service' (although the accompanying submission notes describe B3 as a 'Light' cruiser). D'Eyncourt noted that with displacement under 7,500 tons armour was limited to 4in + 3in, compared to the narrow 6in belt on B1 and B2 of 8,000 tons. He replaced their flat deck protecting the steering with a 1.25in turtle deck. A memo from the Third Sea Lord dated 3 July 1913 which accompanying DNC's formal submission stated of B3:

> This design got out as the result of rumours that the new German protected cruisers would be armed with guns of at least 6.9" calibre (probably larger). The First Lord was anxious to have a design ready in case these rumours were true. [Mid-1913 gave time for rumours to grow about ship designs under the German 1912 Naval Law].

B3 had a heavy armament: eight 7.5in guns in single turrets (three down each side), 'which means power

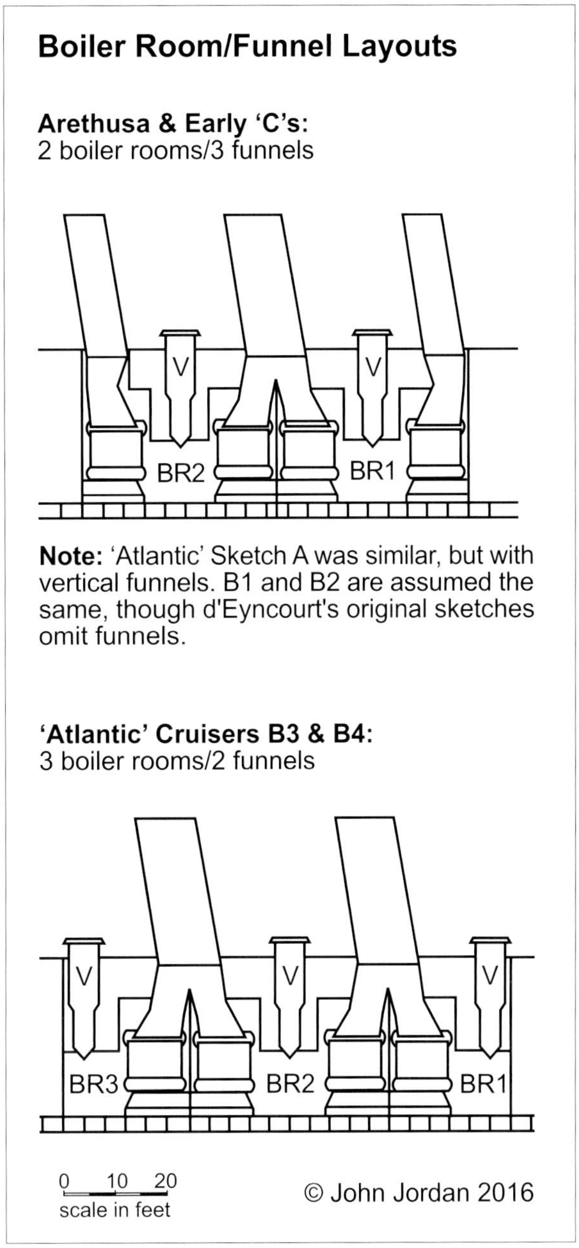

Boiler Room/Funnel Layouts

Arethusa & Early 'C's:
2 boiler rooms/3 funnels

Note: 'Atlantic' Sketch A was similar, but with vertical funnels. B1 and B2 are assumed the same, though d'Eyncourt's original sketches omit funnels.

'Atlantic' Cruisers B3 & B4:
3 boiler rooms/2 funnels

0 10 20
scale in feet

© John Jordan 2016

Fig 3: Atlantic Cruisers: Boiler and funnel layouts.

working that at once runs up the cost of the ship.' Machinery to deliver 30,000shp, 100 tons heavier than the 1912 machinery, kept costs down but made B3 two knots slower. Mean draught was 20ft, with excellent sea-keeping expected from the 26ft freeboard forward. Third Sea Lord commented: 'compared with the County class it is remarkable what a powerful ship she would be for her size.' But B3 would cost £700,000 excluding guns.

The First Lord was less impressed. Disliking the description 'for Atlantic Service' he 'questioned whether [B3] does not go beyond anything required by German cruiser construction'. Exercising his close interest in the details of warship design, on 4 August 1913 Churchill asked d'Eyncourt for a smaller design with six 6in and

Table 3: 'Atlantic' Cruisers 1913

Class	B3	B4 oil	B4 mixed
Displacement	7,400 tons	6,500 tons	7,000 tons
Length, pp	500ft	510ft	510ft
Beam	52ft	53ft[1]	54ft
L/B ratio	9.6	9.6	9.4
Draught, mean	20ft	17ft 6in	17ft 6in
Power	30,000shp	30,000shp	28,000shp
Speed	26 knots (load)	27.5 knots (load)	26.5 knots (load)
Fuel	1,250 tons oil	1,100 tons oil	1,600 tons mixed
Armament:	8 – 7.5in (8 x I)	2 – 7.5in (2 x I)	2 – 7.5in (2 x I)
		6 – 6in (6 x I)	6 – 6in (6 x I)
	4 HA (12pdr?)	4 HA (12pdr?)	4 HA (12pdr?)
	4 – 21in TT sub	4 – 21in TT sub	4 – 21in TT sub
Protection:			
Side	4–3in	4in + 3in ends	4in + 3in ends
Deck	1.25in steering	1.25in steering	1.25in steering
Cost (excl guns)	£700,000	£548,000	£588,000
Weights:			
Hull	3,400	3,200	3,360
Protection	1,000	855	900
Armament	1,090	700	700
Machinery (+Eng's stores)	900+40	830+40	980+40
General	400	360	400
Board margin	70	65	70

Note:

1 Measured at 50ft on the plan, giving an L/B ratio of 10.2.

Table 4: Large Armoured cruisers with 9.2in Guns

	E	E2	E3
Designed	Watts	d'Eyncourt	d'Eyncourt
Date	1907	1913	1913
Displacement	15,750 tons	15,500 tons	17,850 tons
Length, pp	525ft pp	560ft pp	580ft pp
Beam		75ft	76ft
Draught		26ft legend	26ft legend
Power	40,000shp	55,000shp	60,000shp
Speed trial	25 knots	28 knots	28 knots
Fuel tons	2,500 tons coal + 1,000 tons oil	2,500 tons oil	2,500 tons oil
Range	8,000nm		
Armament:	8 – 9.2in (4 x II)	8 – 9.2in (4 x II)	8 – 9.2in (4 x II)
	16 – 4in (16 x I)	8 – 6in (8 x I)	8 – 6in (8 x I)
	2 – 18in TT sub	4 – 21in TT sub	4 – 21in TT sub
Protection:			
Belt	6in/4in upper	6in	6in/4in
Ends	4in–2in	4in/2in	4in/2in
Deck	1.5in/2in slopes	2in/1in slopes	2in/1in slopes
Complement	730	n/a	n/a
Cost	£1.28m	£1.35m	£1.50m
Weights:			
Hull	5,890	5,500	6,000
Protection	2,985	3,560	5,070
Armament	690	1,900	1,900
Machinery	3,400	2,700	3,000

Note:

For more details of the E2, E3 armour schemes, see Figure 6.

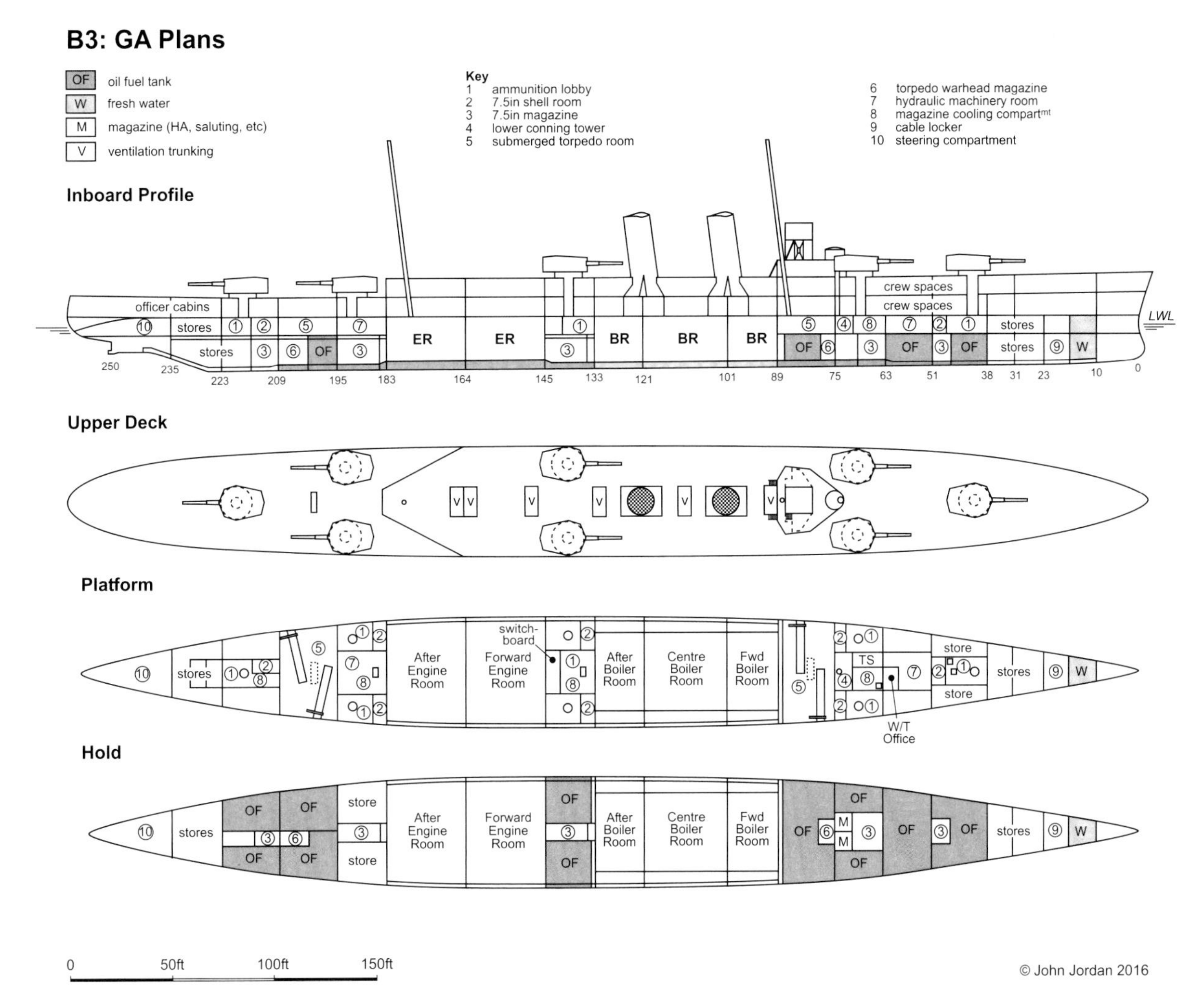

Fig 4: B3 General Arrangement plans.

only two 7.5in, all on the centreline, examining all-oil and mixed coal/oil alternatives. The 1912 designs and B3 were all oil-fired, but Churchill (though negotiating guaranteed supplies of Persian oil) was concerned about the maintenance of oil stocks on distant stations. Often accused of amateurish interference, Churchill's comments on these designs, the 'Colonials', the *Arethusas* and the *Queen Elizabeth* class battleships all show a secure grasp of modern requirements. Next day DNC passed slightly expanded instructions to Berry, who produced two legends for design B4.

B4 was 10ft longer than B3, had 2ft less draught, and mounted a mixed armament with 7.5in single turrets fore and aft (like A), plus six 6in guns (Mk XII with 3in shields); four on the centreline and two sided by the bridge gave a broadside of five and three guns firing forward. Oil firing generated 30,000shp for 27.5 knots. The figure of 1,600 tons of mixed fuel rather than 1,100 tons of oil seems to explain 'mixed' B4's legend displacement, which was 500 tons higher than the 6,500 tons for B4 'oil' (though at the quoted displacements the ships carried only 550 tons of fuel or 450 tons of oil respectively – see Table 3 for the distribution of weights). The War Staff wanted at least half of those 1,600 tons to be coal; B4 'mixed' would have been a knot slower with 2,000shp less power. Savings compared to B3 were £110–150,000 and, assuming that these figures excluded guns, there would have been further savings in replacing six 7.5in turrets with the standard 6in QF gun.

Designs E2 and E3

The largest designs from this period were E2 and E3, which occupied two double pages illustrated by a single (composite) sketch in d'Eyncourt's notebook, dated October 1913.

Why such designs were being considered in 1913 is unclear, unless to revive the Fleet Unit concept with a rather less expensive and more modern design. The 9.2in guns, all on the centreline, would have resulted in a cheaper ship than one armed with 12in guns but one sufficiently fast and powerful to sink all cruisers and any fast German liners armed as corsairs, while the 6in secondary QF battery was adequate against smaller

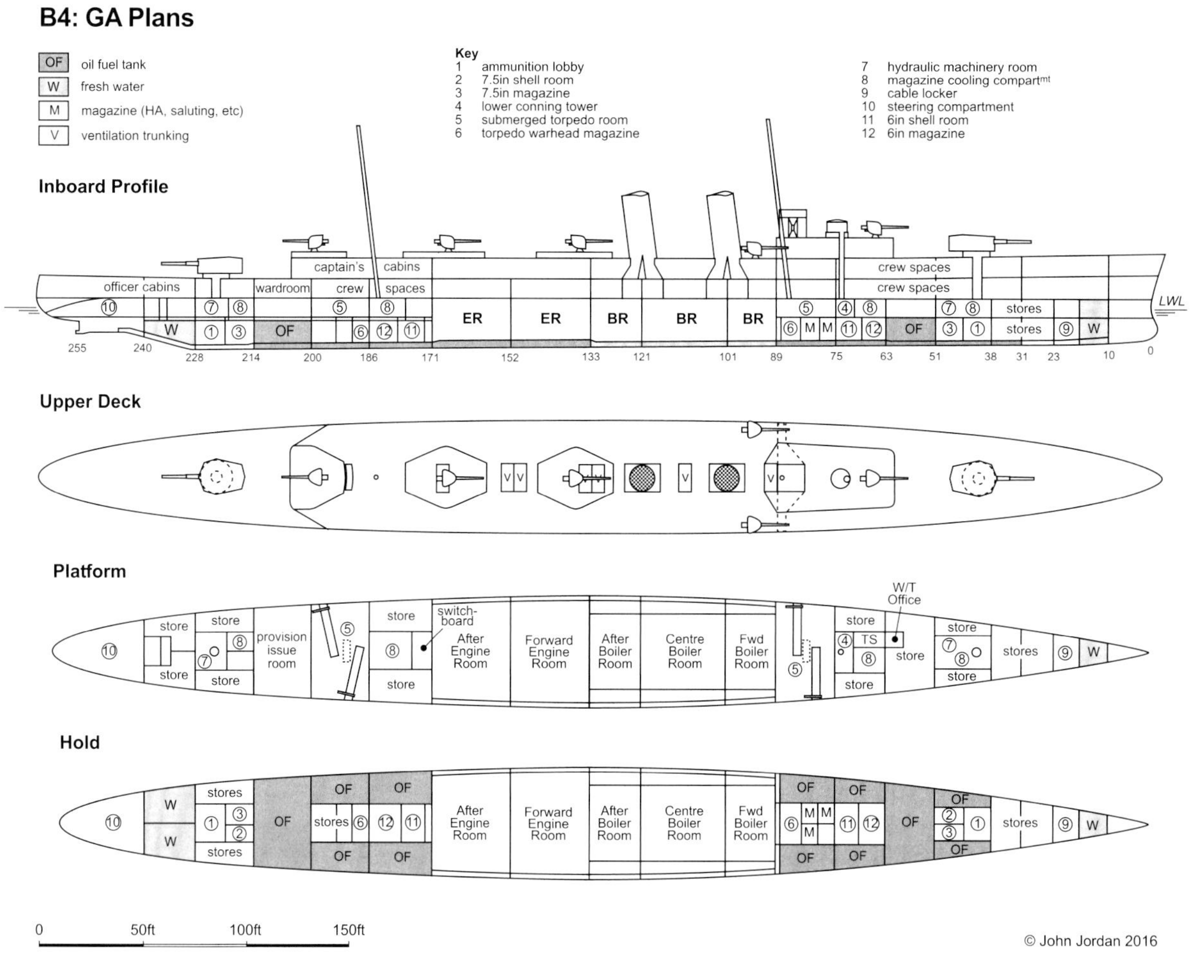

B4: GA Plans

OF oil fuel tank
W fresh water
M magazine (HA, saluting, etc)
V ventilation trunking

Key
1 ammunition lobby
2 7.5in shell room
3 7.5in magazine
4 lower conning tower
5 submerged torpedo room
6 torpedo warhead magazine
7 hydraulic machinery room
8 magazine cooling compart^mt
9 cable locker
10 steering compartment
11 6in shell room
12 6in magazine

Inboard Profile

Upper Deck

Platform

Hold

0 50ft 100ft 150ft

© John Jordan 2016

Fig 5: B4 General Arrangement plans.

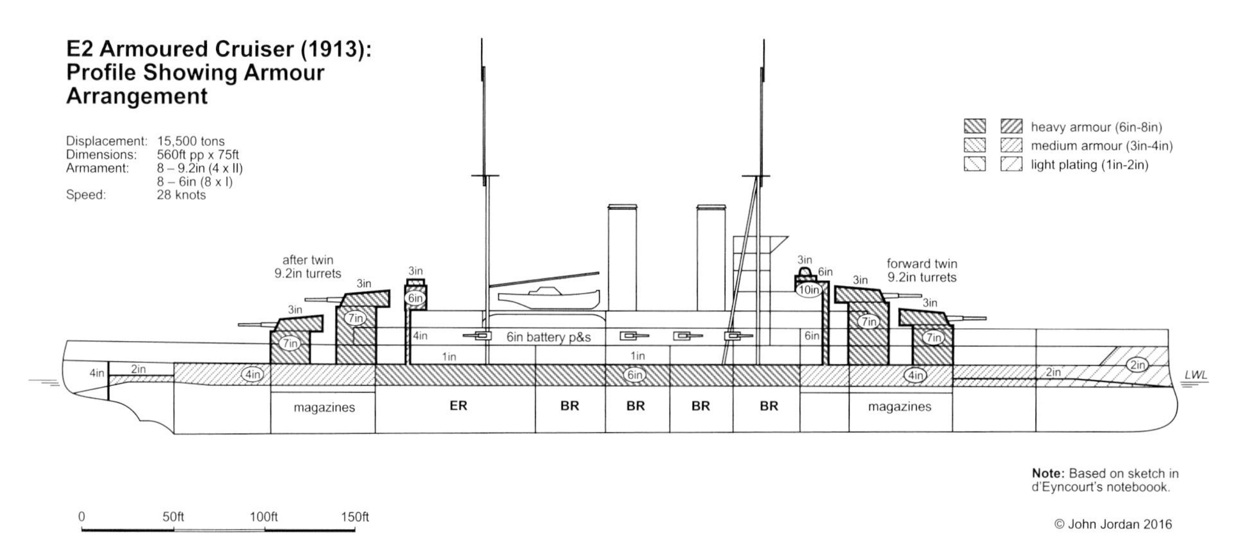

Fig 6: Armoured cruisers E2 and E3, 1913.

E2 is shown. E3 was somewhat larger (displacement 17,850 tons, 580ft pp). D'Eyncourt showed E3's more extensive armour scheme on the same diagram: the 6in gun battery had 4in armour, with a 4in strake between the battery and the 6in belt; there was 1.5in vertical protective plating around the funnel uptakes and in way of magazines and shell rooms, 1in protective plating on the forecastle deck over the battery and 1.25in on the upper deck amidships; on the lower deck there was 2in fore and aft and 1in amidships. E2 had only 1in on the main deck amidships and 2in on the lower deck fore and aft.

ships. Alternatively, the rationale for the designs may have been simply to compare the costs of armoured cruisers with eight 9.2in guns with the mixed 6in and 7.5in batteries of the smaller 'Atlantic' cruisers. The notebook gives the fates of the designs: E2 and E3 were marked 'Dropped', implying abandoned prior to submission to the Admiralty Board; other designs were marked 'Not accepted' or 'Submitted [date]'.

Improved *Birminghams*

Had Germany attacked commerce in 1914 more successfully, the RN would have needed more new, faster cruisers. On 14 Oct 1914 d'Eyncourt wrote:

> Commerce protection on the ocean routes needs vessels somewhat larger and faster than the *Birmingham* with a good radius of action, burning coal and oil. Better protection than the *Birminghams* [would give] a fair chance against 6" gun attack but thicker side armour would add greatly to size and cost. [speed and armament unchanged].[7]

D'Eyncourt was almost certainly advocating some boilers for oil and some for coal in unstated proportions, rather than merely spraying oil on the coal. Nor was he necessarily advocating ships as large as the *Hawkins*, later referred to as 'Improved *Birminghams*' because their designs had some common features.

Third Sea Lord, supporting Design B3, had observed 'if 6in guns were enough, a lengthened *Birmingham* with oil firing was hard to beat', suggesting a knot faster with two more guns or torpedo tubes – a proposal on the conservative side. HMS *Chester*, designed for Greece and completed in 1916, made 26.5 knots on 31,000shp from twelve Yarrow oil-fired small-tube boilers. Her sister *Birkenhead*, in which only the four boilers in the central boiler room burned oil, made 25 knots. Both mounted ten guns, albeit the lighter 5.5in, on the standard *Birmingham* hull.

The Royal Australian Navy's first three 'Towns' had twelve boilers giving 25,000shp for 25 knots with forced draft, 22,000shp with normal draft. Australian coal had a lower calorific value than Welsh. In February 1914 Britain was approached for material for the first of two further ships to be built in Sydney. Geared impulse turbines were specified, with ten boilers (four oil-fired) for 25,000shp with normal draft, and longer range for trade protection duties.

In 1915, when the RAN's flagship battle cruiser *Australia* was with the Grand Fleet, heavier guns were sought. In February the RN Director of Naval Ordnance provided weights of 9.2in and 7.5in guns, including shields, ammunition and spares. Only two 9.2in could be mounted, quite inadequate for ranging on shell splashes, but *Adelaide*'s Cover contains DNC's brief estimates and $\frac{1}{32}$in = 1ft outline tracings of versions with four or five centreline single turrets mounting 7.5in BL Mk VI L45 guns (see Figure 7). A speed of 25.5 knots was estimated, helped by increased length, assuming the current scale of armour protection and 27,500shp from machinery with ten boilers. This was reasonable; several standard 'Towns' exceeded rated power on trials. A 1910 alternative 'Colonial cruiser' scheme (27–30,000shp for 25.5 knots for 8 hours and 26 knots on 2-hour trials instead of 22,000shp for 24.75 knots) differed only in having machinery 110 tons heavier, a displacement of 5,500 tons, and an increase in beam of 6in.

Assuming the same HA guns and torpedo tubes for the four- and five-gun designs, the 111-ton difference between their armament weights gives the weight of the fifth turret. But then four and five main guns would absorb all the quoted armament weight. DNO gave the empty turret weight as 50.75 tons plus a 44.5-ton shield, which with the desired 380 rounds gives a figure of about 145 tons. Adding only 50 tons for HA and torpedo tubes would give armament weights of 630 tons and 775 tons respectively, taking the total weights to 5,754 tons and 6,118 tons.

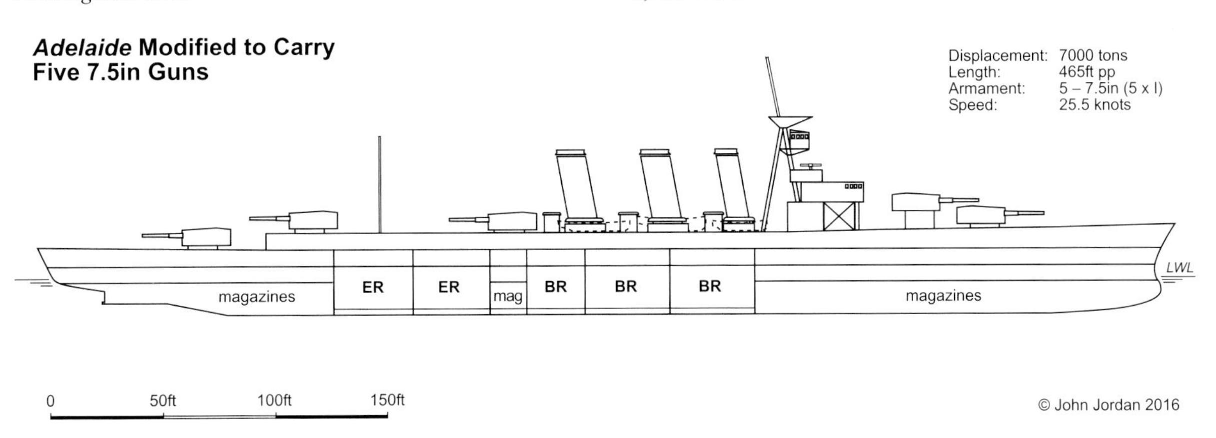

Fig 7: *Adelaide*: 7.5in gun proposals.
Two versions were proposed: one with four 7.5in guns. the other with five, shown here. The four-gun ship was similar but slightly shorter and without the middle turret; its boiler rooms were reversed, the shorter one being forward. The narrow funnel was at the after end of the third boiler room, with the broader funnels still over the inter-room bulkheads, so the forefunnel was farther from the bridge. (In the 5-gun ship, the after funnel so placed would have fouled the turret amidships.)

A projected 95.25-ton empty turret weight was less than the 147 tons estimated in 1911 when considering replacing the twin 6in of the *Monmouths* with single 7.5in Mk I L45 guns as on the *Devonshires*, and significantly less than the 193-ton single turret for the Mk II L50 guns fitted in *Minotaur* (those weights include barbette and tube). The estimates in the table use DNO's figures, but with 380 rounds for high shell expenditure at long ranges instead of his 180. Had these 7.5in variants gone ahead, 30-degree elevation CP mounts as in the *Hawkins* class (design initiated June 1915) might have lengthened gun ranges.

Closer to the original 'Towns' in conception than to *Hawkins*, these 'Improved *Birminghams*' went no further. With building scheduled to start in August 1915, a modified *Birmingham* design with nine 6in guns, two-shaft geared impulse turbines and twelve boilers had to be adopted. A September 1915 Admiralty blueprint in the Cover shows first and third boiler rooms slightly longer than the central one, with the usual four funnels of the 'Towns'. Legend displacement became 5,557 tons with the hull lengthened by 4 feet to maintain trim, rather than by filling out the hull lines forward. (Legend displacements for *Brisbane* were 5,400 tons, for *Birmingham* 5,440 tons.)

Adelaide was laid down in November 1915. The

Table 5: *Adelaide* with 7.5in Guns

	Adelaide 7.5 A	*Adelaide* 7.5 B
Displacement	6,540? tons	7,000 tons
Length, pp	450	465
Power (shafts)	27,500shp (2)	27,500 (2)
Speed	25.5 knots	25.5 knots
Fuel	650? tons	650? tons
Armament	4 – 7.5in (4 x I)	5 – 7.5in (4 x I)
Protection:		
Side	2in aft/3in–2.5in fwd	
Deck	0.75–0.375in	
Weights:		
Hull & protection	3,124	3,391
Armament	448 [630]	559 [775]
Machinery	1,049	1,049
General	301	303
Total weights given	4,922 [+650 fuel]	5,302 [+650 fuel]
[Total weights revised]	[5,754 incl fuel]	[6,118 incl fuel]

Notes:

1 *Adelaide* 7.5in estimates from Cover 331 lack some data usually quoted: beam and draught, fuel (neither legend nor full load), secondary armament, Board margins; author's estimates are in square brackets.

2 The Legend displacements quoted are ca 1,000 tons greater than the total given weights plus a reasonable Legend fuel load, and make more sense if full load.

Supplementary table of revised weights (tons):

	B'ham	**7.5 A**	**7.5 B**
Displacement	6,050	6,540	7,000
Total of weights	5,223	5,754	6,168
Hull + protection	2,914	3,124	3,391
Hull/displacement	0.482	0.478	0.484
Hull/weights	0.558	0.543	0.550
Length, pp	430ft	450ft	465ft

Notes on supplementary table:

1 HMS *Birmingham* has more detailed data than are available for *Adelaide* herself; quoted displacement for *Birmingham* is full load; Legend displacement was 5,273 tons, equal to her total of weights plus an assumed 50-ton Board margin as other 'Towns'.

3 The ratios of hull weights to displacement are all close to 0.48, suggesting that all three displacement figures, not just that of *Birmingham*, are full load. Hull weight/total weights all around 0.55 suggest Legend weights for the 7.5in gun designs should be close to their respective totals of weights.

4 The hull is essentially a hollow shell, not a solid, so for similar ships the ratios of hull weights should approximate to the ratios of the squares of their lengths (slightly distorted by the weight of protection [ca 600 tons] contributing to hull strength and being grouped with hull weights): 2,914 x (450/430)2 = 3,191; 2,914 x (465/430)2 = 3,407; 3,124 x (465/450)2 = 3,336.

HMAS *Adelaide*, a light cruiser of the *Birmingham* class. Her completion was much delayed and alternative armaments suggested; compare the ship as completed with the 7.5in versions projected. (Allan C Green collection, State Library of Victoria)

forward and after boilers burned coal and oil, with four oil-fired boilers between. A total of 750 tons of coal and 600 tons of oil was carried – *Brisbane*'s twelve boilers all burned coal and oil, with 1,240 tons of coal and 260 tons of oil. A new bridge, improved fire control, searchlights and W/T, and a stronger hull with HT steel worked longitudinally for the outer thickness of protection all reflected latest practice. A maximum elevation of 30 degrees for her 6in guns was considered, 20 degrees provided. Wartime losses of parts and slow deliveries delayed completion until 1922, when 'latest practice' could embody wartime experience. Oil firing only, centreline 6in guns and a two-funnel profile like the 'D' class were planned in 1924. In 1938 oil-firing simply removed the forward boilers and funnel. *Adelaide* served throughout the Second World War. Light shields first considered in 1915 replaced 3in armoured gunhouses on her 6in guns in 1943.

Hawkins Class

Reports of German raiders with 6.7in or 8.2in guns persisted. When in 1915 'C' class cruiser numbers were expected to match German construction, designs for six large, fast trade protection cruisers began. An Admiralty memo of the time states:

> The general type should be a 30 knot cruiser... ten or more 6in guns mounted as far as possible on the centre line... ⅕th power by burning coal only... The question of 5.5in guns for these ships has been fully discussed... [but as] German cruisers carried 5.9in it is inadvisable to go below 6in.[8]

The *Hawkins* class (renamed after *Raleigh* was wrecked in 1922) had a long but chequered history described in detail by Friedman and Raven & Roberts (qv). The overall budget allowed only five ships. Three ran aground, with two wrecked completely. *Cavendish* completed first as the aircraft carrier *Vindictive*, was reconverted to a cruiser then, in turn, to a repair ship, destroyer depot ship and training ship. All were later converted to oil firing. Reconstructions with three twin 8in (1925); four twin 6in (1930) and three twin 5.25in DP (1938) were considered, *Effingham* being rearmed with nine single 6in guns in 1937.

These 'Improved *Birminghams*' had the long forecastle and short quarterdeck of the *Birminghams* and the unbuilt 'Atlantics' between. However, *Hawkins* displaced and, at £750,000, cost about twice as much as a *Birmingham*. This was rather more than d'Eyncourt's 'somewhat larger': length 565ft compared to 430ft, broadside 1200lb compared to 500lb. A comparison of the length/beam ratios and the balance of weights for versions with eight main battery guns (see Table 7) shows *Hawkins* and *Chatham* to be closer in conception than to B3 (the supposed 'bridge' between them), despite *Hawkins*' very different hull, which was derived from the

Table 6: *Hawkins* Preliminary Designs 1915

The seven-column table from d'Eyncourt's notebook is here recast, omitting some details of armour and ammunition stowage (generally 300 rounds per gun) to make clear where the variants differ. Deep draught rises with armament weights (and some other weights for D). Fuel load was reduced to give the same nominal Legend displacement.

Version	A, B, C, D	E, F	Approved
Displacement	9,100 tons	9,750 tons	9,750 tons
Length, pp	560ft	565ft	565ft
Beam	64ft	65ft	65ft
Draught (deep)	A 19ft 3;in B 19ft 4in; C 19ft 5in; D 19ft 8in	19ft 4in	19ft 3in
Horsepower	60,000shp	60,000shp	60,000shp
Speed	30 knots	30 knots	30 knots
Fuel:			
coal	500 tons	500 tons	800 tons
oil	A/B/C 1,750 tons D 1,600 tons	1,750 tons	1,580 tons
Armament:	A/B/C 10/12/14 – 6in; D 2 – 9.2in, 8 –6in 2 – 3in HA 2 – 21in TT sub	E 2 – 9.2, 8 – 6in; F 8 – 7.5in E/F 2/4 – 3in HA; 2 – 21in TT sub	7 – 7.5in 6 – 3in LA 4 – 3in HA 2/4 – 21in TT sub/aw
Armour:			
Side	3in–2.5in–1.5in	3in–2.5in–1.5in	3in–2.5in–1.5in
Deck	1in	1in	1in
Complement	A 440; B 460; C 480; D 470	470	480 (as flagship)
Weights:			
Hull	A/B/C 4,600; D 4,650	E 4,750; F 4,780	4,900
Protection	A/B/C 850; D 900	E 920; F 940	810
Armament	A 400; B 460; C 520; D 680	E 680; F 600	560
Machinery	1,880	E 1,900; F 1,930	1,950
General	400	400	430
Board margin	90	100	100
Total (excl Bd m)	A 8,130; B 8,190; C 8,250; D 8,360 [with 150t less oil]	E 8,650; F 8,650	8,650

'large light cruiser' *Furious*. Armour was on light cruiser scale (2in+1in and 1in+1in HT steel), but the sides were inclined 10 degrees from deck to waterline so that shells would strike the belt obliquely, thereby increasing its effective thickness. D'Eyncourt's recently-devised light waterline bulge theoretically reduced hull damage from torpedoes. Oil fuel provided 60,000shp for 30 knots, which made the ships markedly faster than either the *Birminghams* or the 'Atlantics', and 20 per cent power from a few coal-fired boilers insured against shortage of oil on distant stations.

Initial designs had ten, twelve or fourteen 6in guns. D'Eyncourt's notebook lacks gun layouts, but obvious parallels were the 11,000-ton *Diadems* with sixteen 6in and the 10,000-ton *Monmouths* with fourteen, mostly sided port and starboard. However, the condition 'mounted as far as possible on the centre line' could be met with two guns forward, one superfiring, three at the after end of the forecastle and one on the quarterdeck, leaving four, six or eight in sided pairs for designs A, B, and C. Legend armament weights show 60 tons for each extra pair of 6in. 'C' class-type light shields rather than the 5-ton 3in armoured shields of the 'Towns' would allow more rounds per gun. More guns meant larger complements.

A 1911 proposal to replace the *Monmouths'* rather ineffectual twin 6in turrets ('A' and 'X') by single 7.5in or 9.2in turrets implied a similar layout to the 1898 armoured cruisers of the *Cressy* class. The latter had

Table 7: Breakdown of Weights

	Chatham	B3	*Hawkins* F
Displacement	5,400 tons	7,400 tons	9,750 tons
Hull	43%	46%	49%
Protection	12%	13.5%	9.5%
Armament	6%	15%	6%
Machinery	20%	13%	20%
General	5.5%	5.5%	4%
Length/beam ratio	**8.6**	**9.6**	**8.7**

HMS *Effingham* in her China Station paint scheme of white hull and buff funnels/masts. One of a class of five, she and her sisters were completed too late to see service in the First World War. They were the only cruisers built by the Royal Navy for commerce protection between 1905 and 1920. (Allan C Green collection, State Library of Victoria)

single 9.2in fore and aft plus twelve 6in sided in casemates; these were plated over around 1915–16, with the 6in guns moved to upper deck single mounts. *Hawkins* Design D could have replaced the proposed forward pair and the after group of centreline 6in with single 9.2in turrets, leaving eight sided 6in (six if a 6in superfired over the forward 9.2in, with another well aft on the quarterdeck if blast from the after 9.2in allowed, the extra 5ft length of design E perhaps helping somewhat). A 9.2in at the after end of the forecastle would be clear of the propeller shafts – which precluded a fourth twin 8in turret on Hawkins' low quarterdeck in 1925 rearmament proposals.[9] A twin 9.2in turret would affect trim and fire more slowly, with both guns disabled if hit.

Design F used the larger hull to carry a uniform armament of eight 7.5in of a new model, the BL Mk VI L45, developed from the L45 Mk I of the *Devonshires* and with slightly higher muzzle velocity. Central pivot mountings with open shields and electrical training and elevation were much lighter than armoured turrets. Elevation was doubled to 30 degrees, range increased from 15,000 to 21,000 yards. This increase in range, allied to a weight of shell twice that of the 6in, increased the chance of an early hit disabling a raider, though hand-loading such shells required two men, a special grab, and the mounting of the guns in shallow circular pits to reduce the lift.

The approved seven-gun design had a centreline 7.5in on the forecastle with a second superfiring, one at the after end of the long forecastle, two well separated on the quarterdeck and two sided amidships – the layout suggested by designs A to E. Six low-angle 3in were added to the four 3in HA of the earlier designs, although these were soon removed (1921). Two of these and two sets of twin torpedo tubes were mounted close to the 7.5in gun at the after end of the forecastle deck, so there was room for a sided pair of 7.5in guns in Design F, the hull narrowing farther aft.

'D's and 'E's

The 3,500-ton small fleet cruisers of the *Arethusa* class grew through several 'C' variants of around 4,000 tons to the 'D' class of 4,750 tons with six centreline 6in guns approved in May 1916 – nearly as big as the prewar 'Towns' and therefore potentially useful on the trade routes. Three were ordered in September. DNC was then asked for a modified variant with 7,000nm range with minimal extra tonnage and loss of speed. AJ Holmes' revised design (25 October) carried an additional 325 tons of fuel oil between the engine rooms and boiler rooms, forward of the 6in magazines for 'Q' turret abaft the funnels. 'Peace' tanks (with crowns above the waterline) held 100 tons of this fuel, to be used first on any voyage.

Drawings were ready for DNC by 18 November. The legend (submitted 8 December 1916 and approved 3 January 1917) increased range from 5,000nm to 7,000nm at 13–14 knots. The design was 15 feet longer (460ft pp, 487ft oa) and had greater beam: 47ft 6in compared to 46ft for standard 'D's as completed. Displacement at load draught would be 5,100 tons (an increase of 350 tons). Redesigned machinery (10 tons heavier) lowered fuel consumption at cruising speed. Triple torpedo tubes replaced twin; searchlights, torpedo and gun control positions and W/T offices were rearranged as in the *Calypso* class.

Another version at least 467ft long, with increased beam and draught and displacement perhaps 5,250 tons, aimed to reduce the silhouette seen from angles abeam. No 3 gun was trained aft before the bridge, below the superfiring No 2 gun. A screen, kept small by ensuring Nos 2 and 3 guns engaged on the same side, protected No 3 from blast but restricted its elevation on forward bearings. The bridge was (desirably) farther from the bow, but amidships the 3in HA guns, the boats and torpedo tubes were somewhat congested. Drawings showed a standard 'D' bow. Either design would have been valuable when the hard-worked 'Towns' were withdrawn in the late 1920s, but both needed complete new building drawings, thereby delaying construction. Instead, three more ordinary 'D's (with higher 'trawler' bows) were ordered in July 1917 and six more in March 1918; four were cancelled when the war ended. *Diomede* and *Dunedin* would serve in the Pacific for several years from the mid-1920s in the then-RN New Zealand Division.

In October 1917 two German minelayers sank most of a convoy and its destroyer escort off Norway. Mistakenly seeking to match their particular size and speed (greatly over-estimated at 36 knots), Project A was a 465ft, 4,500-ton design with 60,000shp for 32 knots but only four 6in guns. Preliminary design B (six guns and 34 knots) required 6,700 tons, while the final 7,550-ton 'E'

Standard 'D' Class **Proposed Modified 'D' Class (1916)**

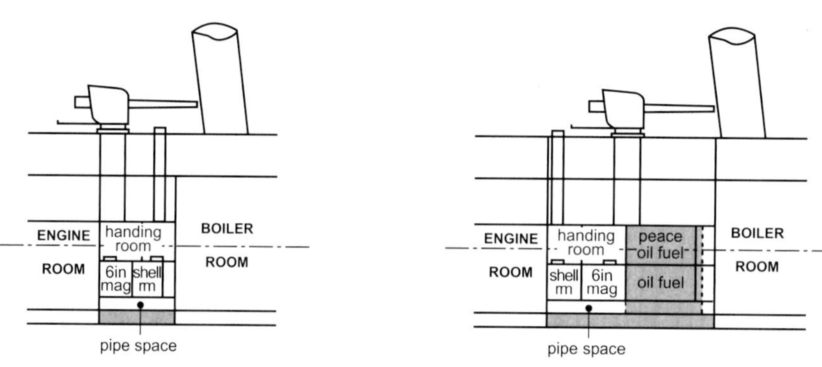

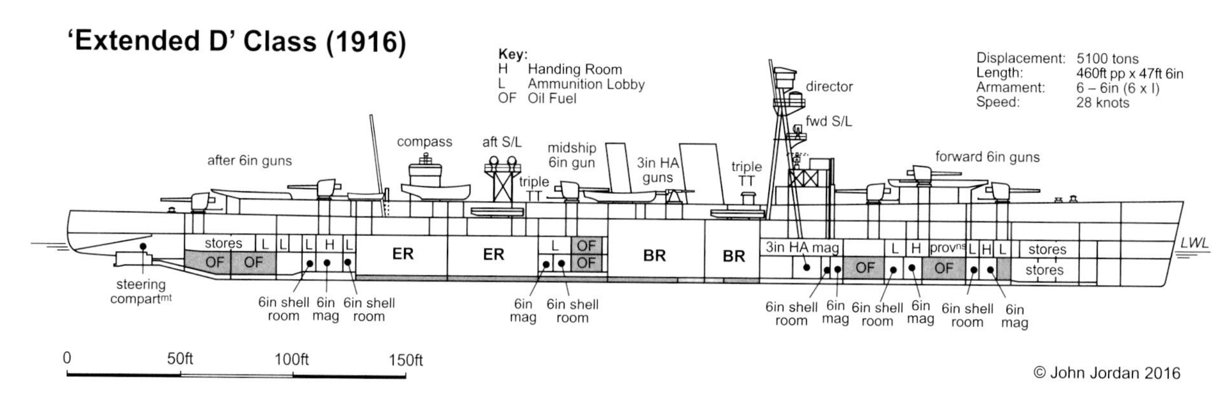

Fig 8: Proposed modifications to 'D' class to extend range.
Two schemes were proposed, referred to here as Extended 'D' to indicate the nature of the modifications. The first simply lengthened the standard 'D' by 15ft to accommodate oil tanks inserted before the shell rooms below the midships 6in gun (see upper drawings). The second, (lower drawing) aimed to reduce the silhouette seen from an angle to abeam by bringing the third gun before the bridge, though the ship was longer still. The documentation and 1/16in =1ft plans of both versions are (surprisingly) in Cover 319, not the 'D' class Cover.

The 'D'-class cruiser *Diomede* served in the Far East from her completion in 1922; in 1926 she was assigned to the RN's New Zealand Division. She was the only ship of her class to have the experimental Mk XVI fully-enclosed 'weatherproof' 6in gun mounting. (Allan C Green collection, State Library of Victoria)

The 'D' class cruiser *Danae* during a visit to Melbourne in early 1924. *Danae* was part of a Royal Navy task force led by HMS *Hood* which was despatched on a world tour during 1923–24. (Allan C Green collection, State Library of Victoria)

Table 8: 'E' Classes

	Project A	Project B	*Emerald*	'E' 7.5 A	'E' 7.5 B
Displacement:	4,500 tons	6,700 tons	7,550 tons	7,700 tons	8,850 tons
Length, pp	435ft	510ft	535ft	535ft	570ft
Beam	43ft	52ft	54ft 6in	54ft 6in	58ft
Draught, mean	15ft 3in	16ft 6in	16ft 6in	16ft 9in	17ft 3in
Power (shafts)	60,000shp (4)	80,000shp (4)	80,000shp (4)	80,000shp (4)	80,000shp (4)
Speed:	32.5 knots	34 knots	33 knots	33 knots	32.75 knots
Fuel:					
Legend	n/a	n/a	650 tons	650 tons	800 tons
Deep	915 tons	1,250 tons	1,600 tons	1,600 tons	[2000 tons]
Armament:					
Main	4 – 6in	6 – 6in	7 – 6in	4 – 7.5in	5 – 7.5in
Secondary	2 – 3in	2 – 4in	2 – 4in	4 – 4in HA	4 – 4in HA
TT	2 x III 21in	4 x III 21in	4 x III 21in	4 x III 21in	4 x III 21in
Protection:					
Side	3in machinery, 1.5–2.5in fwd (all)				
Deck	1in magazines (all)				
Weights:					
Hull + protn	2,250	3,740	3,820+700	3820+700	[4600+750?]
Armament	160	330	355	[420]	[490]
Machinery	1,245	1,640	1,590	1,590	1,640
General	315	315	360	360	360?
Board margin	15	70	75	75	–
Total weights	3,985	6,095	6,900	–	–
Wts incl fuel	4,350	6,595	7,550	7,540	8,640

Notes:

Data for 'E' 7.5s is from 'Proposed Designs for Light Cruisers with 7.5in guns' filed with ADM 1/9223 at TNA. These do not include distributions of weights. Estimates [square backets] for 'E' 7.5 A are closely based on *Emerald* – increased armament weight plus a small addition to hull weight arising from the changed boiler layout. Estimates for 'E' 7.5 B to match a displacement of 8,850 tons are more speculative, the biggest increase being the extra weight of the larger hull.

design had two sets of standard 40,000shp destroyer leader machinery for 33 knots, occupying much centre-line space and needing a sided pair of 6in amidships to maintain a six-gun broadside. Approved in May 1918, two 'E's were eventually completed in 1926; a third was cancelled. Fitted with four triple 21in torpedo mountings, they were well-suited to use as fleet cruisers but their roots were in trade protection. Early in the Second World War they were so employed on distant stations because of their limited horizontal protection and vulnerability to air attack.

A summer 1918 Grand Fleet meeting, which probably discussed doubtful reports of German light cruiser raiders armed with two 8.2in guns, suggested more gunpower to Controller. On 15 August DNC produced preliminary designs with four (A) or five (B) 7.5in guns, which were forwarded to C-in-C Grand Fleet next day. Their legends (see Table 8) compared them with *Hawkins* as well as *Emerald*, thereby confirming interest in protecting trade. They were short-lived. C-in-C responded on 2 September in detailed question-and-answer form (see Raven & Roberts, 102). In order to build in sufficient numbers, it was stated that the requirements for wartime fleet cruisers were a displacement of 4,500–5,000 tons and high speed, implying fewer guns – 6in to avoid power operation. The *Hawkins* type might be needed later – it is possible that a November armistice was unforeseen. Senior Admiralty officers exchanged memos and concurred, CNS deciding (26 September 1918) to discount the German cruiser reports, so C-in-C was not consulted again. These designs (final date stamp 30 October) reached neither d'Eyncourt's notebook nor the Admiralty Board.

DNC's 1/16in = 1ft preliminary plans, redrawn by William J Jurens for McBride's article 'E's and Super 'E's',[10] showed *Dragon*'s unsatisfactory aircraft hangar under the bridge, as in *Emerald*'s first plan.[11] *Emerald* had four boiler rooms, the forward and after rooms being longitudinal, well inboard and less exposed to torpedo attack. A and B had the foremost boiler room transverse, probably to maintain trim with the heavier 7.5in forward. The forefunnel, now the broadest of the three, served four boilers, and was set slightly aft over the bulkhead between the two forward boiler rooms. Design A was very like *Emerald*, but with 7.5in on *Hawkins*-type CP mounts in 'A', 'B', 'X' and 'Y' positions. Two 4in HA on each side with magazines below replaced *Emerald*'s wing 6in, the seventh 6in being omitted. Design B had a new 35ft section abaft the second funnel for her fifth 7.5in gun, its magazine and the 4in HA magazine being grouped together below it. 'X' 7.5in gun was restricted to 30 degrees before the beam to avoid blast on the after control tower, which was relocated abaft the third funnel. The HA guns repositioned nearer that funnel were farther from their magazine but less subject to blast from the midships 7.5in, which gained a clearer field of fire. A rotating flying-off platform aft like *Emerald*'s would have been problematic; four 7.5s would probably have sufficed.

Postwar Poverty

After the war old and worn-out ships were scrapped and much planned construction cancelled. Earlier unbuilt trade protection designs retained features such as mixed armament, sided guns and mixed fuel; later designs were variations on the fleet cruisers given priority.

When *Raleigh* was wrecked on 8 August 1922, *Hawkins* was the RN's only large modern trade protection cruiser until her sister *Frobisher* commissioned on 20 September 1924. The Post-war Questions Committee, drawing lessons from the war,[12] advocated a high-speed cruiser of about 7,000 tons with four twin 6in turrets (later implemented as *Leander*, after other abortive design studies in the 1920s[13]). The Committee of Imperial Imperial Defence produced suggestions[14] akin to the 1909 Fleet Unit scheme, Director of Gunnery Division proposing five new designs: 7,500–10,000 tons with four, five or six 8in guns and 80,000shp, oil fired for up to 35 knots. The Dominions could build the smaller designs, which might not tempt other navies to larger ships, but numbers estimated for a worldwide convoy system were very high. A 15,000-ton cruiser with 10in guns and magazine armour to resist 8in shells was suggested to back them up and deter larger foreign ships.

Workload in the DNC's department on battleships and other vessels precluded detailed designs, but DNC estimated the '15,000 tonner' would be nearer the 18,000 tons of *Glorious* while a *Hawkins*-size ship with 8in guns making 33 knots would top 11,000 tons. Australia planned building *Hawkins*-based hulls with four twin or three triple 8in turrets at Cockatoo Island – ideas bearing out DNC's estimate.

The Washington Treaty of February 1922 limited cruisers to 8in guns and 10,000 tons standard. Designs over 30 knots based on *Hawkins* again proved fruitless, leading to the adoption of a new hull design for the '*Kents*'. Admiralty hopes for 40 such ships – the commerce protection component of Jellicoe's 70-cruiser navy – were soon dashed. Australia purchased two of only thirteen 'Counties' built before the 1930 London Treaty limited total cruiser tonnages. The Royal Navy's emphasis on the fleet cruiser before and during the First World War would be responsible for a long-lasting dearth of modern cruisers for commerce protection.

Acknowledgements:

Thanks to Andrew Choong, NMM Plans and Photographs, for especial help with finding details, and to John Jordan, for careful interpretation of data into drawings, detailed helpful editorial discussions and suitable photographs.

Sources:

Published Sources

Brown, David K, *The Grand Fleet*, Chatham (London 1999).

Friedman, Norman, *British Cruisers: Two World Wars and After*, Seaforth (Barnsley 2016).

Hezlet, Sir Arthur, *The Electron and Sea Power*, Peter Davies (London 1975).

Marder, Arthur J, *From the Dreadnought to Scapa Flow*, Vol IV, Seaforth (Barnsley 2014).

Parkinson, Roger, *The Late Victorian Navy*, Boydell (Martlesham 2008).

Raven, Alan & Roberts, John, *British Cruisers*, Naval Institute Press (Annapolis 1980).

Sumida, Jon Tetsuro, *In Defence of Naval Supremacy*, Routledge (Abingdon 1993)

Watson, Graham, 'From Imperial Policeman to North Sea Battle Fleet: British Naval Deployment 1900–1914', www.naval-history.net/xGW-RNOrganisation1900-14.htm.

Archival Sources

Jellicoe, John, 'Submission on cruisers for 1907 programme', and 'Subject of the armament of cruisers for the 1908–09 programme', with Watts, Sir Phillip, Legend Particulars, all Fisher Papers, Churchill Archives, Churchill College, Cambridge.

Tennyson d'Eyncourt, Eustace, 'Design Particulars', NMM Caird Library MSS93/011.

ADM 1/8397/365 'Warship Design 1914–1922' (TNA).

ADM 116/1100A Cmnd 4948 Papers re Imperial Conference 1909 (TNA).

Ship's Cover NMM No 319 'Atlantic Cruisers'.

Ship's Cover NMM No 331 'HMAS *Adelaide*'.

Notes on sources:

The article relies chiefly on these Archival Sources, citing some subsidiary sources in the text. Cover 319 contains reports and plans for 1913 'Atlantics', Colonial cruisers, modified 'D' classes and some early *Adelaide* papers. The Published Sources provide background, their titles indicating their scope. Sumida has an extensive further bibliography.

Endnotes:

1 Brown, *op cit*, 61.
2 *ibid*, 61.
3 *ibid*, 14.
4 Friedman, *op cit*, 35, 336.
5 Brown, *op cit*, 67.
6 Friedman, *op cit*, 35.
7 ADM 1/8397/365.
8 ADM 1/8424/171.
9 ADM 1/8674/8.
10 *Warship International*, 1996, 257.
11 ADM 1/9223.
12 ADM 1/8586/70.
13 ADM 1/9260.
14 ADM 1/8605/81 & CAB 34/1/6 Empire Naval Policy and Cooperation, Feb 1921.

AMATSUKAZE: A DESTROYER'S STRUGGLE

Captain Tameichi Hara's account of the early service career of the IJN destroyer *Amatsukaze* has become a classic. **Michael Williams** follows on from that account and tells the less-well-known story of the destroyer's service from 1943 to 1945, when Japanese fortunes were on the wane.

On 7 January 1945, 25-year-old Lieutenant Tomoyuki Morita (Japanese Naval Academy Class 68), then serving as torpedo officer on board the Japanese fleet destroyer *Kasumi* at Singapore, received orders transferring him to the *Kagero* class destroyer *Amatsukaze*, also at the Seletar facility. He was subsequently formally appointed as her commanding officer with effect from 10 February 1945. Destroyers, and even smaller escorts such as *Kaibokan* were usually under an experienced full Commander or at least a Lieutenant-Commander; Morita's appointment as CO of *Amatsukaze* therefore had to involve profoundly unusual circumstances, which are the subject of this article.

The exploits of the Imperial Japanese Navy (IJN) destroyer *Amatsukaze* under Commander Tameichi Hara from 19 April 1940 up to 10 January 1943, when he was reassigned upon promotion, is a period well documented in Hara's own memoir 'Japanese Destroyer Captain' (USNIP 1961). Hara's narrative covers the ship's extensive and diverse duties during the initial Japanese expansion phase of the Pacific War: the invasion of the Philippines, the East Indies campaigns (and her participation in the Battle of the Java Sea), the Battle of Midway, and the Battle of the Solomon Islands.

Hara's published memoir is one of the best-known personal memoirs of the Pacific War. However, in his otherwise comprehensive account, *Amatsukaze*'s fate is consigned to a brief passage on page 248:

> She had been torpedoed 250 miles north of the Spratly Islands in January 1944, and ... 80 members of her crew had been killed. Although badly damaged the destroyer was able to reach Saigon. Emergency repairs there enabled her to be moved to Singapore for further work which was not completed until March 1945, when she was again seaworthy.

The author felt compelled to discover exactly what had happened to *Amatsukaze* following the damage sustained off the Spratly Islands. Japan had begun the Pacific War in December 1941 with 111 destroyers, while during the conflict she commissioned a further 52. Her cumulative

Amatsukaze on her machinery trials 17 October 1940. She featured a new experimental type of boiler, which operated at a pressure of 40kg/cm^2 and a maximum temperature of 400°C; the respective figures for other units of the class were 30kg/cm^2 and 300°C. IJN Department of Naval Aeronautics)

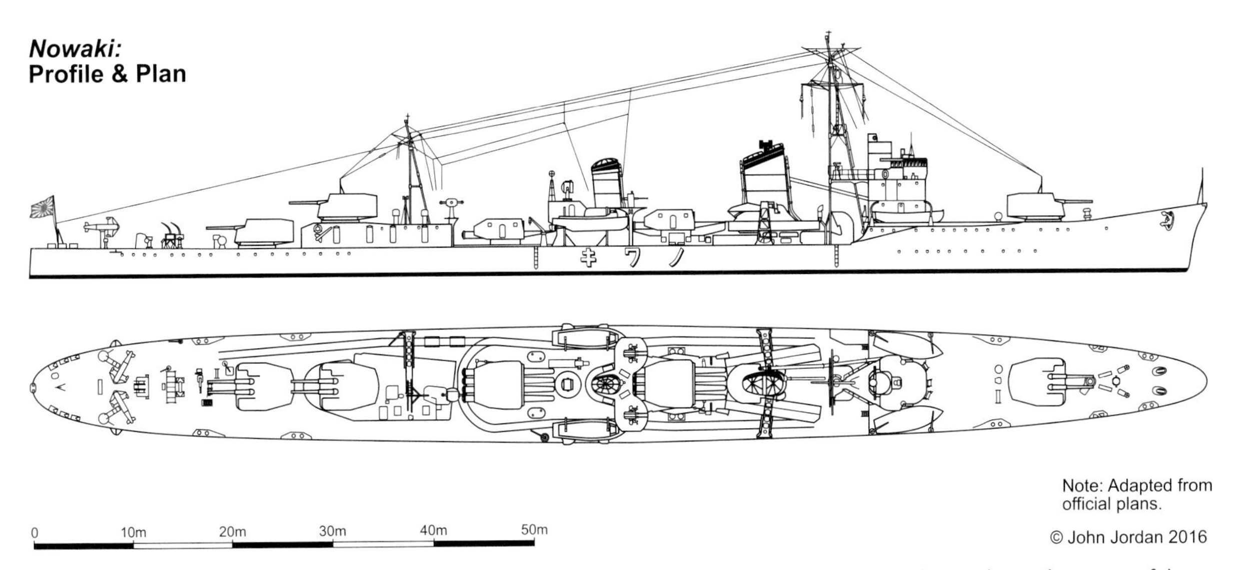

Profile and plan of *Nowaki,* as she appeared upon completion in April 1941. At the time of their completion, the 18 destroyers of the *Kagero* class were among the most powerful afloat. Displacing 2,530 tonnes under battle conditions, they were propelled by two Kampon impulse turbines with a nominal rating of 52,000shp, and were capable of achieving 35.5 knots. Armament as completed was six 127mm/50 Type 3 dual-purpose guns in three twin mounts, four 25mm Type 96 anti-aircraft guns in twin mounts, eight Type 92 torpedo tubes in two quad mounts with sixteen 61cm Type 93 'Long Lance' torpedoes, and eighteen Type 95 depth-charges. The designed complement was 9 officers and 219 men.

losses in this one type of warship up until August 1945 were 135, leaving 28 in various states of damage or repair to be surrendered to the Allies. The story of *Amatsukaze* which follows is in many ways representative of the late-war service of the IJN's surviving destroyers.

New Beginnings

When Commander Masao Tanaka took over *Amatsukaze*, she was at Kure undergoing repairs to damage received on the night of 12/13 November 1942 at the First Naval Battle of Guadalcanal. As a member of the 16th Destroyer Division she had been assigned to escort Admiral Abe's Bombardment Force. During the battle she torpedoed and sank USS *Barton* (DD-599), but was herself hit in return by 5in gunfire from the cruiser USS *Helena* (CL-50), sustaining serious damage to her upperworks. Although not all the shells detonated, those which did knocked out her hydraulic system, disabling her turrets and rudder and leaving 43 of her crew dead.

While *Amatsukaze*'s record of achievements from December 1941 up to this battle had been one of supporting Japanese expansion and offensives, the turn of the year under her new captain saw a marked change in her deployment. When she returned to service in February 1943, she embarked on a series of routine escort and transport runs while based at Truk, sailing to Palau, Wewak, and Hansa Bay. These involved the occasional rendering of assistance to damaged consorts and regular periods of maintenance.

On 11 November Commander Tanaka was promoted to Captain, and in December *Amatsukaze* returned to Kure for refit following her prolonged period of high-mileage escort duties. The entire year of 1943 had been fully committed to support duties without a single encounter with the enemy. However, the following year would see a complete shift both in deployment and in fortunes.

Convoy HI-31

The routine relief of Captain Tanaka by Lieutenant-Commander Akiji Suga was notified on 10 January, but did not take place due to Suga's death in the sinking of

A view of the after part of the *Kagero* class destroyer *Shiranuhi* docked at Maizuru after losing her bow to torpedoes from USS *Growler* (SS-215) off Kiska Harbour on 5 July 1942. Note the Type 94 depth charge thrower and its reload rack (under canvas), and the six individual depth charge chutes above the stern; prominent between them are the paravanes used for minesweeping. The upper hull has been fitted with a prominent degaussing coil. (IJN Department of Naval Aeronautics)

Sazanami by USS *Albacore* (SS-218) on 14 January – there were 153 casualties, leaving 89 survivors to be rescued by the destroyer *Akebono*. Captain Takanobu Sasaki was then appointed as Suga's replacement to *Amatsukaze* on 16 January 1944, but on that day she would herself be very nearly lost.

On 11 January *Amatsukaze*, still under Tanaka and with the commander of the 16th Destroyer Division, Captain Furukawa Bunji, on board, along with her sister *Yukikaze* supported the light carrier *Chitose* in escorting the high-speed tanker convoy HI-31. The latter was designated a 'breakthrough fleet' and comprised the empty *Omurosan Maru*, *Tatekawa Maru*, *Itsukushima Maru*, and *Genyo Maru*, which were to take on oil at Singapore and return to Japan. Together with the fully-loaded military transport *Hokuroku Maru*, the ships departed Moji (Japan) at 0830. The mission would take this valuable fast convoy through the submarine-infested South China Sea.

Another tanker, *Kuroshio Maru* (often erroneously claimed to be part of Convoy HI-31), had sailed earlier with HI-29, finally arriving at Singapore on 16 January, to load amphibious troops, military cargo, and 16,000 cubic litres of crude oil – perhaps indicative of the cargo the returning Convoy HI-32 would have carried. *Kuroshio Maru*'s inclusion in the return convoy was undoubtedly responsible for the error. Every one of these irreplaceable modern fast tankers would be lost in 1944, evidence of the fatal fuel haemorrhage that was to cripple not only the IJN but also Japan's mercantile arm, leading directly to her eventual defeat.

The first few days of Convoy HI-31's transit of the Yellow Sea, then the South China Sea were uneventful, but on the evening of 16 January this was to change dramatically. The protective combat air patrol of 'Kates' from *Chitose* had to return to the carrier before sunset, and this vital aerial cover disappeared just as *Amatsukaze*, which was stationed astern of the convoy, detected a surfaced enemy submarine 10km away and immediately increased speed to close, forcing the submarine to dive and break off contact. Arriving at the general position, the destroyer did not drop depth charges through the lack of a definite contact and, conscious that he had left a gap in the rear of the convoy screen, Tanaka decided to return to his station.

The *Gato*-class submarine *Redfin* (SS-272) was on her first war patrol. Commander Robert D King, in his subsequent report, relates how he had tracked the convoy on the surface until dusk, when *Amatsukaze* was sighted overtaking *Redfin* steadily from astern. The destroyer then apparently lit all boilers, resulting in a large plume of exhaust, and began to close rapidly, forcing him to dive.

There is a pronounced difference in the Japanese and American accounts of this early encounter, with King noting that *Amatsukaze* opened fire at 1949 (adjusted to JST), while no such direct action appears in the Japanese version, which simply states that the destroyer forced the enemy to dive, thereby disrupting its attack on the convoy for that night, and that this was the primary objective, with not even a depth charge attack carried out. However, as the destroyer turned to starboard to return to the convoy, a cry went up from the bridge

Convoy HI-31, designated a 'breakthrough fleet', comprising four fast tankers and a troopship, was escorted south by the fleet destroyers *Amatsukaze* and *Yukikaze* and the light carrier *Chitose*, seen here shortly after her conversion from a seaplane carrier at the Sasebo Navy Yard. The attached 653 Naval Air Group was based around the Nakajima B5N2 ('Kate') single-engine three-seat carrier-borne bomber, which was to prove useful in the anti-submarine warfare role. *Chitose*'s inclusion in the escort force for Convoy HI-31 was a crucial factor in its success, as she was able to provide aerial cover throughout the daylight hours. (IJN Department of Naval Aeronautics)

watch warning of a torpedo approaching. The torpedo was too close to avoid and struck *Amatsukaze* amidships.

Cornered, King's *Redfin* had retaliated with her four stern tubes. One of the torpedoes struck the destroyer's port side at 1952, causing the forward magazine to explode and producing a powerful underwater concussive effect that led the submarine crew to believe that all four torpedoes had struck. The torpedo hit about Frame 77, just abaft the fore-funnel and square in No 2 boiler room. The explosion blew the forward torpedo mounting overboard and tore a huge hole in the hull. For a brief time the ship held together, but with boiler rooms Nos 1 and 2 flooded the destroyer began to buckle very quickly amidships. The forward section began to list, twisting to port. The officers on the bridge hastily made their way to the starboard side, then the forepart lay over in the sea at 90 degrees, snapping free at Frame 84 from the rest of the ship and drifting away, flooding slowly.

Night was falling, and already the stern section was barely visible in the gloom of twilight. Captain Tanaka and some 35 men in the fore-section had a hasty conference, and decided to swim for the stern section before it was too late. One by one they jumped into the sea and struck out across the gap which separated the two halves of the ship, now some 300m apart, through the rising seas. They looked back and saw the bow rise vertically as it plunged into the seas; about 30 of them made it across. Once aboard Tanaka organised a roll-call and determined that 76 officers and crewmen were missing, including the commanding officer of Destroyer Division 16, Captain Furukawa.

The stern stayed afloat due to the rugged construction of the forward bulkhead of the after boiler room. Although the break was farther forward, at frame 84, this bulkhead was severely warped and leaking. Progressing flooding would eventually extend to No 3 boiler room, but it held for a sufficient time to allow the crew to shore it up and establish a flood boundary.

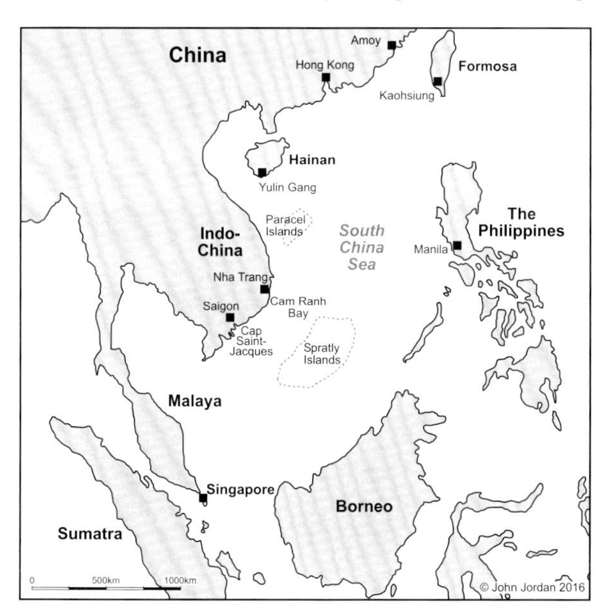

Map of the South China Sea, through which all Japanese mercantile shipping had to pass in order to re-supply Japan with oil and precious metals.

In the face of the submarine threat to this vital convoy all ships, including *Yukikaze* (the only vessel which could have offered any practical immediate assistance), rapidly left the scene. It was believed that *Amatsukaze* had been sunk with all hands in the massive explosion observed. Abandoned to their fate, the crew began their struggle to survive; the after section, still afloat and with up to 160 survivors, was crippled and cast adrift.

Because of the danger of further enemy attacks, the surviving after torpedo mount and the aft pair of 12.7cm gun mountings were manned. The night was long and tense, but with no follow-up attack and a steady improvement in the weather throughout the 17th, there was some restoration of power thanks to the considerable efforts of the damage control and engineering teams. This was not only essential for the remaining pumps, but it also enabled the reserve W/T installation aft to be used. A member of the communications team forward had survived and he had memorised the emergency cipher, but the full set of codes, along with all charts, had been lost with the bridge.

Survival and Rescue

The only map available was an overall general map of the Greater East Asia Co-Prosperity Sphere in the appendix of a book, from which the survivors had to estimate their position. This was duly passed on to the Kaohsiung shore station on Formosa (now Taiwan) they had raised; however, it proved to be very inaccurate, and no search plane launched that day from Saigon (Air Group 936) found them. They could not freely use the W/T again in case the transmission was also detected by prowling enemy submarines, which would be fatal in their crippled state. However, on the 18th a revised position estimate was transmitted despite the risk. The Takao and Saigon stations both acknowledged and assured them a new search would begin at dawn the 19th.

However, dawn of the 19th was cloudy and squally with poor visibility, and they were informed that the air search had been suspended. It had been intended to renew the search as soon as weather permitted, but at noon on 19 January an event occurred that would fully occupy the attention of the Japanese shore command. The escort carrier *Unyo* was badly damaged by submarine torpedoes (fired by USS *Haddock*, SS-231) some 140 miles east-southeast of Guam, and had to make her way slowly to Saipan. A safe path had to be cleared for the crippled carrier, and this incident generated considerable communications traffic and fully stretched Japanese resources.

On board the destroyer, the deterioration in the weather between the 18th and the 20th aggravated an already serious situation. All aerial searches were aborted or turned back, leaving *Amatsukaze*'s crew to struggle on alone in the face of severe conditions. (In a poignant reflection on the ship's abandonment, Japanese records state that on the 20th the otherwise intact Convoy HI-31 arrived safely at Singapore, and that this crucial operation had been concluded with the 'acceptable' loss of just one escort.)

On the 21st a mast was sighted on the distant horizon by *Amatsukaze*'s alert crew, but she was unable to communicate with the vessel, while her greatly reduced freeboard aft left her extremely low in the water, apparently hiding her from view in the poor conditions prevailing. What edible food had not been lost or contaminated with oil or water had been consumed by the 22nd, and the stamina of the crew began to suffer in consequence. Constant vigilance on the forward bulkhead had to be maintained; given its crucial importance to the ship's survival it had been shored up, but the constant pumping and other physical demands were taking their toll. The situation was alleviated only by harpooning some of the sharks that were now swarming around, smashing their heads with iron bars, then hauling them aboard to cook them in sea water.

To make matters worse, on the 23rd Tanaka was warned that enemy submarines were gathering in the area, presumably vectored in due to the projected path of *Unyo* and radio intercepts, and he was advised to suspend transmitter use. A desperate Tanaka finally decided to use the wireless transmitter again on the 24th, as there was no other option in their present dire situation. Medium-wave broadcasts were used as a directional beam. Their luck held, and at 1500 a patrolling aircraft was sighted approaching them directly; it circled over them again and again as the crew cheered wildly. As it flew away it dropped a communication tube and, despite the risks from sharks, a crewman swam to pick up the message. Once retrieved it was discovered that their estimated position had shifted 100 nautical miles compared to their own calculations on the 17th, but it was stated that relief ships would now be sent to this revised position.

Later that day the veteran (1923) relief destroyer *Asagao* and the new *Kusentei* (subchaser, 1941) *CH-19* finally arrived on the scene. Ready-prepared emergency food (rice balls) was immediately sent across by line, to the great joy of the crew, as towing preparations were made throughout the night. It was intended to head initially for the nearest base facility at Cap Saint-Jacques in the former French colony of Indo-China for emergency repairs.

The tow got underway, with *Asagao* towing *Amatsukaze* slowly stern first, so as to not strain the bulkheads forward. Tanaka stood atop No 2 gun mounting to navigate his ship, and messages were passed from there to *Asagao* as necessary.

Meanwhile, on 25 January, *Chitose* and the destroyer *Yukikaze* successfully escorted the returning Convoy HI-32, comprising the same ships as HI-31 above plus

On 6 September 1944, while screening Convoy TAMA-25, the destroyer *Hibiki* was torpedoed by USS *Hake* (SS-256), losing a section of her bow. After emergency repairs at Takao, she returned to Yokosuka where she received a new, modified bow. The photo shows her in dry dock. She was subsequently assigned to the 2nd Destroyer Squadron of the Second Fleet on 25 January 1945 and was later assigned to escort duties with the First Fleet. *Amatsukaze* would likewise receive a new bow, although in her case only the after part of the ship survived the torpedo. (IJN Department of Naval Aeronautics)

some additions, from Singapore back to Japan without loss, arriving at Moji on 4 February.

For *Amatsukaze* on the 25th the slow passage to Saint-Jacques continued, as *Asagao* towed and *CH-19* circled, offering protection to the vulnerable group. Fate now smiled on the ship, as signals intelligence advised the American wolf-packs on the 26th that *Unyo* was starting north from Saipan, and all submarine attention between Saipan and Formosa was focused on the wounded carrier until 2 February. *Unyo* started out north on the 27th, crawling at 12 knots.

Between the 26th and the 28th the tow line parted on several occasions and had to be reset. On the 29th the group finally reached Saint-Jacques, where *Asagao* passed the tow to a French tug. The devastated after part of *Amatsukaze* had to be carefully manoeuvred into harbour, where an emergency inspection was undertaken. It was clear that she needed to be docked for a full inspection, and the only suitable dry docks were at the former French naval dockyard at Saigon.

At 1500 on the 30th *Amatsukaze* was towed (stern-first) by tug the 80km up-river to Saigon, finally arriving at her berth at 2200. The prolonged period of repair and restoration which followed was to be undertaken by the ship's crew and local labour from the 11th Special Base Force. Over the subsequent months skilled personnel was siphoned off to serve on board active units, thereby slowing the progress of the work, but a cadre remained to service *Amatsukaze*.

The principal IJN close-range anti-aircraft gun was the Type 96 25mm (*Kyuroku-shiki nijyugo-miri Kokakukiju*) automatic cannon in single, twin, and triple mounts. The triple mounting seen here had a nine-man crew. Ammunition was fed using 15-round magazines and the gun had an effective rate of fire of 110rpm; normally one tracer round was added to every four or five rounds to aid aiming. The Type 96 has been described as a mediocre AA gun, with slow training and elevating speeds, excessive vibration and muzzle flash, and inferior fire control. Nevertheless, in interviews conducted by the US Naval Technical Mission to Japan after the end of the war, Japanese naval personnel cited it as the most reliable Japanese anti-aircraft weapon. Most effective at ranges of 1,000m or less, it required an average of 1,500 rounds to down an aircraft at a height of 1,000m and a range of 2,000m; fire beyond that range was deemed ineffectual. (IJN Department of Naval Aeronautics)

Repairs

Amatsukaze entered dry dock at Saigon on 2 February 1944. The first task was a major clean-up; every compartment was affected by seawater contaminated with oil from her breached tanks. Repairs to the ship's residual structure and renovation of her machinery subsequently met significant delays due to a lack of spare parts and material, as Japanese commitments elsewhere took priority. Throughout this period the crew was housed in a rented part of the Naval Hospital in Saigon, pleasant lodgings which fostered a good working relationship with the remaining Vichy French naval personnel.

On 15 March, Captain Sasaki Takanobu, who had initially been appointed to *Amatsukaze* on that fateful day of 16 January 1944 and was currently serving as the Executive Officer of the 11th Special Base Force, was assigned extra duty as her commanding officer, formally relieving Tanaka. Two weeks later, on 31 March, *Amatsukaze* was formally removed from the 16th Destroyer Division and reassigned to the 1st Southern Expeditionary Fleet, under Vice Admiral Fukudome Shigeru, for administrative purposes.

On 1 April *Amatsukaze* left the Saigon dry dock, but on 14 April was towed back in for more lasting work. While there, between 2237 and 2400 on 5 May, *Amatsukaze* had a close call when enemy B-24 Liberator bombers struck the Saigon port area, but she received no further damage.

While the ship was still at Saigon, on 2 September Captain Hasebe Ichizou of the old (1924) fleet tanker *Hayatomo*, based as a hulk at Singapore, was assigned extra administrative duty as *Amatsukaze*'s CO, formally relieving Captain Sasaki. Finally on 4 September 1944, *Amatsukaze* was able to leave dry dock.

Suitably patched up between 8 and 14 November, *Amatsukaze* left under tow of the 3,520-ton auxiliary gunboat *Eifuku Maru* for the 1,100km transit from Saigon to Singapore. There were no escorts to spare, due to the need to reinforce the Philippines – convoys left Manila for Ormoc Bay in the wake of Leyte Gulf. Even the valuable damaged heavy cruiser *Myoko* had to make do with an escort comprising only one destroyer (*Kishinami*) and a minesweeper on her dangerous voyage from Brunei to Singapore between 30 October and 3 November, while between the 8 and 12 November the damaged heavy cruiser *Takao* had to make a similar voyage, again with just one destroyer.

Upon arrival at Singapore on the 15th *Amatsukaze* was placed in the capable hands of Seletar's No 101 Repair Facility. Here what has been described as a substantial new bow (*Karikanjubi*, or 'wave-cutter') was fitted, to enable her to return to Japan for full repair and restoration as a fleet destroyer. The hull was cut at the former location of No 1 torpedo-tube mount, and a temporary open bridge and snub-nosed bow fitted. Her original waterline length of 116.2m was reduced to just 67.5m,

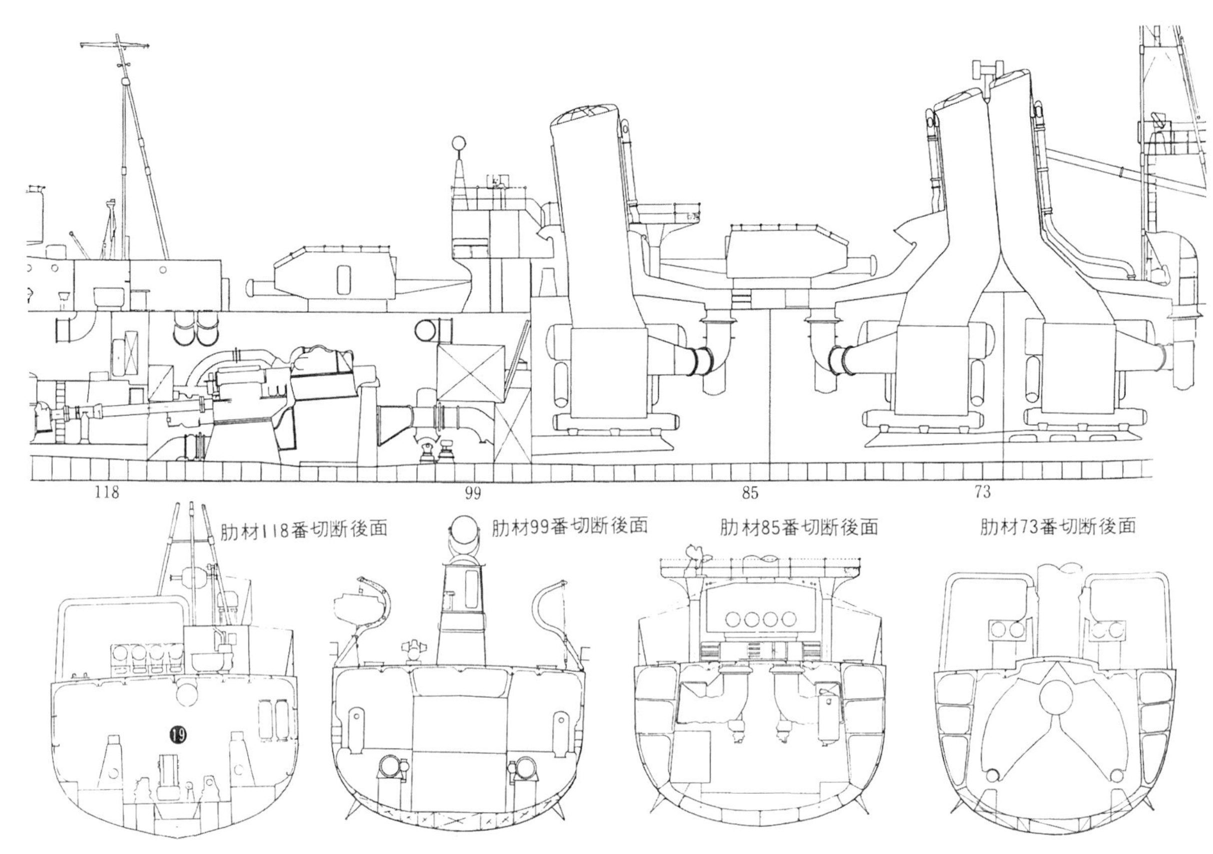

The midships section of a *Kagero* class destroyer. The bulkhead between the centre and after boiler rooms at Frame 89 was crucial to the survival of *Amatsukaze* following the torpedo hit of 16 January 1944. (*Mechanism of Japanese Warships No 4 - Destroyers*)

but the fitting of the new provisional bow **increased this to 72.4m (see contemporary sketch).** No 2 funnel, just abaft the temporary bridge, housed her only boiler uptake. She retained both twin 12.7cm mountings aft, along with an increased anti-aircraft outfit which initially comprised one triple 25mm mounting abreast the funnel to port, together with four single 25mm and five single 13mm MGs on deck, while an anti-submarine outfit of 36 depth charges was retained aft. There is no specific mention of No 2 torpedo mounting, which appears to

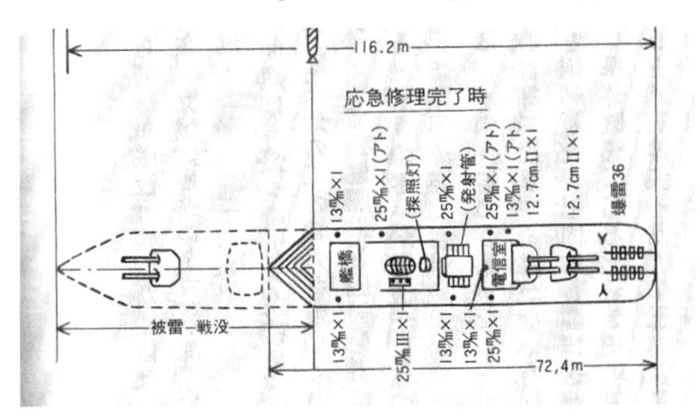

A crude sketch of *Amatsukaze* following her reconstruction at Singapore, showing the new *Karikanjubi* ('wave-cutter') bow, the new depth charge racks above the stern, and the 25mm and 13mm single MGs scavenged to enhance the ship's AA capabilities. Note the triple 25mm MG to port of the remaining funnel.

have remained in place; it is believed this carried four torpedoes but no re-loads.

One practical aspect of the reconstruction which has not been highlighted, but was obviously crucial, was crew accommodation; it seems likely that new quarters replaced the old torpedo reload lockers aft, while further accommodation may have been built around the funnel and beneath the new bridge forward.

The memoirs of Constructor Lt-Cdr Shizuo Fukui, the naval architect who supervised the temporary reconstruction of *Amatsukaze* at Singapore, contain some interesting information about the final refit plans envisaged once the ship was back in Japan. There was to be a new bow of mild steel and a greatly augmented depth charge outfit, requiring the stern to be significantly modified to a configuration similar to a Transport Ship No 1 (*Dai 1 Go-gata Yusokan*) type vessel. The ship was to be capable of a maximum 30 knots using two boilers instead of three, and *Amatsukaze* would become a unique type of anti-submarine destroyer which would serve as the prototype for a new design. Since her surviving after boiler room housed only a single boiler, a new boiler room would have had to be provided in the rebuilt forward section.

At Singapore in late 1944, *Amatsukaze* was a greatly reduced unit with limited capabilities, but in these desperate times the Japanese were prepared to employ her locally in an active role as a convoy escort, and her

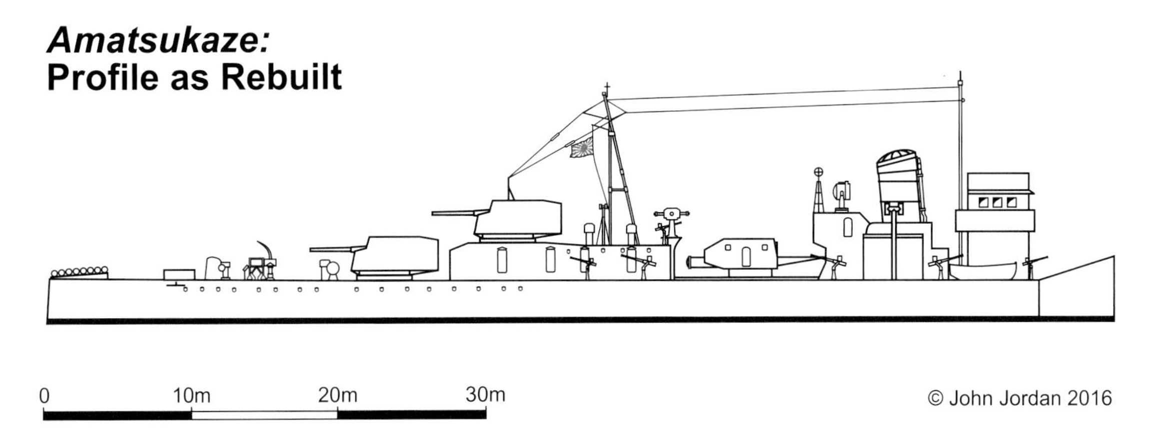

Amatsukaze as she would have appeared in late 1944, following her reconstruction at Singapore; the notable Japanese naval architect Fukui Shizuo was personally involved in her reconstruction. The drawing has to be regarded as provisional, being based on the crude sketch published opposite plus the aerial photos of *Amatsukaze*'s destruction; many details are unclear, and drawings published elsewhere have major errors which include the configuration of the new bow and the bridge, the omission of the simple pole foremast carrying the W/T aerials, and an unrealistic AA armament comprising several triple 25mm MG mountings.

remaining crew underwent an intensive work-up period to prepare them for active service again after their prolonged period ashore. It was considered that with the work done to date *Amatsukaze* could comfortably sustain 12 knots – adequate for convoy work – with 20 knots the absolute maximum available in an emergency. However, Singapore was no longer regarded as a safe location after recent bombing raids by B-29s flying from Kharagpur in India, commencing late October. The desire of the IJN High Command that *Amatsukaze* should return to Japan, allied to her ability even in her current state to serve as escort for a convoy, effectively decided her fate.

Convoy HI-88J

With Lieutenant Tomoyuki Morita's appointment to *Amatsukaze* on 7 January 1945, the final phase in her career began. On 10 February Captain Hasebe officially handed over command to Morita, and by the end of that month all repair work had been completed, with trial runs undertaken. Also appointed around this time was Lieutenant Haruo Ogawa (Academy Class 70), as executive cum communications and navigation officer.

On 1 March Lieutenant Nakayama Yamamoto (Academy Class 72), a survivor from the super-battleship *Musashi*, came on board. Yamamoto has left some interesting notes concerning his appointment to *Amatsukaze*. He describes how the veteran destroyers *Kamikaze* and *Nokaze*, stopping over at Mako with the south-bound Convoy HI-91, were used to transfer Yamamoto and fifty other men to Singapore. On passage *Nokaze* was torpedoed with the loss of 209 of her crew. It is not known how many of the 21 survivors picked-up by *Kamikaze* were from Yamamoto's party. Eventually the latter ship arrived at Singapore on 22 February, and the personnel transferred brought *Amatsukaze*'s final complement up to 200.

Morita thought he appreciated the situation regarding *Amatsukaze*. In his considered opinion, basing the cut-down *Amatsukaze* at Singapore for local defence would be a sound decision. He was not aware of the intentions of the authorities in Japan, who wanted her to be repatriated to a fully-equipped home base for a comprehensive reconstruction.

Vice Admiral Fukudome told Morita that in recent sailings most ships had failed to reach Japan, and that his fleet was very weak. Unusually, Fukudome suggested that if Morita wanted *Amatsukaze* to remain at Singapore he would gladly discus this matter further with his superiors on his behalf. It seems likely that Fukudome had recommended the appointment of Morita to command the destroyer, and given the extreme dangers of such a return voyage to Japan, he not only hesitated to despatch his *protégé* to Japan through such deadly waters, but was concerned about the effect of the removal of even such a crippled unit from his already depleted command.

Morita replied he would talk over his situation with his officers before responding – an unusual course of action in the tightly-disciplined IJN, where any command was obeyed without discussion. He had 200 men under him, including eight other officers. Two had been his juniors at the Navy War School, one was from the Merchant Ship School, one from the accelerated officer programme for college students, and four were *tokumu* officers, promoted from the lower deck. This was a very young leader and officer team, supported by a core of seasoned ratings, but all of them expressed the desire to return home despite the poor chances of success.

Intriguingly, only mentioned later in Morita's narrative is a brief note of a special cargo: ten tons of 'rare metals' to have been embarked at Singapore. The precise nature of this cargo has not been identified, but its embarkation aboard the convoy's sole destroyer escort, albeit one with much-reduced capabilities, suggests its importance. If High Command entrusted this command and responsibility to Morita, it could be seen as confirming Fukudome's high opinion of him; in recognition of his

The *Kaibokan* Type C escort was designed in the Spring of 1943 as a simplified version of the Type B (*Mikura* and *Ukuru* classes) to facilitate mass production. Altogether 132 were ordered (all odd-numbered), and their construction continued until the end of the war. Hull design was simplified, and electric welding was used extensively in their construction. Their diesel engines were capable of 16.5 knots, and they were armed with two older-model 12cm HA and between six and twelve 25mm guns, an 80mm trench mortar and 120 depth charges. They proved to be effective antisubmarine escorts. The photo shows *CD-17*. (IJN Department of Naval Aeronautics)

The *Kaibokan* Type D was similar to the Type C but had single-shaft turbine propulsion; the turbines gave these ships an extra knot of speed at the expense of higher fuel consumption and decreased range. The exhausts were trunked into a tall slim funnel placed farther forward than that of the Type C. Altogether 143 units (all even-numbered) were ordered. The photo shows *CD-8*. (IJN Department of Naval Aeronautics)

new elevated status and responsibilities, Morita was promoted to Lieutenant-Commander at some undisclosed date during this period at Singapore.

Morita received a telegram from First Fleet on 7 March ordering him to join Convoy HI-88J, comprising three tankers and one merchantman loaded with strategic materials from the southern resources area. The convoy would have to break through a strong Allied submarine and air blockade in what all appreciated would be a very difficult transit to Japan. Three vessels bound for Saigon would also accompany the convoy on its first leg.

Despite his ship's own precious cargo Morita expressed a desire to be an active member of the convoy escort, *Amatsukaze*'s remaining ordnance being fully operational. The convoy's senior officer, Commander Yasuhiro Hirano, in the *Kaibokan* (coast defence ship) *CD-134* agreed, and *Amatsukaze* was assigned a position covering the rear of the convoy. The seven merchantmen in HI-88J were *Kaiko Maru*, *Asokawa Maru*, *Honan Maru*, *Sarawak Maru*, *Araosan Maru*, *Tencho Maru*, and *Saigon Maru*, an assortment of elderly and modern, small and large general cargo ships and tankers. The protective screen would comprise *CD-18*, *CD-26*, *CD-84*, *CD-130* and *CD-134* together with *Amatsukaze*.

The convoy would have been disposed in two columns 1,000 metres apart, with the ships at intervals of 600 metres. Two of the *Kaibokan* would be on either beam 1,500 metres out, with the flagship *CD-134* leading the convoy, leaving *Amatsukaze* to bring up the rear. The escorts would have zig-zagged in what could be loosely described as a defensive ring formation.

Convoy HI-88J, the last oil convoy in Operation *Minami-Go* (South), departed Seletar Naval Harbour, Singapore, on 19 March bound for Moji, sailing via Saint-Jacques. However, *Sarawak Maru* soon struck an air-dropped mine and lost way. *Araosan Maru* was ordered to assist her, disrupting the formation for about one hour. The convoy then resumed its formation and proceeded at 7 knots. *Sarawak Maru* was beached in shallow water off Bintan Island. Salvage efforts failed and she was to eventually sink on 21 March; the loss to the Empire included 4,400 tons of oil, 690 tons of rubber and 116 tons of tin. On the evening of 22 March, Convoy HI-88J made an overnight stay at Kamao, southern Indo-China, leaving at 0800 the following morning.

The experiences of Convoy HI-88I, which departed Saint Jacques on the same day, illustrate the hazards Convoy HI-88J was about to face. This convoy comprised the merchantmen *Hosen Maru*, *Fushimi Maru No 2*, *Motoyama Maru No 1*, *Nanshin Maru No 21*, *Nanshin Maru No 30*, and *Takasago Maru No 6*; it was escorted by *CH-9*, *CH-20*, *CH-33*, the auxiliary subchaser *Kainan Maru*, and the cable-layer *Tateishi*.

On 20 March, some 50 miles south of Cam Ranh Bay (Indo-China) at about 0100, convoy HI-88I was attacked by USS *Baya* (SS-318), which torpedoed and sank *Kainan Maru*. *CH-9* counter-attacked *Baya* and dropped 21 depth charges. Despite sustaining damage *Baya* remained on patrol. That same afternoon at about 1720, the convoy was attacked by USS *Blenny* (SS-324), which torpedoed and sank *Nanshin Maru No 21* and *Hosen Maru*.

Off Nha Trang (Indo-China), about 500km north-east of Saigon, at about 1120 on the 21st the convoy was attacked by B-25s belonging to the 345rd Bomb Group of the US 5th Air Force, which bombed, strafed, and sank *CH-33*, *Tateishi*, *Motoyama Maru*, and *Fushimi Maru*, while also damaging *CH-9* and *Takasago Maru*. Finally *CH-9*, accompanied by *CH-20*, took refuge at Nha Trang. Convoy HI-88I had effectively been halted, while Convoy 88J still proceeded undetected and unmolested towards these heavily contested waters between 22 and 25 March.

Convoy HI-88J arrived off Cap Saint-Jacques on the 26th. *Araosan Maru*, *Tencho Maru* and *Saigon Maru* then left the convoy, destined for Saigon; however, there was a welcome addition in the form of *CH-20* which now joined the escort. The following day, at about 2000 off Nha Trang Bay, the remnants of Convoy HI-88I, *Nanshin Maru No 30* and the escorts *CH-9*, *Manju*, and *CD-1* all joined the main convoy. Shortly after this junction the convoy was finally discovered by a US reconnaissance aircraft, and all units immediately prepared for an enemy attack as they continued to push north.

Aerial and Submarine Attacks

The anticipated air attacks began at 1040 on 28 March. *Asokawa Maru* was hit in the engine room by a B-24 Liberator and sank; 92 passengers and 42 crewmen were killed, survivors being rescued by *CD-84* and *Manju*. At 1220 USS *Bluegill* (SS-242) torpedoed *Honan Maru*; her captain finally ran her aground later that day off Cap Varella but she was lost, along with five gunners and 44 crewmen. *Amatsukaze* dropped ten depth charges and found some oil, but *Bluegill* continued her patrol. A party from *Bluegill* subsequently boarded the hulk of *Honan Maru* on 5 April, and completed her destruction manually with demolition charges and incendiaries. *Nanshin Maru No 30* also became detached from the convoy at some point, finally being sunk by air attack on 3 April.

At 1124 the submarine USS *Blackfin* (SS-322) was about to attack when she was sighted, forced down to the shallow bottom and nearly destroyed by a 79-depth-charge attack. The convoy was then subjected to aerial attacks. Although the target of a number of aircraft, it is possible that the great bow wave produced by the blunt provisional bow of *Amatsukaze* deceived the attackers as to her actual speed, throwing them off their aim; eyewitnesses claimed that the skip-bombs released all fell forward, where her original bow would have been.

At 0200 on the 29th the depleted convoy headed for Hainan Island. At 0710 USS *Hammerhead* (SS-364) torpedoed and sank *CD-84*, killing 191 crewmen together with six survivors from *Honan Maru*, leaving *Manju* to rescue the few survivors. At 1130 another submarine attack coincided with an air attack in which *Kaiko Maru* was bombed and sunk; losses were 22 passengers, 4 gunners, and 19 crewmen killed. Twelve B-25s then arrived. Luckily for *Amatsukaze* a bank of squally weather was nearby, and she hid under the low-lying clouds. That afternoon at 1300 a further twenty B-25s arrived and sank *CD-18* at the rear of the convoy; the escort was seen to rear up and plunge bow first beneath the waves. No fewer than fifteen B-25s singled out *CH-9*, which was fortunate to survive with moderate damage.

That evening a Martin PBM-5 Mariner flying boat made a night radar bombing attack and sank *CD-130*. At 2230 a second aircraft, probably another flying-boat, located the convoy and damaged *CD-134*, the flagship of Commander Hirano, during its two attacking runs. The *Kaibokan* claimed to have emerged victorious after shooting it down; her after magazine was completely flooded and she was left with a list to port and 30 men wounded, but she could still make 10 knots. The crew of *Amatsukaze* also claimed to have succeeded in shooting down a shadowing flying boat as it attempted a bombing run that night, but it is possible that this was the same aircraft.

By now Convoy HI-88J comprised only escorts. On the 30th, at 1000 off Yulin Gang harbour on the south coast of Hainan Island, B-25s sank the auxiliary subchaser *Shinan Maru*, which was coming to their aid. The convoy had barely arrived in Yulin Gang when twelve more B-25s ambushed them at 1045 by approaching from the landward side of the anchorage over the hills of Hainan. Morita had anticipated this possibility and had moored very close to the shore; this meant that the B-25s had insufficient space for effective skip-bombing. The first bomb hit a stay while others fell harmlessly into the sea; only *CD-26* was hit, but she was disabled with the loss of eight men.

The destruction of Convoys HI-88J and HI-88I marked the end of the Singapore–Empire convoys. Even in Yulin Gang there was a threat of relentless enemy air attacks, and Commander Hirano was compelled to order the survivors to flee to Hong Kong; they departed Yulin Gang harbour on 31 March, leaving the damaged *CD-26* behind (she survived the war).

On 2 April *Amatsukaze* and her five battered consorts finally dropped anchor in Hong Kong, which was far from a safe haven, having been subjected to recent enemy air attacks. While laying over in Hong Kong her crew fitted three additional single 25mm anti-aircraft guns,

The B-25H variant of the Mitchell bomber was specially fitted for attacks on maritime targets. Armed with eight forward-firing Browning AN/M2 0.5in (12.7mm) heavy machine guns – some were also fitted with heavy, slow-firing 75mm cannon – it had a 1,350-mile (2,174km) operational range. There were under-wing hard-point mountings for eight 5in (130mm) High Velocity Aircraft Rockets (HVAR), plus an internal bomb load of 3,000lb (1,360kg). The 345th Bomb Group was the first full Air Force combat group sent to the Pacific, where it perfected the art of anti-shipping strikes and skip-bombing. A stick of two to four bombs, usually 500lb (230kg) or 1,000lb (450kg) devices, with four/five-second time delay fuses, were released at a distance of 20-90m from the side of the target ship; the bombs would either 'skip' over the surface of the water directly into the side of the ship and detonate, submerge and explode under the ship, or bounce over the target and explode as an air burst. (USAAF)

salvaged from a vessel which had run aground. Following their recent trials the crew naturally desired to fit as much anti-aircraft ordnance to their destroyer as possible.

Convoy HOMO-03

At Hong Kong some merchant ships were lacking escorts, and with the serendipitous arrival of *Amatsukaze* and her worn-out consorts a last-ditch convoy, HOMO-03, was hastily formed by the command at Hong Kong, comprising the merchantmen *Kine Maru* and *Dai 2 Tokai Maru* escorted by *Amatsukaze*, together with the *Kaibokan CD-1*, *CD-134* and *Manju*, and the subchasers *CH-9* and *CH-20*.

While Convoy HOMO-03 was making final preparations at Hong Kong, fifty USAAF Far East Air Force B-24s arrived over Hong Kong on 3 April. In this air raid *Manju* was severely hit by two bombs forward, leaving her grounded with 52 fatalities (she remained there until the end of war). Also hit and sunk in this raid were the cargo ships *Heikai Maru* and *Shozan Maru*, possibly slated for Convoy HOMO-03. Another forty B-24s came in on the 4th, but their target was mainly the port facilities.

Amatsukaze departed Hong Kong with convoy HOMO-03 at 1730 on 4 April; the escort, minus *Manju*, was again under the command of Commander Hirano in *CD-134*. The two merchantmen were carrying many civilians who were desperate to return to Japan. On board *Amatsukaze* were 201 officers and men, including one unidentified passenger.

CD-130 under attack by US bombers on 29 March 1945 and in a sinking condition; all 178 on board were lost. Note the depth charges strewn around the quarterdeck. (USAAF)

The operation got off to the worst possible start when *Dai 2 Tokai Maru* was bombed and crippled that same night, presumably by B-25s belonging to the 345th Bomb Group, which was to devastate this beleaguered group of ships. The 5 April saw *Dai 2 Tokai Maru* sink before dawn; *CH-9* had also been damaged in the night bombing and was forced to return to Hong Kong with a few survivors from the lost merchantman. That afternoon at 1500 one B-24 Liberator and three B-25 Mitchells, accompanied by five P-38 Lightnings, arrived and *Kine Maru* was sunk, with many cast into the hostile sea. *Amatsukaze* and *CH-20* attempted to rescue as many as they could, and the subchaser then returned to Hong Kong with 400 survivors from the two merchantmen.

Amatsukaze, however, would not return to Hong Kong for two reasons: Lt-Cdr Morita was determined to deliver the cargo of precious metals embarked at Singapore to Japan; and all on board wished to return home for the full restoration of their ship. *Amatsukaze* continued northwards alone, trailing her two *Kaibokan* consorts because she had been engaged in helping survivors of *Kine Maru*.

It is unclear why *CD-1* and *CD-134*, still the convoy flagship, continued on a northerly course while leaving the rest of Convoy HOMO-03 behind. One possible reason for this is the damage inflicted upon *CD-134* on 29 March, when her after magazine was flooded and her speed cut to 10 knots, greatly reducing her offensive and manoeuvring capabilities. It may also be that Hirano felt the survivors of *Kine Maru* were being well served by *Amatsukaze* and *CH-20*, and that his other two ships could best serve the overall objective of a return to Japan by proceeding separately.

Morita recalled that the waves were high and the wind strong that night, slowing the passage north. That passage was at least uncontested, as the USAAF seems to have temporarily lost contact in the poor weather. Morita contacted Commander Hirano, warning him that the three ships were so dispersed that they might be attacked and overwhelmed in detail, but Hirano did not slow his speed (estimated at 10 knots) to enable *Amatsukaze*, which was making 12 knots, to catch up. Morita speculates that Hirano may have feared that if *Amatsukaze*, with her distinctive and highly-visible wake, were in company with his own two ships she would attract an aerial attack onto them.

The Last Battle

As the day dawned on 6 April the two *Kaibokan* were well ahead, fatally exposed. Eventually they were located by the enemy while some 30km distant from the trailing *Amatsukaze*; Hirano's decision saved *Amatsukaze* but resulted in the destruction of his own command. At 1150 a single B-25 was sighted from *Amatsukaze*, quickly followed by four others, but they held a course passing to the rear of the destroyer out of range. Her crew were greatly relieved, as in her significantly reduced and isolated condition *Amatsukaze* was in no position to fight off such an enemy formation. From the direction ahead from which the planes had come, columns of smoke were sighted.

The crews of twenty-four B-25s of the 345th had volunteered for the long over-water flight from their base at San Marcelino, Luzon, to the Japanese ships off the Chinese coast. The B-25s were in four six-plane squadrons each comprising two flights of three aircraft. The 501st and 499th squadrons attacked *CD-134* and *CD-1*, while the 498th made a pass at the sinking *CD-1*. The 501st Bombing Squadron went on south to hit *Amatsukaze*, later joined by most of the 498th after *CD-1* went down.

While the precise fate of Commander Hirano and the crew of *CD-134* has not been uncovered, the fate of *CD-1* under Commander Arima Kunio has been recorded, and can be regarded as matching that of her consort. All told there were 176 on board *CD-1* that day, which was quickly overwhelmed in the B-25 onslaught; survivors struggling in the water were then strafed, with total losses of 155 dead. It is unclear which vessel picked up the 21 survivors; the only other vessel nearby was *Amatsukaze*, which was herself fighting for her life and was soon to limp away from the area. A Japanese report states that four wounded crewmen subsequently died in Hong Kong.

The formation of B-25s which passed over *Amatsukaze* earlier had apparently expended all of its ammunition on the two hapless *Kaibokan*; it was therefore unable to attack, but almost certainly broadcast the location of this valuable isolated target. Soon, around 1230, a formation totalling some eighteen B-25s gathered around the beleaguered destroyer. For *Amatsukaze* a desperate battle for survival began.

The B-25s divided into six distinct groups, each of three bombers, seemingly intent upon executing a concerted series of close-in skip-bombing attacks to swamp and overwhelm the crippled destroyer's defences from all directions. However, the anti-aircraft fire of the manoeuvring *Amatsukaze* broke up the first wave before it was in a position to release its bombs. The bombs of the second wave were also avoided, but not the accompanying strafing.

The third wave, comprising six B-25s of the 500th Bomb Squadron, approached from astern, In the face of a fierce AA barrage put up by the Japanese destroyer, the leading B-25 closed to land a bomb between the main 12.7cm mountings aft, putting both out of action; the second aircraft achieved only near-misses, and the third was downed by *Amatsukaze*. The second flight of three then move in for the kill.

Amatsukaze suffered severe damage in these attacks. The rear bridge collapsed, topside damage was extensive, all power was lost, and the rudder failed, leaving *Amatsukaze* adrift with serious fires raging aft. Morita ordered a fire of oily rags to be lit on deck to persuade the circling fliers that the damage was terminal, and heavy smoke was emitted from her single funnel. However, these ploys failed to deceive his opponents for long; the combustible material was soon exhausted and the aerial attacks were resumed after a brief lull of around four minutes, during which time engineering personnel were called up to replace fallen gunners.

The fourth wave to attack was the 498th Bomb Squadron. Two more bombs hit home, destroying the radio room and the after auxiliary machinery room, and setting off uncontrolled fires. The fifth and sixth waves attacked in unison; they dropped no bombs but instead unleashed an eviscerating strafing attack, apparently concentrating upon the exposed 25mm positions and devastating the light superstructures. *Amatsukaze*'s end seemed inevitable.

Suddenly two 'friendly' fighters appeared, and at this point the B-25s, which were low on ordnance and presumably fuel, broke off their attack. The identity of the Japanese pilots and their unit, and whether they were responsible for any of the B-25s lost or damaged, has not been recorded, but their appearance undoubtedly saved

The D-Type *Kaibokan CD-134*, flagship of Commander Hirano, under aerial attack from the B-25s belonging to the 345th Bomb Group off Amoy on 6 April 1945. She sank the same day. (USAAF)

Two photos of the C-Type *Kaibokan CD-1* taken during her sinking by B-25s. (USAAF)

Amatsukaze. They may have been from the IJN's 1st Escort Squadron (901st Air Group based on Formosa), which included shore-based fighters within its 170-aircraft strength, and which was tasked with carrying out offensive anti-submarine sweeps of the seas between Shanghai and Formosa.

Photographs of *Amatsukaze* under attack had been taken by a crewman on board one of the B-25s. Often used as a symbol of a Japanese convoy's annihilation, many viewing it assume it caught a devastated vessel being sunk. However, the battle for survival had not yet ended. *Amatsukaze*'s situation was certainly critical: she had taken three direct hits, along with intense strafing, and was seriously damaged, with flooding and fire resulting. The conflagration aft at the base of the two crippled 12.7cm mountings was critical, as it threatened the magazines. The valves had been wrecked so no controlled flooding could be undertaken, but as the ship settled flooding caused by her extensive hull damage had the effect of quenching this blaze before it overheated the magazines.

The flooding had to be checked, and all efforts were now fully committed towards this task. Another problem arose: sea-water contamination of the lubricating oil system for the shaft bearings; if they were to seize the ship's ability to manoeuvre would cease, leaving *Amatsukaze* at the mercy of further enemy aerial attack. The starboard shaft was soon restored to full operation while work continued on the port shaft. The damaged rudder was under manual operation, and *Amatsukaze* slowly headed at 6 knots towards the Japanese-occupied enclave of Amoy (now Xiamen) on the coast of China, some 30nm distant across the Formosa Strait (now Taiwan Strait). The despair on board gradually dissipated with the prospect of reaching a safe harbour.

At 1930, as *Amatsukaze* arrived off Amoy, the shaft bearings finally gave out and she came to a halt. The crippled destroyer had to wait 20 minutes adrift outside the local defensive minefield before being laboriously towed by the small local patrol boat through the protected channel between Amoy Island and the mainland. With no motive power *Amatsukaze* was at the mercy of the prevailing wind and tidal currents; she finally came to rest gently against a reef on the southern shore (mainland) at around 2020, opposite the Japanese naval base. Despite the best efforts of her crew in trying to secure their ship with her light reserve anchor and the endeavours of the small patrol boat, she was hard aground. The only option was to await high tide to move the ship to a safer mooring the following day.

Amoy

Additional guard boats from the base arrived shortly after the grounding to stand by. High water on the 7th did not alter the situation; pushed by the wind and tide, *Amatsukaze* was by then firmly aground on the mainland, and there did not appear to be the facilities on Amoy Island to salvage her. Her fires had been mastered by mid-

Amatsukaze under attack on 6 April 1945, some 30 nautical miles off Amoy on the coast of China. Note the foreshortening of the new bow fitted at Singapore and the distinctive bow wave in the prevailing heavy seas. At this point the ship is relatively undamaged. (USAAF, 345th Bomb Group)

morning, but she was still unable to use her engines.

Ominously, local residents began looking curiously at the new arrival from the shore. Even though the ship was grounded close to an important Japanese base, local Chinese communist guerrillas were active on the mainland opposite, and the ship was in a precariously exposed position, as was soon confirmed by a burst of light machine gun fire from the shore. Her remaining heavy 25mm cannon quickly responded and silenced the concealed foe – although one member of her crew was killed in this surprise attack. It was clear that the mainland territory was in enemy control, and the captain had to consider scuttling the ship to prevent her capture. There was also the real threat of another air attack if they were found by a reconnaissance sweep off the coast in this exposed position. Every effort was then committed towards refloating the ship, but without success, and Morita was compelled to accept that the end had finally come. With all options now exhausted, he would have to abandon ship; he formally communicated this to the local HQ, which concurred with his decision.

Amatsukaze was abandoned by most of her complement on the 8th, leaving a skeleton crew to ward off any attack. It appears that 44 of *Amatsukaze*'s crew had been killed in the action on the 6th and the days which followed, together with one unnamed passenger; 156 of those on board had miraculously survived, although many were wounded.

The evacuation of the ship decreed by Morita was now methodically undertaken. All consumables and materiel useful to Japanese forces were removed by her remaining crew, loaded aboard the local patrol boat and lighters, and transported back to the Japanese base at Amoy. Items salvaged included food, fuel, ammunition, four single 25mm and five single 13mm mounts, and even the triple 25mm mount weighing 1,800kg, while seven large bags of sugar from Singapore were especially welcomed by Army personnel. Not specifically noted, but certainly

Three views of *Amatsukaze* under heavy air attack. The damage to the after 12.7cm gun mountings is clearly visible and the lattice mainmast has collapsed. In the circumstances, the survival of the ship and most of her crew is remarkable. (USAAF, 345th Bomb Group)

undertaken given its importance, would have been the transfer of the 10 tons of precious metals loaded at Singapore.

Salvage work continued throughout the 9th alongside preparations to scuttle *Amatsukaze* with the assistance of a special base party from Amoy. Depth charges were to be used as explosives; these were presumably late-war versions of the ubiquitous Type 2 device, each with an explosive charge weighing 162kg, with an estimated 26 remaining on board. While these onerous duties were being undertaken, Morita performed a funeral service to honour those lost.

Some sources state that the hulk slid back into the sea and sank on the 8th, but the author can confirm that *Amatsukaze* was finally scuttled by demolition charges at sunset on the 10th, resulting in massive internal explosions. On 10 May the hulk was officially relegated to Reserve Ship 4th class status, and is said to have been later expended as a target for Japanese aircraft, although what remained of the hull after her scuttling is difficult to determine. Finally, on 10 August 1945, just prior to the Japanese unconditional surrender on the 15th, *Amatsukaze* was formally removed from the Navy List. The ship had served for a total of 58 months.

Although *Amatsukaze* had gone, her efforts to reach Japan had achieved at least one objective: the majority of her crew who boarded in Singapore managed to survive the war, eventually to be repatriated. Morita was appointed by Japanese HQ in Shanghai to command of the Sijiao Island (Shengsi group) garrison, China Theatre Fleet, and remained there until after the war, finally returning to Japan in 1946; he subsequently rose to the rank of Vice Admiral in the postwar Japanese Maritime Self Defence Force.

Sources:

Books

Goralski, Waldemar, *The Japanese Destroyer Kagero* (*Kagero Oficyna-Wydawnicza*), Kagero Publishing (Lublin 2013).

Tameichi, Hara, with Fred Saito and Roger Pineau, *Japanese Destroyer Captain*, US Naval Institute Press (Annapolis 1961).

Fukui, Shizuo, *Japanese Naval Vessels at the End of WWII*, Greenhill Books (London 1991).

Jentschura, Jung & Mickel, *Warships of the Imperial Japanese Navy, 1869-45*, Weidenfeld Military (London 1977).

Watts, Anthony, *Japanese Warships of World War Two*, Ian Allan (London 1966).

Mechanism of Japanese Warships No 4 – Destroyers, Maru Mechanic/Kojinsha (Japan 2000).

Internet

Tully, Anthony: Overall History, Imperial Japanese Navy Page - http://www.combinedfleet.com/amatsu_t.htm.

Destroyer Wind Tianjin: Amatsukaze - www.jam.bx.sakura.ne.jp/dd/dd_career_amatsukaze.html.

Tomoyuki, Morita's war diary: http://forum.axishistory.com/viewtopic.php?f=65&t=184197.

Nakayama, Yamamoto's account: http://www5f.biglobe.ne.jp/~ma480/senki-1-amatukaze-yamamoto1.html.

Kaibokan CD-1 sinking: http://blog.goo.ne.jp/toranokobunko/e/731442a1b0c4b0a89b2cc50f4519cac0.

Yoshiaki, Harada, *Naniwa-kai*: http://www.naniwa-navy.com/index.htm.

USS *HUNTINGTON* (EX-*WEST VIRGINIA*)

Huntington was one of ten large armoured cruisers authorised for the US Navy in the wake of the Spanish–American War of 1898. **A D Baker III** tells her story and provides detailed plans of the ship as she appeared in 1920.

The United States Congress, riding the wave of nationalistic expansion begun in the late 1890s, authorised two classes of large armoured cruisers between 1899 and 1904: six ships of the *Pennsylvania* class, and four larger and more heavily-armed units of the *Tennessee* class. All ten of the armoured cruisers were initially named for US states, but had their names changed to those of cities and towns in those states between 1912 and 1920 to allow their state names to be used on new-construction battleships. With the exception of *Maryland* (renamed *Frederick* in November of 1916) the ships had their military foremast replaced by a cage mast after 1910.

Ordered on 21 June 1901 from the Newport News Shipbuilding Co in Virginia, *West Virginia* (Armoured Cruiser No 5) was the second unit of the *Pennsylvania* class. Her keel was laid on 16 September of that year, and

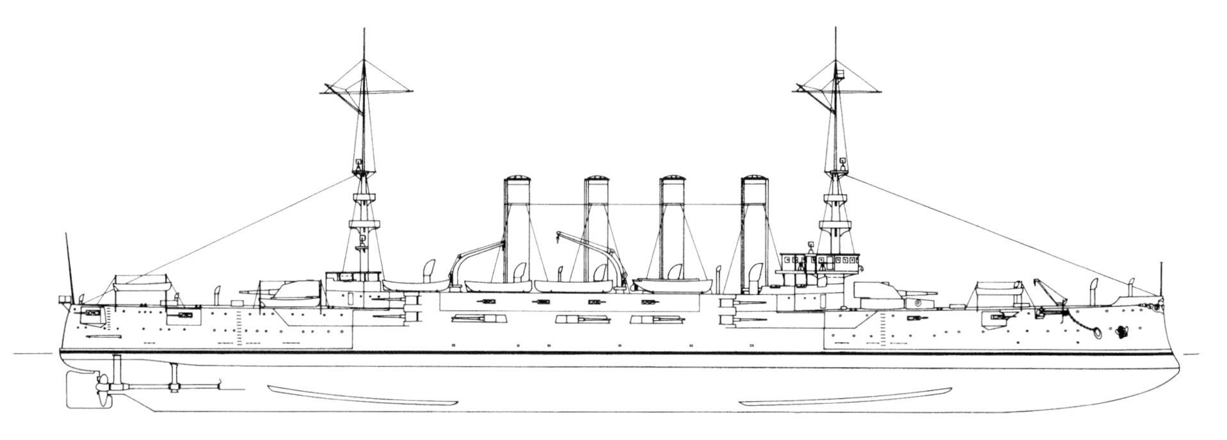

Profile of the name-ship *Pennsylvania* (renamed *Pittsburgh* in 1912) as completed. This ship remained in active service until 10 July 1931, her forward funnel and boilers having been removed in 1922. (Drawing by A D Baker III, from John Alden's *The American Steel Navy*)

USS *Huntington*: Characteristics Table

Displacement (1922):	13,720 tons normal; 15,138 tons full load
Dimensions:	503ft 11in length oa x 69ft 7in beam x 24ft 1in mean draught
Armament (1920):	2 twin 8in 45-cal; 14 single 6in 50-cal; 18 single 3in 50-cal; 2 single 3in 50-cal AA; 1 3-in field gun; 4 single 3pdr saluting; 2 submerged 18in broadside TT
Protection:	
Belt	5in belt amidships, 3.5in ends Frames 1–35 and Frame 95 to stern; 2in sides to 3in gun batteries
Decks	4in main deck (1.5in amidships)
Turrets	6.5in face and sides, 6in backs; 6in barbette at main deck, 4in at gun deck, 3in at berth deck; 2in roof
Conning Tower	9in sides; 2in roof; 5in tube
Engines:	Two 4-cylinder triple-expansion, inboard-turning engines; 26,477ihp maximum
Boilers:	Sixteen Babcock & Wilcox (265psi)
Speed:	22.44 kts at 13,810 tons
Coal:	2,005 tons maximum
Electrical:	Three 100kW and four 50kW dynamos (all 125 volt)
Crew (1922):	41 officers, 815 enlisted, 72 Marines

West Virginia as completed in 'white and buff' paint scheme. Her name was changed to *Huntington* in 1916. Note the military masts fore and aft, the elaborate scrollwork at the bow, and the inclusion of four gooseneck cranes for her then-large complement of boats. (US Navy)

the ship was launched on18 April 1903. The US Navy commissioned her on 23 February 1905, and the ship was initially assigned to operational duties in the Pacific.

As a unit of the Reserve Force she participated in patrol duties off the Pacific coast of strife-torn Mexico during September 1916. The ship was then overhauled at the Mare Island Navy Yard and equipped with an aircraft take-off rail system aft that was theoretically capable of accommodating up to four small floatplanes or flying boats. On 11 November of that year, her name was changed to *Huntington*, commemorating the name of an important railroad junction in West Virginia and, incidently, the name of Collis P Huntington, the railroad magnate who had founded her building yard in the 1890s.

At the completion of her yard period and conversion – one that was also made to two half-sisters of the *Tennessee* class, *Seattle* (ex-*Washington*) and *North Carolina*, whose name was not changed until 1920, to *Charlotte* – the renamed *Huntington* was sent to the Navy's flight training and experimental centre at Pensacola, Florida, detaching from the Pacific Fleet in May of 2017. Intensive trials were made off Pensacola over that summer with the aircraft launching system, which not only proved unwieldy and suitable for launching only very light aircraft but also prevented the after 8in gun turret from being trained. That August, *Huntington* departed for New York City, from where the ship escorted a six-ship convoy carrying troops to

Taken on 26 June 1917 off Pensacola, Florida, this overhead view of *Huntington* shows her with a single-float Curtiss JN-4 mounted on the take-off rails aft. Note the aircraft stowage rails running forward on either side of the ship nearly as far forward as the fourth funnel; the ship's starboard boat crane was equipped with a lengthened boom to recover aircraft after landing and had a capacity of 15 tons, while the port crane could lift only 10 tons. (US Navy)

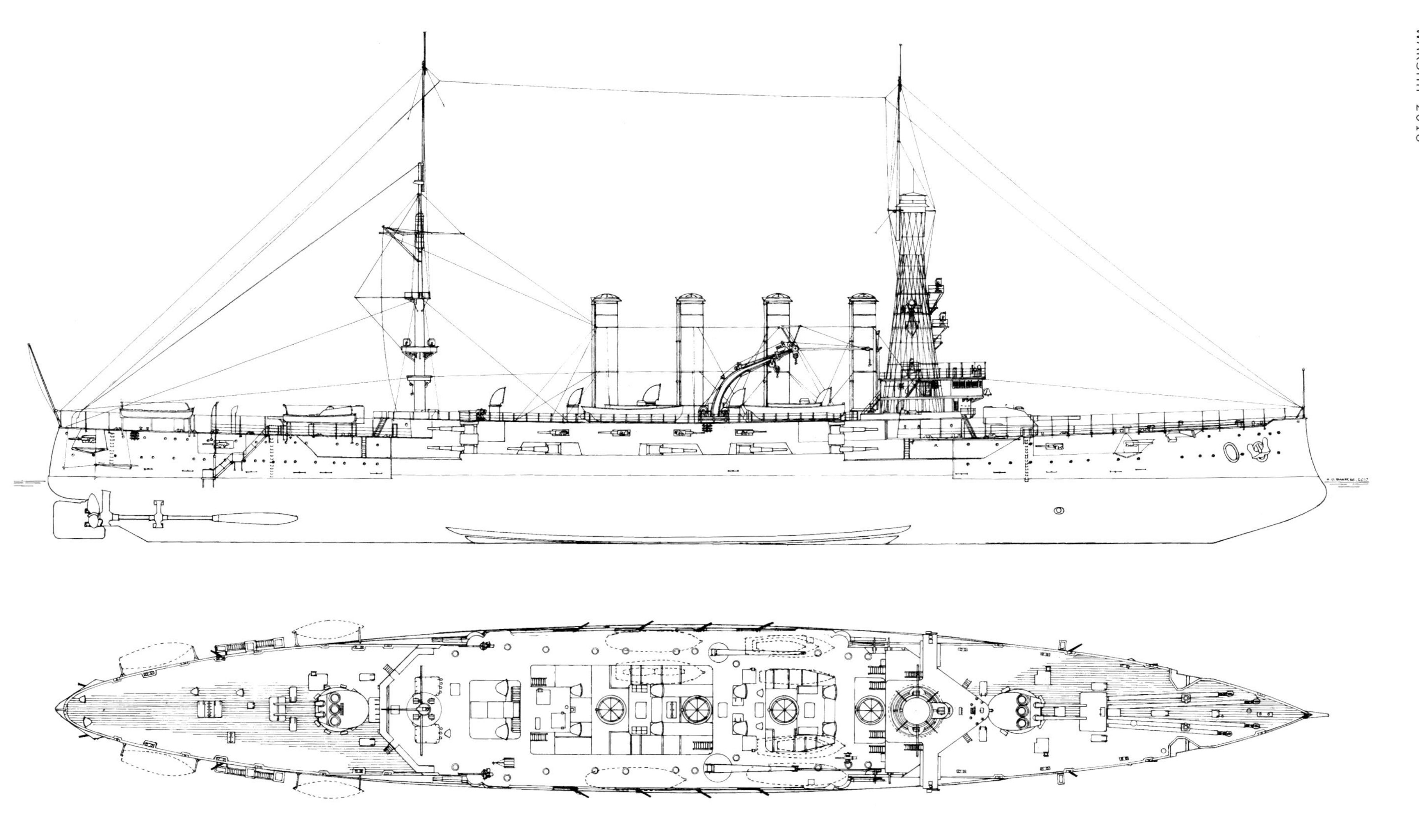

Huntington in 1920 with her 6in and 3in guns restored and all evidence of her aircraft-launching career removed except for the extended boom to the starboard boat crane. The stocked anchor pockets port and starboard and the starboard hawsepipe seem never to have been used. The boats shown on davits aft would have been stowed on skids amidships when the ship was underway. (Drawing by A D Baker III)

Building Data and Disposals

Pennsylvania Class

Number[1]	Name as built	Builder[2]	Laid down	Launched	Commissioned
ACR 4	*Pennsylvania*	A	7 Aug 1901	22 Aug 1903	9 Mar1905
ACR 5	*West Virginia*	B	16 Sep 1901	18 Apr 1903	23 Feb 1905
ACR 6	*California*	C	7 May 1902	28 Apr 1904	1 Aug 1907
ACR 7	*Colorado*	A	25 Apr 1901	25 Apr 1903	19 Jan 1905
ACR 8	*Maryland*	B	29 Oct 1901	12 Sep 1903	18 Apr 1905
ACR 9	*South Dakota*	C	30 Sep 1902	21 Jul 1904	27 Jan 1908

Tennessee Class

Number[1]	Name as built	Builder[2]	Laid down	Launched	Commissioned
ACR 10	*Tennessee*	A	20 Jun 1903	3 Dec 1904	17 Jul 1906
ACR 11	*Washington*	D	23 Sep 1903	18 Mar 1905	7 Aug 1906
ACR 12	*North Carolina*	B	21 Mar 1905	6 Oct 1906	7 May 1908
ACR 13	*Montana*	B	29 Apr 1905	15 Dec 1906	21 Jul 1908

Notes:

1 Prefixed ACR (for Armored Cruiser) until 1920, when the prefix changed to CA, which, however, continued to be described as 'Armored Cruiser' until well into the 1920s, when it came to stand for Heavy Cruiser. *Maine* (briefly) was ACR 1, *New York* was ACR 2, and *Brooklyn* was ACR 3.

2 A = William Cramp & Sons, Philadelphia, Pa.
B = Newport News Shipbuilding & Dry Dock, Newport News, Va.
C = Union Ironworks, San Francisco, Ca.
D = New York Shipbuilding Co, Camden, NJ.

Name changes	Disposals
Pennsylvania became *Pittsburgh* 27 Aug 1912	Sold scrap 21 Dec 1931
West Virginia became *Huntington* 11 Nov 1916	Sold scrap 30 Aug 1930
California became *San Diego* 1 Sep 1914	Mined (?) 19 Jul 1918
Colorado became *Pueblo* 9 Nov 1916	Sold scrap 2 Oct 1930
Maryland became *Frederick* 9 Nov 1916	Sold scrap 11 Feb 1930
South Dakota became *Huron* 7 Jun 1920	Sold scrap 11 Feb 1930
Tennessee became *Memphis* 25 May 1917	Aground 17 Dec 1917, scrap 17 Jan 1922
Washington became *Seattle* 9 Nov 1916	Sold scrap 3 Dec 1946
North Carolina became *Charlotte* 7 Jun 1920	Sold scrap 29 Sep 1930
Montana became *Missoula* 7 Jun 1920	Sold scrap 29 Sep 1930

Europe. The aircraft facilities were removed at New York that October, and the ship resumed her convoy escort duties until November of 1918.

During their service in the Great War, the large armoured cruisers proved unable to employ the 6in and 3in guns mounted on the gun deck in any sort of seaway, and the weapons were removed. When *Huntington* was refitted to act as a troop ship herself at the end of the war, the remaining 6in and most of the 3in guns were also removed, and the space they had occupied was employed for troop berthing, with the ship bringing home some 1,700 troops on her first transport voyage. The *Huntington* repatriated 12,000 more soldiers on five additional trooping trips lasting into July of 1919. Reassigned as flagship of the cruiser Flying Squadron in July 1919, the ship had her full gun armament restored prior to her decomissioning on 1 September 1920. Despite several attempts to draw up a modernisation programme for the surviving large armoured cruisers – *Memphis* (ex-*Tennessee*) was grounded by a tidal wave at San Domingo on 28 August 1916 and could not be

The camouflaged *Huntington* in European waters during 1918, when she was serving with the Cruiser & Transport Force of the US Atlantic Fleet. Note the cage mast replacing the military foremast, the reduction from four to two boat cranes, and the suppression of the 6in guns on the gun deck, leaving only four on board; a rangefinder has been added atop the after twin 8in turret and (not visible here) two 3in HA guns installed. (US Navy)

The larger *North Carolina* (renamed *Charlotte* in 1920) in a close-up showing the take-off ramp with a take-off trolley on the starboard-side stowage rail, which was considerably shorter than that on the *Huntington*. At the after end of the take-off rail, a plunger can be seen between the rails to stop the trolley as the aircraft left the rails. Note that the after turret could not be trained with the rail system installed. (US Navy)

saved, while *San Diego*, ex-*California*, was lost to a suspected German mine or torpedo off Long Island on 29 July 1918 – *Huntington* never operated again. Only *Pittsburgh* (ex-*Pennsylvania*) remained in service through the 1920s, as flag and station ship on the China Station, shorn of her forward funnel and four of her 16 boilers. *Huntington* remained in reserve until stricken from the Navy on 12 March 1930 and was sold for scrapping that August, a fate that was shared with seven of her sisters and half-sisters. *Seattle* (ex-*Washington*) lingered on as a stationary Receiving Ship at New York, serving in that capacity until stricken on 28 June 1946, having been redesignated unclassified auxiliary IX 39 on 28 June 1942.

Conclusion

The US Navy's big armoured cruisers had proved to be of little use as combatants in wartime despite their formidable gun batteries and handsome profiles. Their principal contribution to warship development had been in testing shipboard aviation techniques. In addition to the 1917 experiments, Pennsylvania had briefly had a 130-ft long platform added above her fantail on which civilian pilot Eugene Ely on 19 January 1911 successfully landed a Curtiss biplane equipped with three improvised hooks that caught ropes stretched across the deck. All subsequent US Navy cruisers were designed at the outset to operate aircraft, either fixed-wing or helicopters.

WARSHIP NOTES

This section comprises a number of short articles and notes, generally highlighting little known aspects of warship history.

THE IJN'S 15.5cm GUN & TRIPLE TURRET

To complement his article on the cruiser *Oyodo* published in this year's annual, **Hans Lengerer** has provided this Note on her main guns and turrets, which were removed from the light cruisers of the *Mogami* class when they were converted to heavy cruisers in the late 1930s.

The 15.5cm gun was designed by naval engineer Hata Chiyokichi, who was responsible for most of the IJN's guns and turrets from the 1920s, and is particularly associated with the design of the 46cm main guns of the 'super-battleships' of the *Yamato* class. Adopted on 7 May 1934 as the 15.5cm 60-cal 3rd Year type after a development period of around four years, the gun was judged by the US Naval Technical Mission to Japan (USNTMtJ) in its report O-54 (N) *Japanese Naval Guns* (page 37) as one of three guns[1] with outstanding ballistic features but a relatively short lifespan. In designing this gun, the largest cruiser gun calibre permitted under the London Treaty of 1930, Hata aimed to maximise performance by adopting the longest possible barrel, which would have the maximum chamber pressure and the highest muzzle velocity.

When the IJN officially moved from the German Krupp-type gun to the British Armstrong type at the end of the 1880s, British wire-wound construction methods were also adopted for domestically-produced guns. However, the employment of wire-wound construction in a very long barrel leads to muzzle droop during firing, an increase in dispersion and, in consequence, a lower hit probability.

In 1927 the Japanese Army (IJA) had acquired a licence from the French Schneider Company to manufacture gun barrels using 'autofrettage', and had made the Navy aware of the process.[2] Following experimentation with materials and the production process, a 20cm gun was trialled in 1930 and demonstrated the superiority of the autofretted barrel: there was no droop or muzzle whip, which suggested that a barrel as long as 60 calibres could now be produced. In addition, the gun would have a smaller number of components. The new barrel was built up as follows:

Breech: two tubes and a jacket with a liner
Muzzle: single (monobloc) tube with a liner

The maximum working pressure was 3,400kg/cm^2, which was exceptional for the period, and muzzle velocity was 920m/sec – the highest mv of any low-angle (LA) gun. This value was surpassed only by the later 10cm 65-cal Type 98 HA gun, which was designed for

15.5cm/60 3rd Year Gun: Characteristics

	15.5cm/60 3rd Year Gun	20cm/50 3rd Year Gun
Weight (incl breech)	12,700kg	17,800kg
Length		
breech face to muzzle	9,300mm	10,000mm
overall	9,615mm	10,310mm
Grooves: no, depth x width	40: 1.80mm x 7.51mm	48: 2.28mm x 8.30mm
Length of rifling	8,025mm	8,481mm
Twist	uniform, 1 turn in 28 cal	uniform, 1 turn in 27.56 cal
Chamber		
length	1,128mm	1,348mm
volume	38 litres	68 litres
Muzzle velocity	920m/sec	840m/sec
Maximum bore pressure	3,400kg/cm^2	3,130kg/cm^2
Weight of projectile	55.87kg	125.85kg
Weight of propellant	19.5 kg (one bag)	33.8kg (two bags)
Weight of igniter	0.075kg	0.170kg
Projectile travel	8,172mm	8,652mm
Point of complete combustion	29 cal from muzzle	17 cal from muzzle
Maximum range	27,400m (AP shell) 26,500m (common)	28,895m (AP shell)
Maximum altitude	12,600m	10,058m
Barrel life	250 service rounds	320 to 400 service rounds

Notes:

Data for Type I gun barrel. Other than this there were:

- an experimental Type I2 with: reduced thickness of the layer at the muzzle; a different shape to the jacket; a wire-wound powder chamber and breech block
- an experimental Type I3 with a further decrease in the thickness of the layer at the muzzle compared to the Type I2

The results of the tests, and the reasons for the decision in favour of Type I and the rejection of Types 12 & 13 are not known. In the Annual Report of the Navy Ministry for FY 1930 (*Kaigunshô nenryô shôwa go nendo*) *Ko* (A) and *Otsu* (B) 15.5cm gun barrels are mentioned, but the difference between A and B is not stated. The A gun was produced in series.

15.5cm Triple Turret: Sectional Profile View
(Trial production model, designed for 75° elevation – in practice reduced to 55°, based on classified plans)

3 8-metre duplex rangefinder
4 Gunhouse
5 Rammer
6 Powder hoist container
7 Shell hoist tube for LA fire
8 Shell hoist tube for HA fire
9 Motor room
10 100hp motor
11 Pump
12 Oil pressure cylinder
13 Main distribution pipe
14 Shell room
15 Powder hoist tube
16 Magazine
17 Powder handing room
18 Double blast-tight door
19 Powder hoist gear
20 Powder hoist container

Source: Gakken No 38 *CA Mogami Class* (*Mogami gata Ittô Junyôkan*), 106.

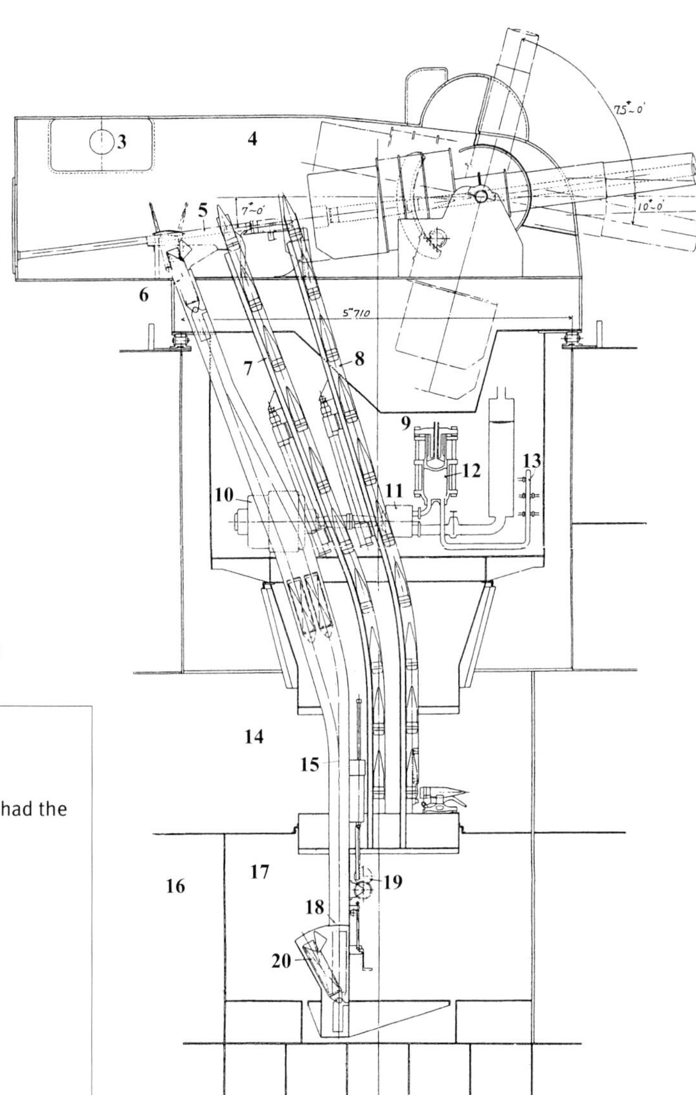

15.5cm Triple Turret: Plan View

1 Turret officer
2 Gunhouse supervisor
3 Elevation angle receiver for LH gun (other guns had the same receiver)
4–6 Gunlayers for LH, centre & RH guns
7 Training angle receiver
8 Trainer
9 Sight setter
10 Target designator
11–13 Breech workers for LH, centre & RH guns
14 Lever for opening/closing breech
15 Powder hoist
16–18 Powder handlers for LH, centre & RH guns
19–21 Shell handlers for LH, centre & RH guns
22 Rammer LH gun (other guns had the same rammer)
23 8-metre duplex rangefinder
24–26 Loading, breech inspection and rammer operators
27–29 Shell & powder hoist operators
30 Shell loading tray for LA fire
31 Shell loading tray for HA fire
32 Messenger
33 Ready-use shell stowage

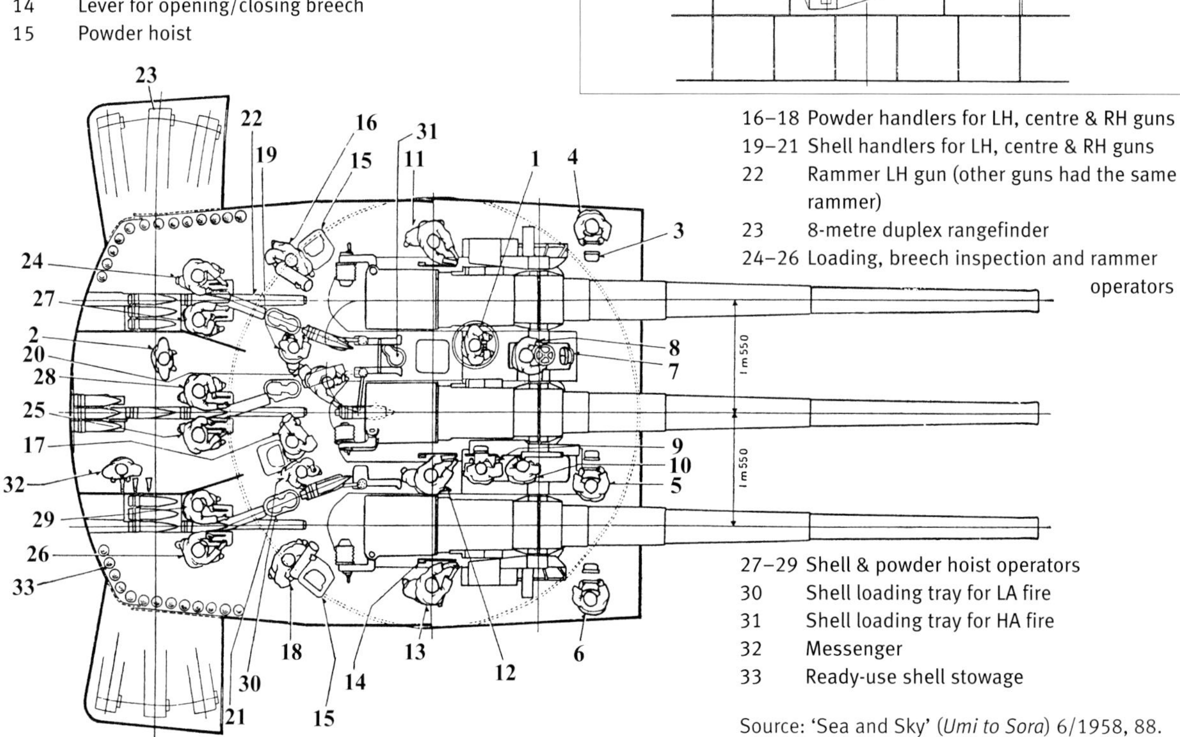

Source: 'Sea and Sky' (*Umi to Sora*) 6/1958, 88.

the anti-aircraft destroyers of the *Akizuki* class but also fitted as *Oyodo's* main HA gun.[3] The maximum range of 27,400m surpassed that of the 6in (152mm) guns fitted in the light cruisers of the US *Brooklyn* class and was comparable to the 8in (203mm) gun mounted in the American heavy cruisers which were seen as *Mogami*'s likely opponents. Naturally the penetration power of the projectile was reduced compared to the 20cm gun, particularly at distances greater than 20,000m, and the barrel had a comparatively short life – a result of the high chamber pressure and the high temperatures generated, which in turn caused the rapid erosion of the liner.

The principal dimensions and ballistic particulars of the 15.5cm 60-cal 3rd Year gun are listed in the accompanying table. Data for the 20cm 50-cal 3rd Year No 2 gun, derived from the No 1 gun (nominal and actual calibre: 200mm) also designed by Hata, are included for comparison.

Hata was also responsible for the design of the turrets, which were the first triple mountings fitted in a Japanese warship. Despite this, the general arrangement of the 20cm twin and 15.5cm triple turrets was similar, with a high degree of commonality. The IJN planned to replace the latter with the former as soon as the arms limitation treaties expired, so the principal dimensions of the 20cm turret, particularly the diameter of the roller path but also the gun well and rotating central trunk, were primary considerations when designing the triple mount.

The gun was mounted in the usual manner in a cradle that slid backwards on a slide[4] when the gun recoiled, until both were brought to halt by hydraulic recoil cylinders mounted below. During this process high pressure (HP) air (100kg/cm^2) in the pneumatic run-out cylinder was compressed using a ram to 125.3kg/cm^2 at full recoil. This increased pressure was then used for controlled run-out at all angles of elevation.

The distance between the gun barrels (measured from gun axis to gun axis) was 1.55m. This was too narrow to allow side hinging of the screw breech mechanism of the centre gun, which opened upwards at a 45° angle, the weight being balanced by springs. The screw breech mechanisms of the right-hand (RH) and left-hand (LH) guns were hinged to the outer side and were operated by power or manually.

The maximum elevation of the gun was 55°, maximum depression -10°.[5] The reason for the latter figure is not evident. The elevation angle corresponded with that used for the guns of the heavy cruiser *Maya*, the last ship of *Takao* class. Other ships of this class had 70° elevation with a view to their use as dual-purpose (DP) guns; however, the 20cm gun with 70° elevation proved ineffectual for air defence, and the angle of elevation was reduced to 55° for *Maya* and the *Mogami* class.[6] At 45° elevation maximum range was 27,400m, and the highest point of the trajectory 9,500m.

The outer diameter of the roller path was 5.71m. The gun mountings were operated by electro-hydraulic pressure using mineral oil. Two 100hp electric motors (B-ends) were installed in the training engine room below

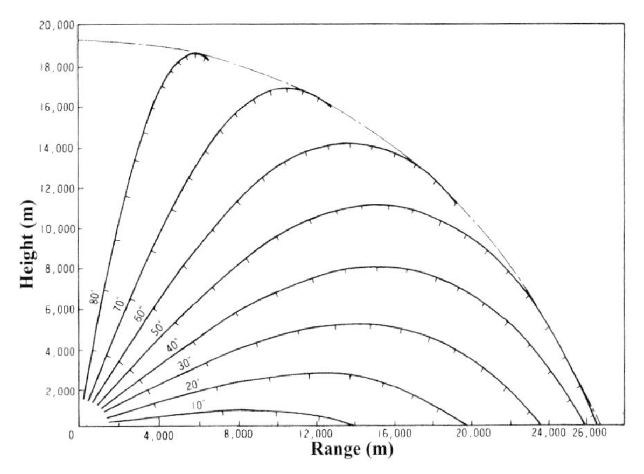

15.5cm 60-cal Gun: Trajectory Graph

Source: *Ships of the World* (*Sekai no Kansen*) 12/1996, 109.

the gun well and each motor drove a hydraulic pump (this was the A-end of the standard transmission gear – or swashplate – that ran at 500–650rpm) which supplied hydraulic pressure to a common ring main via hydraulic accumulators. One motor (B-end) and one pump (A-end) were normally in operation, and a second set was on standby to take over the operation immediately in the event of breakdown. Training was via worm gear, and maximum training speed was 6°/sec, the power being supplied by an hydraulic motor. Elevation and depression of the gun used the same power source, which actuated piston-type gears with a maximum speed of 10°/sec.

The projectiles were stored in bins in the shell magazine and transported to the shell handing room via a roller track which circled the base of the revolving trunk. The shells were then manhandled into pusher-type hoists which raised them to the gunhouse at a maximum rate of six projectiles per minute. The guns had individual shell hoists, so there were three per turret. There were also individual hoists for the bagged propellant charges. These were stowed in a magazine below the shell room and moved to the powder handing room via flashtight doors. The bagged charges were then transferred via double flashtight doors to the bucket-type hoists that brought them up to the gun house. Two powder cars, attached to an endless wire cable, were fitted in each hoist trunk and operated by a complex system using hydraulic pistons, racks and winches. The supply rate was five powder bags per minute for each of the three hoists. The six projectile and powder hoists were hydraulically powered from the common ring main.

Fixed loading was adopted; however, the angle was increased from 5° in the *Takao* class to 7°. The projectile and the powder bag were loaded separately. The tilting bucket for the shells was worked by a simple handle and lever, and the fuse was set manually when the projectile was on the loading tray. The shell was rammed in a single stroke; the powder bag was then placed manually on the tray and rammed. The rammer, which was of the conventional piston and rack type, was hydraulically powered.

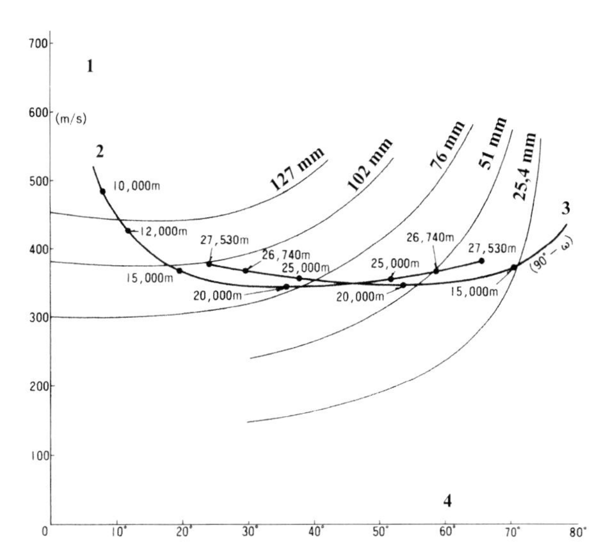

15.5cm 60-cal Gun: Penetration

Using Type 91 AP Shell against New Vickers Non-Cemented (NVNC) Armour: figures for plates of 127mm (5in), 102mm (4in), 76mm (3in), 51mm (2in) and 25.4mm (1in) thickness.

1 Remaining speed (m/s) = impact speed
2 Side (vertical) armour
3 Deck (horizontal) armour
4 Angle of impact (°)

Source: *Ships of the World* (*Sekai no Kansen*) 12/1996, 111.

The theoretical firing cycle is reported to have been seven rounds per minute; in practice it was a maximum 5rpm.

The gunhouse had all-round splinter protection comprising 25.4mm Ducol steel plates – the same thickness as on the 20cm turret. There was internal lagging inside the plates with an air space of 100mm to absorb most of the heat when the ship operated in tropical waters. In order to protect the gun crew against blast, canvas bags were fitted to the guns over the protective plating, and pivoting rods arranged to keep the canvas clear of the gun ports at high angles of elevation.

Evaluation

The performance of the 15.5cm gun was very good, and there were no stoppages or handling problems. Dispersion was much smaller compared to the 20cm gun: the salvo pattern of the latter at 20,000m was 400m, that of the 15.5cm gun around 250m. Some artillery experts questioned the decision to up-gun the *Mogami* class.

However, the noise generated by the hydraulic pump (B-end) in the training engine chamber was deafening, and the modified turrets mounted in *Yamato* and *Musashi* were fitted outside the chamber in consequence. The second problem, identified in report O-47(N)-1, was the damage to the rubber-insulated electric cabling caused by oil, indicating leakage from hydraulic pipes.[7]

Another issue was the weight of the gun. The 6in 47-cal Mk 16 gun mounted in the cruisers of the *Brooklyn* class, which was the US Navy's response to the *Mogamis*, weighed 4,363–6,587kg; its Japanese 15.5cm counterpart weighed 12,700kg. In part this reflected the shorter barrel of the American gun, which had a corresponding 3,000m disadvantage in range but had a life (according to Friedman[8]) of 750–1,050 rounds, three/four times greater than the Japanese 15.5cm gun. The technical data for the American gun indicate more advanced design and production methods, and superior metallurgy.

Arguably the principal defect of the turret was its light splinter protection. US Navy cruisers from the *New Orleans* class onwards had thick armour plating on their main turrets. Yasufumi[9] states that a cruiser of the *Mogami* class in a fight against a US cruiser of the *Brooklyn* class would have been forced to reduce the range to less than 10,000m in order to pierce the front shield of the main guns. He suggests that the decision to up-gun the *Mogamis* was very much influenced by the increased penetration of the 20cm gun, which could pierce the front shield of the US turrets at 15,000m.

A further disadvantage compared to the US gun mount was the lower firing cycle: five rounds per minute for the Japanese gun and 8–10rpm for the gun mounted in the *Brooklyn* class. If all 15 guns were laid correctly and fired uninterrupted salvos for one minute the weight of shell amounted to 4,190kg (55.87kg x 15 guns x 5 salvos) for a ship of the *Mogami* class, and 5,715kg (47.63kg x 15guns x 8 salvos) for a cruiser of the *Brooklyn* class.

Endnotes:

1. The others were the 10cm 65-cal HA gun and the 8cm 60-cal HA gun.
2. The IJN subsequently despatched Vice-Admiral Hishikawa Manzaburô to Schneider. He was allowed to observe the manufacturing and testing processes, and his reports strongly influenced the later decision to adopt this construction method. After acquiring the licence from Schneider the IJN applied autofrettage to all barrels below 20cm in calibre.
3. Data for this gun can be found in *Anti-Aircraft Gunnery in the Imperial Japanese Navy* in *Warship 1991*, 81–101.
4. The trunnions were fixed to the slide and rested on plain bearings which took the shock of firing in the usual manner. However, the members of the USNTMtJ who investigated the Japanese naval guns and mounts found 'the method of reducing the friction for ease of elevating' so unusual that the drawing of a typical trunnion design was added to report 0-47(N)-1 as enclosure K.
5. It is reported that the guns in some turrets could be depressed only to -7°.
6. Maximum range in the LA mode was attained with an angle of elevation around 45° (48° max), so range cannot have been the determining factor for choosing 55°.
7. It is reported that the crew in the hydraulic pump room of the 36cm twin turrets of the battleships of the *Fusô* class had to wear oilskins due to leakage of the pipes, and that ' large quantities of hydraulic water were consumed when firing the large calibre guns in twin turrets'.
8. *US Naval Weapons*, Conway Maritime Press (London 1983), 48.
9. Yasufumi Kunimoto, 'Main Batteries of *Mogami* Class Light Cruisers', *Ships of the World* 12/1996, 108–11.

THE SINKING OF *U-56* IN 1916: AN ENDURING MYSTERY

The U-boat *U-56* disappeared in the Barents Sea in November 1916. In this note **Stephen McLaughlin** uses Russian, British and Norwegian sources to trace her last days.

At 1145 on 2 November 1916 the Russian destroyer *Grozovoi* (Lieutenant MM Korenov) was standing 3nm ESE of Khorne Island, off the Norwegian port of Vardø.[1] She was an elderly vessel, a veteran of the Russo-Japanese War, and had only recently joined the Arctic Flotilla after a long voyage from Vladivostok to Arkhangelsk. Her mission on this occasion was to escort a small convoy from Vardø to Aleksandrovsk (now Poliarnyi); accompanying her were the dispatch vessel *Kupava* (Lieutenant GA Bleze), and the minesweeper *T-13* (*Michman* N Iakovitskii). The British minesweeper *A-516* was in the vicinity as well. The weather was clear and calm, wind force 2 from the WNW, with 'small' waves. The first of the merchant ships to arrive was the steamer *Lomonosov*, coming out of Vardø under the protection of the Norwegian patrol craft *Heimdal* and *Wiking*. The ships all hove to in order to exchange messages by megaphone, and at 1210 *Heimdal's* captain came aboard *Kupava*. But at almost the same moment *Kupava* sighted a submarine running on the surface about 20 cables (4,000 yards) distant to the WSW, and immediately alerted *Grozovoi*. Korenov ordered *Kupava* and *T-13* to protect the vessels coming from Vardø, while he went after the submarine. Bleze in turn ordered *Lomonosov* back to Vardø and went off to assist *Grozovoi*.

The submarine was *U-56* (*Kapitänleutnant* Hermann Lorenz); she was on her first – and, as it turned out, last – war patrol. However, the outcome of the action was by no means a foregone conclusion; the submarine displaced twice as much as *Grozovoi*, her two 88mm guns outclassed the destroyer's two 75mm guns, and she was a smaller and more difficult target (see Table 1).[2] The disparity in gun-power initially led Korenov to opt

Table 1: **The Opponents**

	Grozovoi	*U-56*
Builder:	F C Med, Le Havre	Germaniawerft, Kiel
Laid down:	1900	28 December 1914
Launched:	11 March 1902	18 April 1916
Entered service:	27 June 1902	23 June 1916
Displacement:	346.5 tons full load	720/902 tons
Dimensions:	56.5m oa x 6.3m x 3.4m	65.2m oa x 6.44m x 3.64m
Armament:	2 x 75mm 6 x 7.63mm 2 x 381mm TT	2 x 88mm 4 x 50cm TT
Machinery:	2 x VTE = 5,800ihp	2 x MAN diesels = 2,400bhp
Speed:	26 knots (max)	17 knots
Complement:	57	35

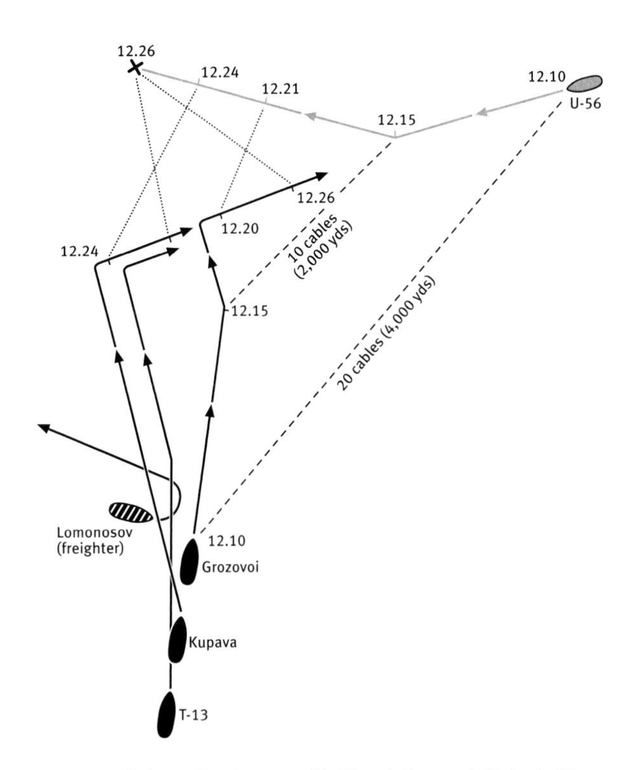

Diagram of the action between *U-56* and *Grozovoi*. (Adapted by Stephen Dent from a diagram in *Gangut* no. 25, page 20)

for a ramming attack; he increased speed and headed straight for the submarine. But *U-56* apparently either did not see *Grozovoi* or incorrectly identified her, and was instead concentrating her attention on *Kupava*, perhaps in the belief that she was a merchant ship – understandable, given that she was the former Norwegian commercial vessel *Sürendalsfjord* (805 tons, speed 10.5kts, 76mm guns). In view of the submarine's apparent distraction, Korenov changed his mind and decided on a gunnery attack, in the hope that he could get to close range before the submarine realised her mistake. His tactics proved successful: keeping the submarine on a bearing of 45° on the port bow, by 1220 *Grozovoi* had closed to 5 cables (1,000 yards), then opened fire with her forward 75mm gun. The second shot appeared to hit the conning tower. As the action progressed, the two vessels passed one another on opposite courses, and *Grozovoi* believed she scored three more hits: two aft below the waterline, one of which seemed to cause an internal explosion, and one on the deck between the conning tower and the after gun. After six minutes the submarine began to sink by the stern, and at 1226 *Grozovoi* ceased fire, having expended twelve rounds.

Kupava had also joined in the action, forming up behind *Grozovoi* and firing on the submarine. *T-13* stayed on *Kupava's* starboard side and did not engage. *A-516* was too far away to join in the action, while the neutral Norwegian vessels observed from a distance.

An hour later, while the convoy was in the process of forming up, *Grozovoi* passed close to the site of the action and observed an oil slick and bubbles rising to the surface. Despite this evidence, however, *U-56* had not sunk. And this is where things get a bit murky.

We can pass quickly over the account in the German official history of U-boat operations, which in describing *U-56*'s fate relied on 'a communication from the British Admiralty'.[3] So what did the British know? In 1919–1920 members of Room 40, the Admiralty's code-breaking unit, compiled a secret history of German naval operations based on intelligence reports. They wrote that the submarine was hit three times by *Grozovoi*, but survived the encounter and proceeded to Tana Fjord; however, she sank later, 'presumably in consequence of damage sustained in action with *Grozovoi*'.[4]

The British obtained this information from the crew of the Norwegian steamer *Ivanhoe*, which *U-56* had sunk the day before her action with *Grozovoi*. The U-boat took the steamer's small crew aboard – operating under prize regulations, *U-56* was obligated to ensure the safety of the crews of the ships she sank, and not a single life had been lost in any of her five attacks (see Table 2).[5] The Norwegian crew was on board the submarine during her battle with the Russians, and after being put ashore at Tana Fjord, they described their experience of the action:

> Suddenly, [the U-boat] was shot at by enemy warships; ten rounds were fired, three of which hit, according to statements by the [German] officers. The U-boat immediately dived below 50 metres ... It remained underwater for approximately two and a half hours.[6]

By the time she surfaced, the Russians had moved on and she set course for Tana Fjord. Apparently *Ivanhoe's* crew subsequently made a more complete statement, for British sources report that the U-boat sustained the following damage: 'An after tank was hit, the radio gear was shot away, and three men were killed'.[7]

Combining the information from Russian, Norwegian and British sources, we may cautiously speculate on the causes of *U-56's* disappearance. When *Grozovoi* observed an oil slick and bubbles an hour after the action, *U-56* was still submerged near the site, and it seems likely that the 'after tank' hit was a fuel tank leaking oil. The fact that the Russians saw her sink by the stern suggests that she had not submerged in the normal fashion, but nevertheless she must have managed to correct the problem – whatever it was – after diving. She was able to proceed to Tana Fjord on the surface and deposit *Ivanhoe's* crew, but after she put to sea something went wrong, probably as a delayed result of damage inflicted by one of *Grozovoi's* shells. Perhaps she dived and was unable to surface again. But that's as far as reasonable speculation takes us, and the exact cause of *U-56's* loss is likely to remain a mystery.

Endnotes:

1. Details of the action from the Russian point of view come from DIu Kozlov and MM Kolyshkin, '*Potoplenie germanskoi podvodnoi lodki U56 minonostsem "Grozovoi"*' [The Sinking of the German Submarine *U-56* by the Torpedo Boat *Grozovoi*] (*Gangut*, No 25, 2000, 16–22).
2. Some sources state that *U-56* had 1 x 105mm and 1 x 88mm.
3. Arno Spindler (editor), *Der Handelskrieg mit U-Booten*, Vol 3: *Oktober 1915 bis Januar 1917* (Mittler, Berlin 1934), 251. The captured German naval records held by the US National Archives indicate that the Germans were still looking in foreign publications for information about *U-56*'s loss as late as October 1940 (T-1022, microfilm roll 19, PG61618). My thanks to Sven Brummack and Tim Lazendörfer who helped me puzzle out a barely legible handwritten note in this file.
4. Frank Birch and William F Clarke, *Room 40: German Naval Warfare 1914-1918*, Vol 2: *The Fleet in Being*, edited by Hans Joachim Koerver (LIS Reinisch, Steinbach 2009), 139.
5. See also Paul Kemp, *U-Boats Destroyed: German Submarine Losses in the World Wars* (Naval Institute Press, Annapolis 1997), 20–21; Dwight R Messimer, *Verschollen: World War I U-Boat Losses* (Naval Institute Press, Annapolis 2002), 77; and the U-Boat Net website: http://uboat.net/wwi/boats/ index.html?boat=56 (accessed 26 July 2017).
6. *Sjøforklaringer over Norske Skibes Krigsforlis*, Vol 1: 1914, 1915, 1916 (Sjøfartskontoret, Kristiania 1917–1918 [?], 304–305), available at https://babel.hathitrust.org/cgi/pt?id=umn.319510021724248;view=1up;seq=9 (accessed 30 August 2017).
7. Robert M Grant, *U-Boats Destroyed: The Effect of Anti-Submarine Warfare 1914–1918* (Putnam, London 1964), 37; my thanks to John Roberts for bringing this passage to my attention. Unfortunately, Grant does not indicate his source, and I have so far been unable to trace it, but it was almost certainly from a debriefing of *Ivanhoe's* crew.

Table 2: The Victims of *U-56*

Date	Name	Displacemt	Nationality	Location
22 Oct 1916	*Theodosi Tschernigowski*	327 tons	Russian	Near Vaida Guba
23 Oct 1916	*Rensfjell*	781 tons	Norwegian	24nm E of Vardø
25 Oct 1916	*Dag*	963 tons	Norwegian	3nm off Berleåg
26 Oct 1916	*Oola*	2,494 tons	British	22nm NEbyN of North Cape
1 Nov 1916	*Ivanhoe*	1,136 tons	Norwegian	30nm E of Vardø
Totals:	5 ships sunk	5,701 tons		

Source: http://uboat.net/wwi/boats/successes/u56.html

POLITICAL NOMENCLATURE IN THE US NAVY

Kenneth Fraser looks at the naming of ships in the US Navy.

The practice of naming warships after recent, or even living politicians has a long history in the United States of America. Thus the first Continental Navy included frigates named after the Revolutionary War leaders Hancock and Randolph, and the original US Navy, founded in 1794, had one named after President John Adams. This situation changed in 1819, when Congress passed an Act regulating the names of warships. It prescribed the names of states for ships of the line, river names for frigates and town names for sloops, and was generally adhered to.

In the late nineteenth century, however, as new categories of warship emerged, new naming systems were also required. It was decided to name torpedo boats, and later destroyers after distinguished naval figures, whether officers or men, and at an early date this was extended to a few Secretaries of the Navy or other politicians who had supported it. The first example was the USS *Fox* (torpedo boat no 13, of 1898), named in honour of Gustavus V Fox, Assistant Secretary of the Navy during the Civil War. Such cases do not make up a major proportion of the names of destroyers and destroyer escorts, but currently nearly all Secretaries of the Navy up to 1900, and many later ones, have had ships named after them.

Aircraft carriers also required a new system of nomenclature, and at first were named either after famous battles (eg CV-3 *Saratoga*), or after the earliest American warships (eg CV-4 *Ranger*). But a significant change occurred in 1945, when CVB-42, which was to have been named *Coral Sea*, was renamed *Franklin D Roosevelt* shortly after the President's death. So high was his reputation that there would have been no objections to this breach of precedent; but it would have widespread consequences later. The next such example was CVA-59 *Forrestal* of 1954, named after the first Secretary of Defense who had died a few years earlier.

A further innovation, the introduction of the ballistic missile submarine, gave rise to a whole class named after prominent Americans, naturally beginning with SSBN-598 *George Washington*. Most of those honoured had lived in previous centuries, but it was notable that much effort was devoted to commemorating members of different ethnic groups, from SSBN-632 *Von Steuben* (German) and SSBN-633 *Casimir Pulaski* (Polish) to SSBN-642 *Kamehameha* (Hawaiian) and SSBN-656 *George Washington Carver* (black). This precedent too would be followed in later years.

By the 1970s a new influence began to make itself felt. While warship names were bestowed by the Secretary of the Navy on the basis of guidelines approved by Congress it was, paradoxically, possible for Congress to advocate a breach of these guidelines by passing a motion urging the Secretary to consider the merits of a particular name, which in practice has nearly always been that of a politician. The Secretary would have it in mind that Congress, unlike the British Parliament, has the power to amend the Budget line by line! Thus, while submarines continued in general to receive the names of marine creatures, there unexpectedly appeared the SSN-680 *William H Bates*, SSN-685 *Glenard P Lipscomb*, SSN-686 *L Mendel Rivers* and SSN-687 *Richard B Russell*, all named after recently-deceased Congressmen with naval interests. Since then, similar cases have occurred sporadically, although Congress's suggestions have not invariably been accepted.

During the same period the aircraft carriers of the *Nimitz* class marked a considerable departure from previous policy. Whereas there had, as remarked above, been occasional carriers named after politicians (CV-67 *John F Kennedy* had been another), this class were all to be named after prominent people. CV-68 *Nimitz* and CV-69 *Eisenhower* were obvious choices as the most eminent naval and military commanders of the Second World War, but CV-70 *Carl Vinson* was remarkable as being named after a long-serving Congressman who had been instrumental in building up the Navy, and was still alive at the time the vessel was named. This precedent too (almost unheard of in other democratic republics) is still occasionally followed. Although it has sometimes been asserted that no ship had been named after a living person since the Congressional legislation of 1819, it has been pointed out that the original US submarine *Holland* (SS-1) had been named after her designer in 1900. No previous writer, however, seems to have noticed that the aircraft carrier *Wright* (CVL-49) of 1945 was so named during Orville Wright's lifetime.

From the *Eisenhower* onwards, it transpired that with one exception, CVN-74 *John C Stennis*, named after a distinguished Congressman, every new US aircraft carrier was named after a President, sometimes of historical importance (eg CVN-71 *Theodore Roosevelt*) but in later times often of recent date (eg CVN-75 *Harry S Truman*). CVN-76 *Ronald Reagan* and CVN-77 *George H W Bush* were so named during the lifetimes of the statesmen honoured. We now know that CVN-75 was originally to have been named *United States*,[1] but when, in 1996, the Republican-controlled Congress called for CVN-76 to be named after President Reagan and the Democrat President Clinton was reluctant, his Secretary of the Navy proposed a successful compromise whereby the name for CVN-76 was agreed on condition that CVN-75 be renamed *Harry S Truman* (another Democrat).

The extent to which politics have dominated US aircraft carrier names is also illustrated by CVN-78 *Gerald R Ford*. Earlier ships commemorated some of the most illustrious figures in US history (eg Washington, Lincoln, Roosevelt), but surely not even his own supporters would assert that President Ford could stand comparison with them. He must have been honoured purely on the strength of having been a recent President.

Meanwhile, political influence continued to make itself felt from time to time in lesser classes of ship. It would be tedious to enumerate them all, from the submarine *Henry M Jackson* (SSBN-730), named after a distinguished

Senator, to the Littoral Combat Ship *Gabrielle Giffords* (LCS-10), named in honour of the Congresswoman who in 2012 had narrowly survived an attempted assassination. But a controversy arose over the amphibious transport *John P Murtha* (LPD-26). As a deceased Congressman with naval interests and a veteran of the Vietnam war, he might have appeared an appropriate choice; but he had been accused of using his position to divert defence contracts to firms in his constituency in return for payments to his election funds.[2] A less objectionable aspect of political nomenclature was the effort, notable in the last few years, to ensure that members of ethnic minorities were appropriately represented in ship names. In fact the first to be named after a black sailor had come into service as early as 1943: USS *Harmon* (DE-678).[3] The most significant examples recently have been T-AKE-13 *Medgar Evers* and T-AKE-14 *Cesar Chavez*, named respectively after a martyr of Civil Rights and a radical trade unionist.

Growing unease at the rise of such non-traditional names for US warships prompted Congress, in 2012, to request the Department of the Navy to draw up a report on naming policies and practices and, in particular, on the question of whether fixed policies for naming each class of ship would be advisable. The report was duly issued, and provides a great deal of explanatory detail on ship naming policies, especially those of recent years. It frankly admits that political considerations often influence ship names. When Admiral Rickover was asked why he was advocating that future SSNs be named after cities rather than the traditional fish, he replied 'Fish don't vote', and commentators claimed that the first twelve of the *Los Angeles* class had all been named after cities in the constituencies of Congressmen who had strongly supported building them. It goes on to describe recent controversies, including that surrounding *John P Murtha*, but does not allude to the specific allegations against him.

The report makes the reasonable point that naming conventions cannot be fixed permanently, as some traditionalists wish, because types of warship in service may differ over time. For example, it had been natural in the late nineteenth century for the United States to give the names of states to battleships, its most prestigious warships; but when they all went out of service, it would have been unthinkable to abandon the names of states altogether, and thus they were later used for SSBNs; as there are now only about a dozen of the latter, they are in future also to be given to SSNs. Having supplied a list of the current naming conventions, the report declares that out of 285 ships in service in May 2012, only eleven had names that departed significantly from them; of these, significantly, six were named after national and congressional leaders. But this comforting conclusion is reached only with the aid of the assertion that the names of aircraft carriers (the largest warships of all) are individually considered, so that the numerous political ones do not count as exceptions. While the report in effect exonerates the present system of nomenclature, the call for it to be compiled must be considered a shot across the

The aircraft carrier *George H W Bush* (CVN-71), commissioned in 2009, was controversially named after the 41st President of the United States while he was still alive. The naming of the ship coincided with the accession of his son, George W Bush, to the presidency. (US Navy)

bows of the Navy Department; and it may be significant that CVN-80 is to be the USS *Enterprise*, in a return to a more traditional style of name.[4] Time will tell whether or not this more conservative principle will continue to be followed.

Sources:

DL Canney, *Sailing Warships of the US Navy*, Chatham, 2001.

Jane's Fighting Ships, 1962/63, 1975/76, 1986/87, 2012/13.

http://www.history/navy/mil/danfs.index.html [Dictionary of American naval fighting ships].

http://en.wikipedia.org/wiki/List_of_United_States_Navy_ships [Provides lists under each type of ship, in their numerical order].

http://en.wikipedia.org/wiki/List_of_US_military_vessels_named_after_living_Americans

http://www.history/navy.mil.faqs/faq37-1.htm [List of Secretaries of the Navy].

http://www.history.navy.mil.faqs/faq63-1 [Ship naming in the United States Navy].

http://www.fas.org/man.dod-101/sys/ship/names.htm [Federation of American Scientists: Military Analysis Network: US Navy ships: Naming ships].

http://navy.mil/navydata/ships/ [Provides links to the individual websites of current USN ships, normally incorporating information explaining the ship's name].

https://www.hsdl.org/?view&did=712827 [R O'Rourke: Navy ship names: background for Congress, Congressional Research Service 2007].

http://www.history.navy.mil/download/Shipnamingreport.pdf [United States Department of the Navy: A report on policies and practices of the US Navy for naming the vessels of the Navy, Department of the Navy 2012].

http://www.fas.org/sgp/crs/weapons/RS22478.pdf [R O'Rourke, Navy ship names: background for Congress, Congressional Research Service 2013 – revised version of his 2007 report, taking into account the Department of the Navy report of 2012].

Endnotes:

1. The name *United States* seems invariably to bring trouble to the ship which carries it. The first, a frigate of 1798, was eventually to be captured at her berth by the Confederates; the second, a battlecruiser laid down in 1920, was aborted by the Washington Treaty; the third, an aircraft carrier of 1949, was cancelled a few days after her construction began, owing to a change of policy.
2. There are many reports on this controversy, for instance: http://www.washingtonpost.com/politics/john-p-murtha/gIQA8baN9O_topic.html [John P Murtha].
3. http://www.history.navy.mil/commemorations/diversity/diversity-index.htm.
4. http://www.navy.mil.submit.display.asp?story_id=70899 [*Enterprise*, Navy's first nuclear-powered aircraft carrier, inactivated].

A's & A's

HACS: Debacle or Just in Time? (*Warship* 2017)

Enrico Cernuschi has pointed out a 'timing' error on page 113 (bottom left column): the concerns about dive-bombing expressed by C-in-C Mediterranean Fleet in March 1936 cannot have been 'based on observing the performance of early-model Ju 87 Stukas in the Spanish Civil War', as the Ju 87 did not make its debut with the Condor Legion until 1937.

The author, Peter Marland, has accepted that this was an erroneous assumption on his part, and has suggested that the focus of C-in-C's concerns may have been either level bombing or dive-bombing techniques currently being trialled by other navies, including the *Regia Marina*.

The US Navy's Last Monitors (*Warship* 2017)

The author, **A D Baker III**, has pointed out that the drawing of the armour scheme on page 161 has been wrongly attributed; it should have been credited to the US National Archives.

The Pen & the Sword (*Warship* 2017)

Enrico Cernuschi has supplied the names of some further Italian ships with literary connections:

Giulio Cesare (battleship): author of *De Bello Gallico* and *De Bello Civili*, widely studied as texts in Italian schools.

Raimondo Montecuccoli (light cruiser): author of *Aforismi della Guerra*, he was widely regarded as the 'European Sun Tzu'; his work predates von Clausewitz.

Marco Polo (armoured cruiser/sail training ship): author of *Il Milione*.

Antonio Pigafetta (destroyer): author of *Relazione del primo viaggio intorno al mondo*; Sir Francis Drake used it for his escape route after the attack on Panama.

Ippolito Nievo (destroyer): a follower of Garibaldi during the *Risorgimento* and a prolific writer.

Giuseppe Cesare Abba (destroyer): another of Garibaldi's 'Red Shirts' (*camicie rosse*), and author of *Da Quarto al Volturno*, the most famous account of the 1860 adventure.

Alfredo Oriani (destroyer): known in Italy as the 'poet of the bicycle'; his political and historical treatises underpinned the rise of fascism in Italy.

After the Kaiser: The IGN's Light Cruisers After 1918 (*Warship* 2017)

Enrico Cernuschi has pointed out that there is an error in Note 24 (page 160) regarding the fate of *Viribus Unitis* and the handing over of the Austro-Hungarian fleet. He writes as follows:

The delivery of the Austro-Hungarian fleet to the newly-created federate state of Zagreb (which included, within the framework of the Habsburg empire, all the territories dating from before the Great War where the Croat, Slovene and Serb languages were spoken) was accomplished at Cattaro only on the morning of 1 November 1918, ie some hours after the loss (at dawn) of *Viribus Unitis*. The latter ship had been handed over, together with the other warships at Pola, to the new state of the Empire at sunset on 31 October, after the Italian warships (a torpedo boat and a MAS boat ferrying the attack craft) bound for Pola had sailed from Venice. (For the record, the Italian TBs and the MAS boats had no radio sets, which were far too bulky for such small vessels, only carrier pigeons.)

The Allies never recognised the new state, which declared its neutrality in a broadcast from a radio station on the morning of 1 November 1918, eliciting an angry reaction from Vienna following the delivery of the 'magnificent present' of its navy. Moreover, the Austro-Hungarian submarines were not informed about the (actually non-existent) truce nor about the unilateral end of hostilities at sea, and the boats went on with their usual war patrols during the following days. So there was no 'nullification' as claimed in the footnote, as that was not relevant on 31 October.

The announced unilateral end of hostilities by the former Austro-Hungarian warships did not come into effect until the signature, on the afternoon of 3 November 1918, of the armistice between Vienna (whose delegates been instructed not to say a word about the passage of the fleet) and Italy; the armistice then began 21 hours after signature, on 4 November. (There was a similar delay, dictated by communications, on the Western front, except that it was a mere 7 hours; this was because the Western Europe battlefields were in flat countryside, not in mountainous country as they were in Italy, where the lack of mobile radio sets meant that the instructions for the end to hostilities, on both sides, had to be delivered by staff cars, horses and mules.)

Russian Cruiser *Aurora* (*Warship* 2017)

Reader **Adam Smigielski** of Gdansk, Poland, has written in to say that the name of this ship should strictly be rendered *Avrora* with a 'v'; the ship was named after the Roman goddess of the dawn.

Meanwhile, regular contributor **Steve McLaughlin** has sent in a photograph (below) he took showing the ship in Kronshtadt naval dockyard during her recent refit.

The Russian crusier *Avorora*, photographed at Kronshtadt on the evening of 27 May 2015 from the cruise ship *Viking Star*, which was departing St. Petersburg after a two-day visit. She is tied up at the Naval Works wharf in Kronshtadt's Middle Harbour. The refit lasted from September 2014 to July 2016 and cost around 840 million rubles (about £15 million). (Photograph by Steve McLaughlin)

NAVAL BOOKS OF THE YEAR

Andrew W Boyd

The Royal Navy in Eastern Waters: Linchpin of Victory 1935–1942

Seaforth Publishing, Barnsley 2017; hardback, 538 pages, B&W plate section; price £30.00.
ISBN 978-1-4738-9248-4

This is not the sort of book one normally associates with Seaforth Publishing. It shares a common format with the books of Geirr Haarr covering the early actions of the Second World War, but the primary focus is not operational: *The Royal Navy in Eastern Waters* is an in-depth analysis of the RN's strategy for a two-hemisphere war as it developed during the late 1930s and early 1940s. At 538 pages of closely-spaced print it is a daunting work, but from the very first pages it commands the reader's attention.

The conventional narrative which has become established over the years is that:

- The Royal Navy's plans for simultaneous conflict in the West and in the Far East were unrealistic and over-ambitious
- Hampered by the naval arms limitation treaties, the Royal Navy failed to build the ships needed for a two-hemisphere war
- It was Winston Churchill who was largely responsible for sending 'Force Z' to its demise.

The author disagrees with all three of these assertions, and presents evidence in support of his views which is always compelling. His own thesis is that during the late 1930s the Royal Navy correctly identified the Indian Ocean as the key to defending the Indian sub-continent and Australasia, which were to provide the manpower and resources necessary to win a global war, against a potentially hostile Japan. The discussions which took place, and the changes of emphasis over time, primarily concerned where the additional naval forces which would be available by 1942 were to be based. If Japan entered the war on the side of the Axis, should the Navy adopt a defensive strategy based on Ceylon, or a forward, offensive strategy based on Singapore? And to what extent could naval forces in the Mediterranean be drawn upon to provide the necessary ships and aircraft to support a credible Eastern Fleet?

The unexpected turn of events during the first two years of the Second World War imposed considerable strains on this strategy and on the forces with which it was planned to implement it. The fall of France in June 1940 opened up the Atlantic to the German U-Boats and compelled the Royal Navy to create Force H to shore up the Western Mediterranean. Britain's position in the Middle East was undermined first by the Italians and Rommel's panzers in North Africa, then by the German assault on the Caucasus, which threatened the vital Persian oilfields. A planned build-up of the fleet in the Indian Ocean was stymied not only by the loss of Force Z in December 1941, but by the disabling by Italian limpet mines of the two modernised battleships *Queen Elizabeth* and *Valiant* during the same month. Nevertheless, throughout this difficult period the Royal Navy never lost sight of what its key objectives were, and adjusted force structures accordingly. Although the fleet which opposed the Japanese raid into the Indian Ocean in March/April 1942, with the exception of the two modern carriers, was a ragbag of ships which could be spared from other duties, by mid-1942 the western half of the Indian Ocean was secure. By 1944 the Eastern Fleet was a force to be reckoned with, and by 1945 the Royal Navy was able to deploy a powerful and well-balanced Pacific Fleet for the final assault on the Japanese islands.

The author's take on the sinking of Force Z is unorthodox but well-argued. The responsibility for the 'offensive' deployment of *Prince of Wales* and *Repulse* to Singapore is seen as that of Pound and the Royal Navy rather than Winston Churchill, who was inclined (for once?) to leave the Navy to its own devices. The deployment was based not on faulty intelligence regarding Japanese intent or available forces, but on false premises. The Navy failed to appreciate the range and capabilities of IJN naval aviation, and over-estimated the deterrent effect on the Japanese of US offensive air and submarine assets in the Philippines and of the US main fleet based on Hawaii, which was effectively neutered by the surprise attack on Pearl Harbor but which had in any case been denuded of some of its most modern units to free up British naval forces in the Atlantic.

This book is a history game-changer in putting the Indian Ocean at the centre of Britain's war strategy as it evolved from 1939 to 1942. It is well-written and well-researched: there are 85 pages of footnotes and the bibliography extends to 27 pages. For a serious academic book it is also extraordinarily good value for money at only £30.00, and Seaforth are to be congratulated on making Boyd's thesis available to a wider public.

John Jordan

PG Rogers

The Dutch in the Medway

Seaforth Publishing, Barnsley 2017; hardback, 192 pages, 12 colour plates, two maps; price £19.99.
ISBN 978-1-4738-9568-3

This account of what contemporary diarist John Evelyn, no doubt reeling from the shock of what had happened, called 'the greatest dishonour to befall an Englishman', is a new edition of a book first published in 1970, shortly after the tricentenary of de Ruyter's attack on Chatham, but which has remained highly regarded in the years since. It comes with a new foreword by JD Davies, the author of a number of recent works on the period, placing the book, and our understanding of the events it describes, into a more modern context.

Philip Rogers was a Man of Kent, and his knowledge of the specific geography of the Medway and its surrounding area strongly informs his narrative. The Dutch feat – of both arms and of seamanship – comes across as all the more impressive, the English response as pretty abysmal. Rogers begins by setting the scene, outlining the reasons for the Anglo-Dutch wars. These can be summed up as a mixture of greed and arrogance on the part of the ruling elites of both protagonists, often supported by the labouring classes, and riding roughshod over the interests of the professional and mercantile sectors who had more to gain from the maintenance of peace and the resulting trade and commerce. There was much hysterical name-calling directed at their recent friends and allies by English pamphleteers, the tabloid columnists of the day.

Once war was under way, better organisation, funding, and (on occasions) leadership enabled the Dutch to overcome many of their inherent drawbacks in size and geography and, while never fully besting their bigger foe, frequently fight them to a standstill and achieve some impressive successes, of which the 'trip to Chatham' (as they have cheerfully called it ever since) ranks as the greatest.

In England, the fearfully uncertain times, with regicide, civil war, interregnum and the accompanying religious and social division all easily within living memory, and a newly-restored monarchy trying to establish how the relationship with parliament was actually going to work, resulted in a society riddled with inefficiency and widespread corruption, with individuals at every level simply looking out for themselves. Nowhere was this more apparent than in the Navy, and in particular the dockyard at Chatham. Despite ample warning of what was coming, the English commanders initially did little, then proceeded to issue a mass of vague, contradictory or, on occasions, pointless orders, resulting in muddle, chaos, panic and general hopelessness on the part of those on the scene, perhaps best epitomised by the taking of the flagship *Royal Charles* by a boatload of just *nine* Dutch seamen.

Rogers emphasises that the losses of materiel, while heavy – though combined casualties on both sides were probably only in the few hundreds, not least because so many English simply fled – were relatively insignificant compared to the shattering blow to the national psyche. There was the inevitable search for a scapegoat; of the two eventually so identified, the Lord Chancellor, Earl Clarendon, was banished for life, while Peter Pett, Commissioner of Chatham Dockyard, was confined to the Tower for a time, then quietly retired. However, overall the whole matter seems to have been quietly brushed under the carpet, as it slowly became evident just how much the entire system was at fault. The subsequent lengthy and shambolic salvage efforts in the wreck-strewn Medway only served to illustrate this more clearly.

In contrast, de Ruyter and the men under him exhibited an impressive sense of purpose. Yet the fruits of victory were distinctly mixed, with widespread national rejoicing but also squabbles about honours, rewards and prize money. The Peace of Breda that followed the raid was short-lived; and of the commanders van Ghent and de Ruyter were later both killed in action, while de Witt was hideously murdered by an Orangist mob.

Rogers writes crisply and economically, and permits himself to speculate only on who was to blame on the English side (pretty much everybody), and whether the Dutch should have gone farther (no – they withdrew at just the right moment). While the subject may be somewhat outside *Warship*'s normal time period, this is still a notable and worthwhile read.

Stephen Dent

Patrick Beesly

Very Special Intelligence: The Story of the Admiralty's Operational Intelligence Centre 1939–1945

Seaforth Publishing, Barnsley 2015; paperback, 296 pages, 8 B&W photographs, bibliography and index; price £14.99.
ISBN 978-1-84832-821-1

In 1974, FW Winterbottom's *The Ultra Secret* broke the wall of secrecy that had hitherto surrounded Bletchley Park. Patrick Beesly quickly seized the opportunity to tell the story of the Admiralty's Operational Intelligence Centre (OIC), which collated and distributed naval intelligence, including the 'special intelligence' from Bletchley Park, to the operations departments and, as 'Ultra' (a term Beesly uses sparingly in its original meaning) to senior commanders at sea. His book was first published in 1977, then republished in 2000 with an introduction by WJR Gardner, an afterword by Ralph Erskine on codebreaking in the Battle of the Atlantic, and an updated bibliography. This edition has now been reissued by Seaforth.

By the end of the First World War, the Admiralty's Room 40 was beginning to perform some of the functions of an OIC, though the codebreakers were then transferred to the Government Code and Cipher School (GC&CS). However, in 1936 the DCNS, Vice Admiral Sir William James (who had himself headed Room 40) was determined that an OIC must be created under the Director of Naval Intelligence; they made the inspired choice of Paymaster Lieutenant-Commander Norman Denning to set up the centre and he quickly established the necessary close cooperation with GC&CS. In July 1939, a U-boat tracking section was added with the lawyer Rodger Winn as deputy head; at the end of 1940 he became its head, with the temporary rank of Commander RNVR and responsibility for the centre's U-boat plot. By that time, Denning had taken on the same role for the plot of German surface ships.

To no small extent, Beesly's book relates the achievements of these two outstandingly able officers. He himself joined the OIC at the beginning of the war and from December 1941 he became 'Winn's deputy'; self-effacingly, for the remainder of the book he refers to himself (though only occasionally) by that title. Yet, notwithstanding his own close involvement, he is by no means uncritical. He accepts that the OIC's performance early in the war was 'inadequate' and he acknowledges that Winn found delegation difficult and was not always easy to work for. Yet Winn's ability to forecast how the Battle of the Atlantic would develop, even when the intelligence was incomplete, was vital, not least in the dark days of 1942 when Bletchley Park could not 'unbutton' the Triton/Shark cipher encrypted with the 4-rotor naval Enigma machine. Unlike later authors, Beesly does not overemphasise the importance of the 'special intelligence', recognising that direction-finding was its 'vital ally' and that the Allied successes from 1943 onwards owed much to material advantages like 10cm radar, long-range aircraft and well-equipped escort groups.

Denning's surface ship tracking was also important, usually giving timely warning of enemy movements. Even when the final outcomes were unfortunate, the OIC did not fail. Denning had warned of the imminence of the Channel Dash and, although he was certain, albeit from negative evidence, that *Tirpitz* had not sailed, his informed conviction could not dissuade Pound from ordering convoy PQ17 to scatter.

This book remains the only full account of the OIC. Although it occupies an early place in the now considerable literature on signals intelligence in the Second World War, it remains a work (to paraphrase Gardner's appreciation) of intimacy, breadth and balance. Seaforth's new edition is most welcome.

John Brooks

Paul E Pedisich
Congress Buys A Navy: Politics, Economics and the Rise Of American Naval Power, 1881–1921

US Naval Institute Press, Annapolis, 2016; hardback, 286 pages, illustrated with 12 B&W photographs; price $39.95.
ISBN 978-168-2470-770

The possession of a large navy has been, and still is the mark of a country wishing to attain Great Power status. Warships are complex to design, build and maintain, and possessing a navy has always made a statement of a nation's power, confidence and industrial capacity. Along with the ships, navies also need men and infrastructure, as well as a national naval strategy, and these too need funding.

Towards the end of his book Pedisich states that the Order of Battle of the US Navy rested (and still rests) on the decisions of the 435 members of the House of Representatives and 100 Senators. That is the book in a nutshell. In 14 chapters, Pedisich has chosen a very tight focus and restricts himself to discussing Congress' role in the funding and development of the Navy. Readers expecting a wider account of the development of the US Navy during the last two decades of the 19th century and the first two of the 20th will be disappointed; the book is about Congress and its role, and this is both a strength and a weakness.

In 1881 the United States was still a young country, having only existed for just over 100 years. During its short life it had fought two wars with Britain plus a civil war, and had grown to fill a continent. Westward expansion ended only when the frontier was officially closed in 1890. Possessing a large navy was not a priority; there were many more pressing concerns, such as stabilising a currency, tariff reform, balancing the rights of states with those of the Federal government – in short, building a nation. As Pedisich makes clear, few Presidents during the period had much interest in developing a navy, and their Secretaries of the Navy likewise often had little interest in (or experience of) navies; they were primarily members of Congress, and looked to the interests of their own constituents rather the nation. Politicians who represented inland states resented paying money for warships at all; Democrats wanted to limit the powers of the Federal government, southerners wanted to limit the growing economic power of the industrial north, and few in Congress wanted to lose power and have it replaced by a strong central government agency. Only those congressmen who represented coastal states and those with navy yards favoured spending on a navy, and even then they preferred to see Federal money spent in their state, maintaining those yards, or on providing funding for State Naval Militias, rather than spending money on warships. There was no clear vision as to why these warships were needed, anyway. What emerges is a story of consistent underfunding of the Navy, and of Congressional authorisation of ships consistently below that requested by the Secretaries of the Navy, the two exceptions being when America went to war with Spain in 1898 and when she entered the Great War in 1917.

The US Navy had a slow and inauspicious start, moving away from coastal defence and commerce raiding to becoming an ocean-going fleet. Naval expansion was slow and frequently piecemeal, although all this changed dramatically at the time of the war with Spain and later during the Great War, when enormous sums were authorised, to be spent on ships, crews, reserves and supporting infrastructure. On both occasions the return of peace was marked by a more or less instant return to the previous parsimonious approach to expenditure, and the USA was a prime mover when it came to the Washington Treaty and subsequent naval arms limitation agreements.

Pedisich's tight focus can be repetitive and frustrating, but he has produced an interesting, authoritative and unique analysis of how navies grow. This is a book for the specialist rather than the general reader, especially since it often assumes a high level of background knowledge about naval developments during the period and, in particular, the works of Mahan. Keeping the focus firmly on Congress and its dominant role, while arguably limiting the book's scope, makes it clear that acquiring a navy is not a matter just for naval men; it requires political consensus, and that without the political will or desire for a navy, no matter how ambitious and expansive the plans for a fleet, nothing will happen. As Pedisich illustrates, time and again between 1881 and 1921 Congress placed other issues as higher priorities than naval expenditure. This remains the same today, not just for the United States, but also for China and for India.

Andy Field

John Roberts
British Battlecruisers 1905–1920

Seaforth Publishing, Barnsley 2016 (revised edition); large format hardback; 128 pages, many B&W photographs, line drawings and colour plans; price £35.00.
ISBN 978-1-4738-8235-5

The first edition of this book (1997) established itself as the major source of information on British battlecruisers from *Invincible* to *Hood*. Antony Preston's review in *Warship* 1999–2000 described it as 'a joy to the technical historian, the modeller and the rank and file warship lover.' This edition improves on that judgement by adding a total of 16 full pages of coloured plans. Four are used for a foldout profile of *Invincible* (though the quoted scale of ¼in = 1ft refers to the original plan!) and three for foldouts of the profile and platform deck of *Princess Royal*, the remaining nine displaying further plans and sections. (The folding plans of *Queen Mary* from the original edition are still provided inside the back cover.)

The second major change is from two columns of text per page to three, with a slightly smaller print size. This actually makes scanning quickly over a page easier (especially if you have done a speed reading course!), but more importantly uses the space saved – a succession of largely empty half columns for captions in the first edition – to good advantage. Many of the photographs are reproduced larger, and the machinery section has extra double page diagrams of the platform deck of *Invincible* and a profile of *New Zealand*.

Both those changes enhance the value of the book considerably (for a price only a little more than 20 years ago). In contrast, the changes to the text are small – perhaps some three columns added in the whole book. The discussion of the reasons for the choice of fire control system (Dreyer or Argo) is expanded, as is comment on British understanding of the German approach to organising and implementing their system – ideas which would have been of value to the RN. The (tiny print) chapter notes extend that discussion usefully. The discussion of the use of Lyddite and the problems of its sensitivity is similarly amplified, while the notes to the 'Battlecruiser Revival' chapter shed new light on the sources of the 15in gun turrets for the *Renown* and *Glorious* groups. Roberts' preface notes too that he has expanded his concluding views contrasting the quality of design with the value of the concept. All these revisions note the influence of what the higher echelons of Admiralty were prepared to have discussed in contrast to purely technical arguments.

If a little negative criticism is expected, the author's list of sources has not added Ian Sturton's article in *Warship* 2010, which described and illustrated the modifications intended for *Hood*'s cancelled sisters so briefly mentioned by Roberts. The chapter on 'War Construction' in Friedman's book *The British Battleship 1906–1946* charts the early history of the *Rhadamanthus*/*Renown* development and the possible introduction in 1916 of the '15in B' (=18in) gun during development of *Hood*'s design. Pages from early in 1915 in Attwood's Workbook at NMM give estimates for a 25,500ton ship with three twin turrets for 15in B guns arranged as in *Renown*. However, Roberts' primary focus is on what was done, rather than merely contemplated. It is good to see his book back in print with this improved edition.

David Murfin

John Roberts
British Warships of the Second World War detailed in the original builders' plans

Seaforth Publishing, Barnsley 2017; large format hardback, 176 pages, many plans (mostly in colour) plus 4-page colour gatefold; price £35.00.
ISBN 978-1-4738-9068-8

This is a new – and heavily revised – edition of a book originally published by Chatham in 2000. Those who have the earlier edition will know that it presents a selection of original Admiralty draughts of ships and craft preserved in the archives of the National Maritime Museum, with explanatory text and captions by John Roberts, one of the foremost authorities on the Royal Navy of the period.

The main body of the text is divided into chapters which provide an outline of the development of the various types of ship and craft, together with a general Introduction and some observations on warship design from the Washington Treaty to 1945. It is punctuated by reproductions of original plans, carefully selected to illustrate a wide range of original designs and conversions, with captions which highlight the key features of the ship and the style of the plans.

In the original edition the plans were reproduced in monochrome. Some were unaccountably 'smudgy', others had not been conserved terribly well and there were prominent creases and torn edges. This aspect has been improved beyond recognition in the new edition, in part because of advances in reprographics. The larger plans, many of which were drawn using coloured inks, are now reproduced in full colour; they are also reproduced wherever possible at a large scale, with the plans of *Warspite* (as reconstructed) and *Ark Royal* meriting large four-page double gatefolds. There remain a number of black-on-white plans and blueprints, but the reproduction of these is clear with much evidence of 'tidying up' where the originals were creased or damaged. The only slight negative feature of the new layouts is that, when reading the text, it is sometimes difficult to find the end of a sentence you have just begun without turning seven or eight pages; where this occurs it might have been helpful to the reader to have 'continued on page ...' at the bottom of the page.

The text is a masterpiece of compression, communicating a large quantity of information in a minimum number of words. The extended captions for the plans are informative and equally well-written. Many of the plans themselves are works of art, although inevitably, given the standard ⅛in = 1 foot scale of the originals, annotations are very difficult to read even with a magnifying glass. John Roberts has attempted to compensate for this in his captions by drawing the reader's attention to the most interesting features of the plans, and in this he has generally been successful. This is a wonderful book to look at, and the standard of the writing is exemplary; it is highly recommended to anyone with an interest in the conventions and execution of warship plans.

John Jordan

James Taylor
Dazzle: Disguise and Disruption in War and Art

Pool of London Press, 2016; hardback, 144 pages, illustrated in colour and B&W, price £20.00.
ISBN 978-1-910860-14-4

The author is a former curator of paintings, drawings and prints at the National Maritime Museum, Greenwich, and now works as a lecturer and art consultant. He is therefore well placed to draw together the artistic and naval elements of the dazzle paint schemes applied to both warships and merchant ships in the latter part of the Great War. His book, with its very eye-catching cover, is printed in colour throughout and covers the subject from a number of aspects; it features a chapter on Norman Wilkinson, the artist generally credited with introducing the concept of dazzle while serving in the RNR, and the work of other artists who also have a claim to have evolved dazzle schemes and the use of dazzle in the United States. The use of models to test the theory in both Britain and the USA is described and there are numerous illustrations of ships that were painted in dazzle schemes. In all, some 4,000 merchant ships and 400 warships were said by Wilkinson to have been dazzle-painted. The Ministry of Shipping stated after the war that the scheme had certain advantages and no disadvantages, although its effectiveness in minimising successful submarine attack was difficult to establish. The author concludes with a description of the influence dazzle painting had on the artistic world after 1919 and the ships recently dazzle-painted to mark the centenary of the scheme's original wartime use.

Taylor concentrates throughout on dazzle's artistic merit rather than its operational value. If photographs of test models viewed through a periscope still exist, an analysis of their ability to impair a U-boat commander's judgement, especially with regard to fire-control solutions and visual estimates of speed and angle on the bow, would have added wider interest to the text, but all the same Taylor has produced a most interesting study. Potential readers need to be aware, however, that the text contains numerous minor errors that should have been corrected at the editing stage of publication: on page 62 'Henry Devonport' becomes 'Henry Davenport' 23 lines later; on page 77 Arthur Balfour is described as the Prime Minister rather the First Lord of the Admiralty; and on page 72 HMS *Carysfort* is referred to as HMS *Caryfort*. There are many others, which overall serve to detract considerably from the value of the text. Those who regard dazzle more as an art form may well find this an interesting read, despite these blemishes, and for them it probably justifies the modest recommended purchase price.

David Hobbs

Conrad Waters (editor)
Navies in the 21st Century

Seaforth Publishing, Barnsley 2016; hardback, 256 pages, numerous colour and B&W photographs, line drawings and charts; price £35.00.
ISBN 978-1-4738-4991-4

It is now more than a quarter of a century since the end of the Cold War. Sufficient time has elapsed, therefore, to allow a measured assessment of the evolution of the world's navies in the intervening period, to consider their current condition and status, and to speculate about their future prospects and ambitions. Conrad Waters has assembled a highly-informed group of writers to address these issues in a beautifully presented large-format book, the style of which will instantly be familiar to readers of Seaforth's annual *World Naval Review*.

The book is divided into nine sections, with a concise and insightful foreword by Geoffrey Till. The first three provide a strategic and operational overview. The longer Section 4, 'Fleet Analysis', is sub-divided into regions of the world. It is a sign of the times that in the case of Asia-Pacific this runs to five separate chapters, each dealing with a specific navy. Sections 5–9 contain overviews of naval shipbuilding, warship design, technical developments, aircraft and personnel. There are numerous photographs, charts, maps and line drawings which all add to the high quality of the overall appearance of the book.

The end of the Cold War served to remove the bi-polar certainties which had existed during the preceding 40 years. Large numbers of specialised warships became surplus to requirements as many welcomed the 'peace dividend'. However, the world's problem areas merely shifted elsewhere and the existing significant naval powers were required to become skilled in new tasks which ranged from expeditionary warfare to constabulary duties and humanitarian relief. The watch-words became communication and co-operation, and unexpected alliances were forged in diverse areas such as the East Asian Archipelagos, the Mediterranean and the Gulf of Aden in the face of unforeseen and often asymmetric threats. This required many navies, not all of them 'emerging' ones, to undertake blue-water operations for the first time.

The US Navy has remained the monolithic superpower, but the early years of the new century have seen a significant shift in industrial and economic power to the Asia-Pacific region. Rightly, much attention is paid to the expansion in size and capability of the Indian, South Korean, Indonesian and in particular Chinese navies and the way in which existing regional players such as Japan and Australia have responded to this changed geo-political scenario.

There is a huge amount to digest in the pages of this book. However, the limited time-span means that while a finite number of factors have shaped events, individual contributors have each felt obliged to mention them, which has resulted in a degree of repetition; perhaps more rigorous editing and a tighter original brief might have avoided this.

Navies in the 21st Century provides a most useful contextual analysis of the post-Cold War period, explaining how technological developments and a range of world events have variously shaped the fleets of today. The 'Significant Ships' and 'Technological Reviews' chapters in the latest edition of the sister publication *World Naval Review* provide highly topical, in-depth analysis in technical yet accessible language which is currently not provided in other annual publications. Where the two books inevitably overlap is in their regional review sections. Do they complement or duplicate each other? The answer to both questions has to be 'yes'.

Jon Wise

Alistair Roach
The Life and Ship Models of Norman Ough

Seaforth Publishing, Barnsley 2016; hardback, 168 pages, illustrated with 25 colour and 88 B&W photographs and 45 plans; price £25.00.
ISBN 978-4738-794- 8

This is the story of perhaps Britain's foremost ship modeller of the 20th century, Norman Ough. Ough produced many models for the Imperial War Museum, the Royal United Services Museum, Exeter Museum and not least for Earl Mountbatten. Alistair Roach has researched his subject in some depth and has also attempted to trace as many of the models as possible, with possibly the most complete collection on public view being that of Mountbatten at his former home at Broadlands, near Romsey in Hampshire. As part of the ground-work for the models, Ough also produced detailed drawings of his subjects, often based on available information when help from the Admiralty was not forthcoming, and then wrote lengthy articles on all aspects of the build process which were published in magazines such as *Model Maker* and *Model Boats*. His drawings were available and sold in large numbers to would-be modellers.

A measure of the quality of Ough's models is that some which were sold at auction have since been squirrelled away in private collections by anonymous buyers, much like an old master's painting. This quality is apparent from the photographs, particularly those in full colour, the reproduction of which is of better quality than the often mediocre black and white images. Roach acknowledges that the reproduction of the drawings is not all it might be. The originals were bequeathed to the SS *Great Britain* by David Macgregor, who took over the sales business following Ough's death on 8 June 1965, aged 67. Copies are still available from this source, and even those published in the book reveal Ough's close attention to detail.

The book is logically laid out with an introduction, a chapter on Ough's life and then ten chapters on such subjects as 'Ordnance', 'Capital Ships' and so on, followed by one on the drawings and plans themselves with a list of those held by the Brunel Institute at the SS *Great Britain* (http://www.ssgreatbritain.org/about-us/press/ship-plans-7000-ship-plans-scan).

Use of a different typeface to distinguish between the author's own account and the writings of his subject is not uncommon in modern publishing; however, the typeface selected here for the many articles written by Ough himself is difficult to read. This *caveat* apart, this is an important biography of a skilled modeller who devoted his life to his art, and the models themselves are inspirational.

W B Davies

Geirr H Haarr
No Room For Mistakes: British and Allied Submarine Warfare 1939–1940

Seaforth Publishing, 2015; hardback, 450 pages, 209 photographs, 7 diagrams, 13 maps; price £35.00.
ISBN 978-1-84832-206-6

No Room For Mistakes is Geirr Haarr's fourth book to be published in the UK by Seaforth Publishing. It follows two books about the German invasion of Norway in 1940, and *The Gathering Storm: The Naval War in Northern Europe September 1939 – April 1940*. In this latest book the author has once again carried out meticulous research to produce a comprehensive and readable book about an often overlooked aspect of the war at sea. While there have been hundreds of books written about the U-boat campaigns, there are very few which deal with the parallel struggle between the British and Allied submarine services against the surface ships of the *Kriegsmarine*.

The title of the book is taken from a 1940 quote by Vice Admiral Max Horton, who later went on to become C-in-C Western Approaches: 'There is no room for mistakes in submarines. You are either alive or dead.' It is a fitting title for a book that illustrates the mixed fortunes of the British submarine service: for every success there was a failure, for every triumph a tragedy.

The book begins with one such tragedy, the loss of the Royal Navy 'T'-class submarine HMS *Thetis*, tragically sunk with the loss of 99 lives during sea trials in June 1939, and it ends in December 1940 with the resurrected submarine (renamed HMS *Thunderbolt*) on her first war patrol in the Bay of Biscay. The author uses these two events to illustrate how the British submarine service developed during that first year of conflict. In 1939 the Admiralty was largely focused on the surface fleet; submarines were technologically misunderstood and tactically misused. However, the experiences of that first year of war led to a significant shift in attitudes, and the submarine service came to play an important role in Allied naval operations.

The book comprises 28 chapters, plus an introduction and six appendices with key data for the different British submarine classes, flotilla information, a list of the British and Allied submarines lost during the relevant period, enemy ships attacked and minefields laid. There is also a useful glossary of terms, a list of equivalent naval ranks between the *Marine Nationale*, Royal Navy and the *Kriegsmarine*, chapter notes, an extensive bibliography and a comprehensive index.

The book is attractively produced in a style matching the author's other titles, with one of Tony Cowland's superb naval paintings on the cover. It is abundantly illustrated with photographs, and Haarr has used a wide range of sources to cover aspects of the subject that others might have ignored, such as submariner training, clothing and escape apparatus. There is an interesting section on the antisubmarine tactics and technologies of the *Kriegsmarine*, a topic which is usually studied from the Allied side. The author takes a fascinating look at submarine versus submarine warfare, and he returns to his earlier work to discuss the submarine service during and after the German invasion of Norway and Denmark, when submarines were used for special missions such as landing agents or Commando units. While the primary focus is on British submarines, coverage extends to French, Polish, Dutch and Norwegian boats.

John Peterson

Philip MacDougall
Portsmouth Dockyard Through Time

Amberley Publishing, 2017; paperback, 96 pages,
136 illustrations in colour and B&W; price £14.99.
ISBN 978-1-4456-6398-2

Philip MacDougall, of the Naval Dockyards Society, is the author of several titles on the histories of Chatham and Portsmouth Dockyards of which this, part of Amberley's numerous and successful 'Through Time' local history series, is the latest. The book comprises a short introduction followed by four chapters which are thematic rather than chronological. A wide range of illustrations, ranging from Victorian plans and engravings from the *Illustrated London News* to contemporary colour photographs, accompanied by detailed captions, demonstrate the author's extensive knowledge of the subject. After a couple of pages in which the author looks briefly at the dockyard's early history – tantalisingly the remains of the original medieval yard may still exist beneath the area just to the east of the modern Portsmouth Harbour railway station – MacDougall covers the period from the early Victorian era, through the tumultuous 20th century, and up to the present-day developments associated with the arrival of the new aircraft carriers of the *Queen Elizabeth* class. Throughout, the handsome dockyard buildings of Georgian and Victorian times are much to the fore, many of them having had a variety of uses over the years, testament both to their inherent quality and their adaptability. A number of structures have even changed location as well as purpose. (It should be pointed out that this is a book about Portsmouth *dockyard*, meaning that associated establishments, such as Whale Island and Priddy's Hard, are not covered.)

The illustrations, many of which are from the author's own collection, include some rare views of ships being launched or fitting out, and the salvage operations associated with the sinking of the cruiser *Gladiator* in 1908. An example of the unusual approach are the two pages covering dredging of the harbour in different eras – about as unglamorous a task as could be, but absolutely vital for the functioning of the base. In some instances images are paired in a 'then & now' approach, illustrating not only the changes that have occurred but also a considerable degree of underlying continuity.

The book is unfortunately let down by the standard of editing: references to ships alternate between 'it' and 'she'; names are in italics one moment, roman the next; there is also the occasional instance of repetition (in one case of an illustration). More serious, given that this is essentially a picture book, is the very poor quality of the reproduction of the photographs, which are extremely grainy and indistinct throughout. This mars what is otherwise a useful introduction to the ongoing story of a naval establishment which is steeped in history.

Stephen Dent

Arthur Nicholson
Very Special Ships: *Abdiel* Class Fast Minelayers of World War Two

Seaforth Publishing, Barnsley 2016; hardback, 208 pages,
8 additional pages of colour plates, many B&W illustrations;
price £30.00.
ISBN 978-1-84832-235-6

The Royal Navy's six *Abdiel* class minelayers were unique ships that gave distinguished service during the Second World War and beyond in a much wider range of roles than was initially envisaged. Despite this, comparatively little has been written on the class beyond Tom Burton's 24-page monograph published in the 1970s as *Warship Profile* 38. American author Arthur Nicholson's new book is therefore welcome in its detailed recording of the exploits of what were, indeed, very special ships.

Encompassing 208 pages and an additional eight-page section of colour plates, *Very Special Ships* comprises eighteen chapters. The book commences with a brief overview of mine warfare and minelayers prior to the advent of the *Abdiel* class, followed by a short description of the process that resulted in their ultimate design. The fact that the latter is accorded just five pages gives some indication of the author's primary focus. An even shorter third chapter provides a useful summary of the Royal Navy's minelaying organisation at the start of the Second World War.

The bulk of the rest of the book – spanning thirteen chapters in total – is devoted to a detailed operational history of the class throughout the Second World War. Adopting a broadly chronological approach, Mr Nicholson is successful in demonstrating the inherent flexibility of the design. Although undoubtedly successful in their primary minelaying role, the capacity provided by the *Abdiels*' capacious minelaying deck saw them pressed into service as fast transports of personnel and supplies, most notably during the siege of Malta. The geographical extent of these varying operations ranged from minelaying off the Norwegian Coast to amphibious operations in the Indian Ocean and Southwest Pacific. There are good descriptions of the circumstances relating to the loss of the three ships – *Latona*, *Welshman* and *Abdiel* – sunk during the war, as well as the torpedoing of *Manxman*.

The final two chapters provide a very short overview of the three surviving ships' postwar careers – *Manxman* being the last to decommission in 1970 – and an epilogue assessing the ships' record and battle honours. An appendix covers plans and models, as well as providing an overview of camouflage schemes. The last-mentioned is supported by a range of excellent profiles by Eric Leon in the colour section. This also contains a general arrangement profile of *Latona* taken from the plans held by the National Maritime Museum.

The book has been very thoroughly researched – as evidenced by ten pages of copious notes and a further three pages on sources – and benefits from personal recollections as well as archival and published material. Lesser-known incidents – for example, *Apollo*'s grounding off the Normandy beaches and *Ariadne*'s service with the US Navy during the Philippines Campaign – are given appropriate prominence. However, it is possibly the detailed descriptions of *Welshman*'s hazardous supply missions to Malta that are the book's highlight.

Criticisms include a relative lack of technical information and analysis, both with respect to the detail of the original design and subsequent modifications. There is no table of technical data nor any conclusive assessment as to the extent to which the class's legendary speed was achieved in practice. The selection of photographs is excellent, but the book's relatively small dimensions (ca 22cm x 26cm) mean that these are not really displayed to best effect. It would also have been good to see more on the class's sometimes interesting postwar careers.

Despite these limitations, *Very Special Ships* can be recommended as a well-researched and well-written operational history of a remarkable class of warship.

Conrad Waters

Dwight Sturtevant Hughes
A Confederate Biography: The Cruise of the CSS *Shenandoah*

US Naval Institute Press, Annapolis 2015; hardback, 272 pages, 20 B&W illustrations; price £34.50/$38.95.
ISBN 978-1-61251-841-1

On 5 November 1865, a Liverpool-built square-rigged clipper capable of cruising at an impressive nine knots limped back into port, surrendering as a stateless ship returning to her city of origin. So ended the story of CSS *Shenandoah*, the ship that fired the last shot of the American Civil War and flew the last Confederate flag to be hauled down. Her story had started in October 1864 when, as the *Sea King*, she was purchased by the Confederate States of America to become a commerce raider. However, as the presence of Union representatives made an open purchase unwise, she sailed as a merchant vessel, rendezvousing off Madeira with a covert supply ship which furnished her with the weapons, stores and an embryonic crew that would allow her become the CSS *Shenandoah* and thus began a successful, thirteen-month-long voyage around the world from Australia and the southern oceans to the Arctic whaling grounds of the Bering Sea.

Under the command of Captain Waddell, an experienced officer with more than 20 years' service in the US Navy, was a crew that was a rich mix of officers and men from the ships they raided and who volunteered to join the Confederate cause – or more commonly to be free men rather than captives. Their mission was to build on the successful cruise of the CSS *Alabama* and carry out a *guerre de course*, destroying as much Yankee shipping as they could. However, there was also a wider diplomatic role for the officers: a visit to Melbourne, Australia, for repairs needed deft handling to ensure the hosts remained congenial despite increasing pressure from the American consul. Waddell was not one to take unnecessary risks, and ironically it was only in the dying days of the war, as Confederate hopes faded on land, that *Shenandoah* had her greatest success amongst the Union's whaling fleet. Overall her *guerre de course* came a year too late to affect fortunes on land.

The author provides a gripping first-hand account of *Shenandoah*'s voyage, drawing on the log of the captain and a rich selection of personal journals of the ship's officers, and capturing the highs and lows as the crew reacted to events on board as well as the changing fortunes of the Confederacy at home. More than just the fight against the enemy, the book portrays the daily battle against the elements and the boredom that a circumnavigation of the globe entails. Perfectly captured are the frustration, solitude and lack of news that accompanied long periods of cruising, sometimes for weeks without even sighting another vessel, a tedium broken by periods of intense activity as United States merchant vessels were captured and plundered not just for their cargos but also for provisions, furnishings and whatever else the *Shenandoah*'s crew deemed necessary (even books to tackle the boredom), and for potential crew members. We also discover how steam-assisted sailing ships were operated. As the South crumpled and *Shenandoah*'s successes rose, the story is interwoven with snippets of news that place the endeavour in context and underline the ultimate futility of her task. By the end of her mission, *Shenandoah* had captured and destroyed more than thirty ships, but the war was lost.

Supporting the narrative are a detailed map of the voyage, plans of the layout of ship and, in an epilogue, brief biographies of each of the key characters. *A Confederate Biography* is a well-researched story about an obscure part of naval history and provides an intriguing closing chapter on the American Civil War. It is a fascinating and engaging read, and highly recommended.

Phil Russell

Richard Osborne
The Watery Grave: The Life and Death of HMS *Manchester*

Frontline Books, 2015; hardback, 244 pages, 32 B&W illustrations, 3 maps & diagrams; price £19.99.
ISBN 978-1-47384-585-5

In the early morning of 13 August 1942 the Royal Navy 'Town' class cruiser HMS *Manchester* slid beneath the waves off the Tunisian coastline. Just one of a number of ships sunk in the successful but costly Operation 'Pedestal' to re-supply Malta, her loss was, nevertheless, one of the most controversial throughout the whole of the Second World War. Damaged by a torpedo hit from an Italian motor torpedo boat, she was scuttled on the order of her commanding officer, Captain Harold Drew. This action swiftly incurred the wrath of his superiors, who considered *Manchester*'s abandonment to be premature. His decision was to result in him being found guilty of negligence at a subsequent court martial and never again being allowed a sea-going command. At the same time, his judgement was supported by many of the ship's crew. These supporters considered he had taken the only possible action open to him in an impossible tactical situation and saved numerous lives.

Author Richard Osborne begins with an overview of the design history of the 'Towns', and provides a chronological account of *Manchester's* early service. This includes the torpedo damage she sustained during a previous Malta convoy, Operation 'Substance', in July 1941. The damage inflicted in this attack – when she was also under the command of Captain Drew – was remarkably similar to that incurred in Operational 'Pedestal'. However, in this earlier case she was less exposed to further enemy attack and brought safely back to port.

The bulk of the book is inevitably dominated by *Manchester*'s loss during Operation 'Pedestal' and the subsequent enquiry and court martial into her destruction. Osborne records in detail the series of events that resulted in Captain Drew's decision to scuttle the ship in the face of opposition from her senior engineer, Commander (E) William Robb, whose team worked successfully to restore the ship's propulsion and steering. The clash of opinion between these two officers during the subsequent court martial hearing is one of the book's main areas of focus.

The court martial itself encompasses five of the sixteen chapters. The majority of the ship's survivors were interned in French North Africa until released by the Anglo-American invasion of November 1942, and it was not until February 1943 that the court martial convened. Proceedings were brought under section 92 of the then Naval Discipline Act, which allowed all the surviving officers and crew of a lost ship to be tried together without specific charge, and effectively served the dual purpose of acting as both a board of enquiry and court martial. The fairness of this process – which had a long history in the Royal Navy but was all but suspended during the Second World War – was subject to much criticism during and after the trial, notably by the trial's prosecutor, Captain AWS Agar, VC.

Osborne's thorough analysis – which is backed by extensive use of surviving archives supplemented by secondary sources – lays to rest many of the inaccuracies and half-truths that have come to surround *Manchester*'s sinking. His ultimate conclusion – that the court martial was fair and its verdict correct – is hard to dispute when considering the evidence presented. However, Captain Drew's actions did ensure the survival of the vast majority of the ship's complement; only eleven died as a result of her loss. After North Africa was liberated, these personnel were again available for Royal Navy service. By this time manpower shortages were becoming a more significant problem for the Royal Navy than lack of materiel: *Manchester*'s sister *Liverpool* was effectively laid up between 1943 and 1945 due to lack of crew. Captain Drew's controversial decision may therefore have been to the Royal Navy's longer-term benefit after all.

Conrad Waters

Steve R Dunn
Blockade: Cruiser Warfare and the Starvation of Germany in World War One

Seaforth Publishing, 2016: hardback, 208 pages, illustrated with B&W photographs and one map; price £19.99.
ISBN 978-1-84832-340-7

Securing the Narrow Seas: The Dover Patrol 1914–1918

Seaforth Publishing, Barnsley 2017; hardback, 288 pages, illustrated with B&W photographs and one map; price £25.00.
ISBN 978-1-84832-249-3

Published within months of one another, these two books describe important campaigns of the First World War in which naval reservists played notable parts. In the north, the 10th Cruiser Squadron, soon formed largely of Armed Merchant Cruisers (AMCs), maintained the distant blockade that did much to bring Germany to her knees. *Blockade* begins with the largely ineffectual attempt by German AMCs to disrupt British trade, while its third part is a detailed account of the action in March 1917 in which the outbound German raider *Leopard* was sunk by the armoured cruiser *Achilles*. There is also a description of the devastating attack by two German cruiser-minelayers on a Norway convoy in November 1917 and a summary of the unrestricted U-boat campaign, both somewhat tangential to the book's main subject.

To the south, the Dover Patrol quickly found itself facing German flotilla forces based in occupied Belgium, which considerably complicated its principal tasks: protecting the cross-Channel military traffic (in which it was almost completely successful); coastal bombardment supporting the left flank of the Western Front and directed at the German naval bases; and preventing German U-boats from passing through the Dover Straits to attack shipping in the Western Approaches. In the course of the war, the patrol was commanded by three contrasting personalities: Admirals Horace Hood, Reginald Bacon and Roger Keyes. Each of these periods of command is accorded a separate section of the book.

The historical introductions to both books are simplistic and marred by an uncritical acceptance of received wisdom: for example, that much of the Royal Navy's thinking was 'ossified' or that it 'regarded "brains" with suspicion' – the second point being directly contradicted by the appointments of the notably intelligent Hood and Bacon. The only subject of recent scholarship that is mentioned is the questionable concept of 'flotilla defence', though it is the case that the presence of flotillas on either side of the exceptionally narrow waters of the Dover Straits soon made them too dangerous for larger warships unless they were heavily screened. Dunn can also be unreliable on technical matters such as armament, ammunition and rangefinders.

Blockade gives no credit to the Admiralty for the three AMCs that joined the blockade before the end of August 1914, nor for the withdrawal of the elderly protected cruisers as soon as sufficient AMCs were available. Thereafter, until the squadron was disbanded in November 1917, these ships were the mainstay of the Northern Patrol, remaining on station even in the worst of weather. We are given little extended testimony of the conditions that their crews had to endure. While Dunn provides a scattering of statistics, he does not give a coherent account of the course of the blockade, nor of the initial conflict between the Foreign Office and the Admiralty on how strictly it should be enforced. The book ends with a short bibliography, although the latter does not include the Navy Records Society volume on the Northern Patrol, and the most recent book specifically concerned with its main subject was published in 1985. There is a list of the archives visited but not the papers consulted, there are no citations in the text, and even direct quotations are often not individually attributed.

The second book is better in this respect, with collections of papers individually identified by archive reference numbers, and the sources of quotations and some textual information given in endnotes. It describes Hood's objections to Churchill's insistence that coastal bombardment continue after it had ceased to be effective, and his reluctance to use the vulnerable old battleships that the First Lord pressed upon him. Hood's dismissal on 13 April 1915 by Churchill – the immediate cause was some erroneous intelligence about U-boat passages – is recounted in considerable detail. Dunn recognises Bacon's technical ability and his popularity with the men of the Dover Patrol, but also his abrasive personality. While not providing any new insights into Bacon's supersession by Keyes, Dunn emphasises that it was Bacon who first proposed the deep mine barrage which, together with the continuous night-time surface illumination demanded by Keyes, at last created an effective barrier to the U-boats. Reasonably, Dunn also accepts Bacon's claim that if he had been able to influence the equipment used in the Zeebrugge raid, it might have been more successful. Throughout, the book recounts the actions fought by the Dover Patrol, its losses in vessels and men and its connections with the local ports. However, operational details are in places sparse or incomplete; thus the description of the German attack on 14–15 February 1918 does not even mention that the British destroyer patrols did nothing to protect the drifters or that, yet again, the British night recognition system failed.

Blockade ends with an extraordinary coda lamenting the demise of a pre-Great War society and a navy dominated by a 'disinterested' aristocracy and of the values they represented: honour, loyalty, obedience, bravery and sacrifice. In contrast, the second book concludes with a balanced assessment of the achievements of the Dover Patrol and its commanders and a survey of its war memorials, some now sadly neglected.

John Brooks

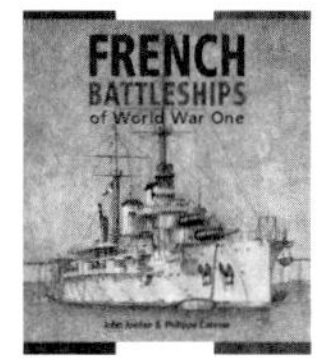

John Jordan and Philippe Caresse
French Battleships of World War One

Seaforth Publishing, Barnsley 2017; hardback, 328 pages, illustrated with 16 maps, 118 line drawings, 16 watercolour reproductions and 240 B&W photographs; price £40.00.
ISBN 978-1-84832-254-7

This book details the conception, design, build and operational service of the eight classes or groups of French battleships which eventually saw service in the Great War. It is divided into two parts: the first covers the technical characteristics, the second the service histories. There are 13 chronologically sequential chapters, as well as the customary index, a list of sources, acronyms and abbreviations, metric to imperial conversion tables and a comprehensive preface from John Jordan, who is also responsible for the very fine line drawings. Added to all this is a collection of superb watercolours from the brush of Jean Bladé, formerly Surgeon General of France, who had served in the *Marine Nationale* as a Medical Officer and whose love of the sea is demonstrated by these paintings. These serve to illustrate the various paint schemes in use over the period, and therefore provide a useful complement to the monochrome photographs which illustrate the chapters of the book.

The photographs make it clear that, in an era that produced some strangely-shaped ships, the French Navy set new standards, the designs being characterised by marked tumblehome (the curving in of the hull from a maximum beam at the waterline) and multiple funnels. Although scientific naval architecture was still somewhat in its infancy, the British would have regarded the tumblehome in particular with suspicion, leading as it does to a severe loss of stability at high angles of heel. In Britain this was the era of Sir William White and his establishment of the Royal Corps of Naval Constructors; centralised design and procurement were producing large classes of ships with identical characteristics and standardised equipment. In France the Navy was still holding design competitions between the chief engineers of both the naval dockyards and the private shipyards; winners were then selected by a committee who seemed always to require constant changes to the successful designs. Nowhere is the difference between the countries better illustrated than by a photograph of the *Gaulois* fitting out in the River Penfeld at Brest. Although part of the Arsenal, it reveals an almost total lack of support infrastructure, with cranage confined to a pair of diminutive sheer legs.

The technical aspects of individual classes are covered in great detail, with clear drawings and detailed period photographs, many of which emphasise the differences between these battleships and those of the Royal Navy. Each class or group of vessels is treated to the same depth of research and their inception, design and construction covered in the same degree of detail, the final part of the first section concluding with details of unbuilt or cancelled projects, involving a particularly interesting four-gun turret. The only uncompleted hull to survive into the post–Washington Treaty era became the sole French aircraft carrier *Béarn*, thereby following the pattern of the other major navies.

The second section deals with the service careers of the completed ships. The period leading up to the Great War is covered in the first chapter, with some overlap as it details the acceptance into service of ships up until 1916. It also covers the various prewar losses, mainly it would seem from unstable propellant, *poudre B*, with both *Iéna* and *Liberté* becoming total losses as a result of internal explosions. The suspect propellant was being removed during 1914 with at least one ship purportedly having no powder charges at all in the magazines in July that year, just over a month short of the outbreak of war. The chapter ends with an appendix covering paint schemes and identification markings.

The chapter on the Great War covers the active service of the ships and further losses; interestingly, all were due to either mines or torpedoes. Those vessels damaged by gunfire, particularly during the Dardanelles campaign, all survived their damage, although drastic steps were needed to improve the stability of some classes, due in part to the tumblehome.

The book concludes with chapters on the interwar period and the Second World War, the careers of the ships being covered up to their ultimate demise, some to gunfire or scuttling, others to the breaker's torch, with the last survivor being broken up in 1953 or thereabouts.

This is a book full of great detail, clearly the result of much research, with the interesting addition of contemporary water colours, and one to be recommended to the serious historian and warship enthusiast alike.

W B Davies

Richard Perkins (with an introduction by Andrew Choong)

British Warship Recognition: Volume III: Cruisers 1865–1939, Part 1

Seaforth Publishing, in association with the National Maritime Museum, 2017; large format hardback, 192 pages, illustrated in full colour throughout; price £60.00.
ISBN 978-1-4738-9145-6

Volume III in this mighty series, 'The Perkins Identification Albums', covers the majority of the Royal Navy's 20th century cruisers completed prior to the Second World War but, following the original author's somewhat individual approach already noted in the review of Volumes I and II (see *Warship* 2017), includes a good number of their 19th century predecessors and omits several cruiser types of the Great War era; the latter are scheduled to appear in Volume IV. In fact, although this division is simply a reflection of the original organisation of the material, two pages of the original Vol III will appear in the new Vol IV so that the *Calliope* class is not divided between the two – there are some limits as to how much is acceptable in terms of pure facsimile accuracy as opposed to actual utility. As an indication of Perkins' often idiosyncratic approach, *Courageous* and *Glorious* feature, in their 'large light cruiser' guise, sandwiched between *Shannon* of 1877 and *Edinburgh* of 1939! A couple of pages bound out of sequence have also been re-ordered, while in addition a dozen sheets of the author's rough notes have been included here – quite how useful these are is open to question.

Reproduction of the original drawings is faithful even down to the odd mark or tear, but of course what really matters is the sheer quantity and quality of information included. The smallest differences between ships of the same class are highlighted in great detail, for example the topmasts, funnel brackets and steam pipes in the numerous armoured cruisers of the late Victorian and Edwardian eras. Funnel bands are given due prominence. Given the greater number of ships in each class in this particular volume, the book's use as an identification aid becomes all the more apparent. The drawings are accompanied by notes highlighting some of the details, as well as where ships were stationed and when they served as flagships. Victorian peacetime paint schemes feature but there is no attempt to depict wartime camouflage, while the older vessels that later went on to serve as harbour training and depot ships again have their colour schemes reproduced.

In a number of instances Perkins' original mechanically copied outline templates appear, unaltered, giving the reader a much clearer idea of how he actually went about creating these collections of drawings. The method used to produce these is not mentioned in Andrew Choong's introduction, but the fact that they are bluish in tone suggests a process such as diazol-printing, an early predecessor of photocopying.

For some the high price will undoubtedly be a deterrent to purchase, but for photograph collectors, modellers, and anyone else interested in the development and appearance of Royal Navy warships during the period, the books are likely to be regarded as indispensable.

Stephen Dent

Robert Brown

ShipCraft: Japanese Battleships – *Fuso* and *Ise* Classes

Seaforth Publishing, Barnsley 2016; softback, 64 pages, numerous illustrations; price £14.99.
ISBN 978-1-4738-8337-6

The 24th in the 'ShipCraft' series, this book is more restricted in its scope than some other volumes, and it is much the better for it. Whereas other volumes have attempted to cover a whole genre of vessels, with the result that coverage is too thin and/or over-simplified, the present volume focuses simply on the four 14in-gun Japanese battleships completed during the First World War, extensively modernised during the 1930s, and sunk during the final two years of the Second World War. The second pair also underwent a remarkable second reconstruction that replaced their after guns with a hangar, aircraft deck and catapults in a vain attempt to make up for the Imperial Japanese Navy's catastrophic carrier losses during 1942.

The first chapter covers the original design and construction of the ships, including a tabulated comparison with their foreign contemporaries; this is followed by a second covering their operational careers. The greatest detail is reserved for the very last years of their service, the first two-and-a-half decades having been distinctly uneventful. Having spent the early years of the Second World War in home waters or in subsidiary roles, in 1944 all four ships were finally pressed into front-line service, the two *Fuso*s perishing in October in the Battle of the Surigao Strait, with some 4,000 dead. Their half-sisters both survived the following day's Battle of Cape Engaño, only to be sunk by bombing the following July in home waters while immobilised for lack of fuel; the wrecks were later broken up.

The third chapter reviews the models available, ranging from 1/1200 to 1/350 scale, together with accessories for enhancing them. Curiously, while there are a number of kits of the ships in both late-1930s and 1944 configurations, none exist of them in their 'as built' forms – indeed, this is true for all Japanese capital ships that survived to serve in the Second World War. One assumes that this is due to a perceived lack of demand for ships without a fighting pedigree – models *are* also available of Japanese ships that served in the Russo-Japanese war.

A section then follows with colour photographs of hand-built models of the ships, many of which are spectacular, including a fascinating one of the half-submerged hulk of *Hyuga*, while the final chapter describes and discusses the modifications made to the ships during their careers. A good level of detail is provided, including images of *Fuso* and *Yamashiro* during reconstruction. The book ends with a bibliography and drawings of the vessels at key phases of their later modification, thereby complementing the 'as built' and 1920s drawings of the *Fuso* class in the first chapter. Oddly, there are no drawings of *Ise* or *Hyuga* prior to 1941 anywhere in the book.

This is definitely one of the best books in the 'ShipCraft' series, and is to be recommended both to modellers and to those interested in the big ships of the Imperial Japanese Navy.

Aidan Dodson

Lawrence Paterson

Hitler's Forgotten Flotillas: Kriegsmarine Security Forces

Seaforth Publishing, Barnsley 2017; 352 pages, 140 photographs, 4 diagrams, 5 maps; price £25.00.
ISBN 978-1-4738-8239-3

Much of the work of a navy at war is not carried out by the glamorous capital ships, cruisers and destroyers, but by such workaday craft as minesweepers, escort and patrol vessels, together with a range of auxiliary units. The German Navy in the Second World War was no exception; indeed, owing to its chronic shortage of destroyers and torpedo boats, such vessels played an even more crucial role. This book tells their story.

The author has previously produced a number of studies of the *Kriegsmarine*, and is clearly familiar with the published and archival sources, which are fully detailed in endnotes. The present book is set out in a straightforward manner, beginning with a most useful glossary of the key German technical and organisational terms, which are employed throughout the book. Background is provided by a chapter on the development of minesweepers and other auxiliary ships by the Imperial German Navy and the *Reichsmarine* of 1919–1935. The latter force not only inherited vessels from its predecessor, but also developed the R-boat motor minesweeper that would become a key workhorse of the Navy during the Second World War. Full size minesweepers also undertook a wide range of roles, often substituting for absent destroyers and dedicated seagoing escort ships.

Once that conflict had begun, the prewar emphasis on mine warfare evolved into the broader task of securing the coasts of Germany, and then of the vastly expanded coastline of German conquests from 1940 onwards. Key vessels were requisitioned fishing trawlers, many of which became *Vorpostenboote* (patrol boats). In addition, larger merchantmen were converted to *Sperrbrecher* ('barrage breakers'), which combined conventional minesweeping methods with the fitting of buoyancy aids to allow them to detonate mines and remain afloat. Other trawlers and small merchantmen (and some ex-enemy warships) became *U-Bootjäger* (submarine chasers), while as the war went on more purpose-built auxiliaries were produced, in particular the ubiquitous *Marinefährprähme* (naval transport barges), designed as landing craft but subsequently modified for a wide range of roles, some mounting two 10.5cm guns.

The book tells the stories of these and other related types through the years of conquest, with the need to set up local security arrangements in newly-occupied territories, through routine patrol work of the middle years of the war, to the withdrawals and desperate evacuations that marked the final phase of the conflict. Much of the narrative is supported by first-person accounts of service in the vessels covered, with many images of members of their crews; these give the account much more colour than a bald operational sketch.

The various types of ship and operation are illustrated by photographs, while excellent maps provide the context for a narrative that ranges from the Bay of Biscay to the Black Sea. Appendices give organograms of the Security Forces at the end of 1940, a list of those who earned Knight's Crosses, and a full enumeration of the relevant units, with dates of formation and disbandment.

This book is an excellent treatment of a side of naval warfare that is rarely given the attention it deserves. It is thus heartily recommended.

Aidan Dodson

Julian Thompson
The Royal Navy: 100 Years of Maritime Warfare in the Modern Age

National Museum of the Royal Navy / Andre Deutsch, 2016; hardback (slip-cased), 80 pages, illustrated in full colour throughout, plus 15 reproduction historical documents; price £40.00.
ISBN 978-0-233-00486-0

This opulently produced publication consists of a full-colour, large-format hardback book, slip-cased together with 15 reproduction historical documents contained in three bound-in folders. Clearly aimed at the mainstream market, it offers a readable if hardly in-depth account of the Royal Navy and Royal Marines operations from the beginning of the First World War up to recent deployments in the Mediterranean and Middle East. The tone of the narrative is often partisan, arguably perhaps even jingoistic, which may not be to every reader's taste. On the other hand, Thompson makes a point of describing just how the grim the consequences of war at sea could be, and has some pithy observations of his own: for example, 'the most successful convoys were those on whose voyages nothing happened'. It should also be noted that the author, a former senior officer in the Royal Marines, places considerable emphasis on their side of the story. Overall the tale told is very much one of victories and successful campaigns; defeats, disasters and controversies, if not exactly glossed over, are often dealt with quickly before moving on. Then again, bouncing back from adversity has long been one of the Royal Navy's defining characteristics.

All this hardly sounds like the usual fare for the average *Warship* reader. However, in two respects this publication does offer something a little different. The majority of the illustrations come from the collections of the NMRN, and as a result various well-known views are offset by, for example, a sketch plan of the Dardanelles operation from the journal of a midshipman on HMS *Agamemnon*, or a dramatic snapshot taken on board a sinking HMS *Hermes* in 1942. Particularly eye-catching are the endpapers, which feature a large and beautifully reproduced plan of the 1911 Spithead Review and a remarkable sight (especially these days) with line upon line of ships extending from the sea forts to past Lee-on-the-Solent. The reproduction documents, while a bit of a mixture, are a definite plus, ranging from excerpts from official reports on actions such as the River Plate, to individual signals and letters/notes by serving personnel recounting aspects of Jutland and the *Bismarck* affair. All provide an unusual and effective way of adding illustration to the rather dry main narrative.

There are a few worrying factual errors: the battleship *Roma* was not sunk 'with all her crew' (many of the 500-odd survivors ended up in the 18th century British naval hospital on the Spanish island of Menorca), while the caption to the photograph of the 2nd Battle Squadron on page 35 repeats the common fallacy that it was taken at Jutland – something disproved in these very pages (see *Warship 2010*, 131). Overall, though, this represents a striking, if notably pricey, way of presenting the last 100 years of the RN's history to a general, non-specialist readership.

Stephen Dent

Barry Gough
Britannia's Navy on the West Coast of North America 1812–1914

Seaforth Publishing, Barnsley 2016; hardback, 408 pages, 35 B&W photos, nine maps; price £25.00.
ISBN 978-1-47388-136-5

This updated expansion of Professor Gough's 1971 publication provides a detailed, scholarly exposition of the influence of sea power on the history of British Columbia, and subsequently on the formation of the Dominion of Canada.

The Royal Navy came in strength to the Pacific Ocean during the War of 1812, and stayed for almost a century. Other nations were interested in the Pacific coast of North America; Russian and British fur traders crossed seas and mountains to build trading posts and forts. The great advance on land would come from countless American wagon trains crossing the plains and the Continental Divide.

The Pacific Station extended from Cape Horn to Alaska, and from Tahiti and the Marquesas to the Hawaiian Islands. For the Admiralty in London the Pacific Coast was the back of beyond, very sparsely populated but forever demanding attention and protection from the few ships on the spot. Gough navigates skilfully through the Anglo-American Treaty of 1846, which divided the Oregon Territory along the 49th parallel, notwithstanding 'Manifest Destiny' and 'Fifty-four forty or Fight'. Command of the seas ensured President Polk's bombast remained just bombast; the Fremont raid that ejected Mexico from Upper California would not be imitated farther north. Before the electric telegraph, local naval commanders had a high degree of autonomy, much depending on their tact and forbearance in adversity.

The Crimean War provides an unexpected break in the long narrative; the inglorious Anglo-French intervention in Siberia brought European-style warfare, described here in detail. Incredibly, local commanders agreed that there would be no hostilities on land between Russian North American and adjacent British territories, although Russian settlements might be blockaded by sea.

In 1858 the Fraser River gold-rush invasion brought 30,000 outsiders to the mainland, more than tripling the existing population. London immediately set up the Crown Colony of British Columbia, which was combined with Vancouver Island (itself a Crown Colony since 1849) in 1866. With the combined colony's accession to the new Dominion (1871), Britain's role began to diminish, although the naval base at Esquimalt continued to grow. The large graving dock (completed 1887) ended reliance on American facilities at San Francisco.

Naval preponderance moved south. In 1890, the most powerful warship in the area was the new armoured cruiser flagship HMS *Warspite*; the only American armoured ship was the Civil War monitor *Camanche*, armed with smooth-bores and laid up 'in ordinary'. The 1895–96 Venezuelan crisis had no discernable effect in the Pacific North-West; the Admiralty had no emergency plans and no ships to spare. By then, a war between the two countries was almost unthinkable. Ten years later, when the *Warspite* was again flagship, the United States, with a first-class battleship and two modern monitors, was far ahead.

It seems almost churlish to find fault with this major historical work. However, there are minor blemishes: the maps are barely adequate and some photo captions are incorrect; the appendices would have benefited from more careful checking, as HMS *Royal Arthur* is listed as both an armoured and a protected cruiser, while the iron-clads *Triumph* and *Swiftsure* and the un-armoured frigate *Shah* are unhelpfully grouped together as iron screw/sailing ships.

Ian Sturton

Robert Brown & Steve Backer

Shipcraft: German Battlecruisers

Seaforth Publishing, Barnsley 2016; 64 pages, numerous illustrations; price £14.99.
ISBN 978-1-84832-181-6

Number 22 in the Shipcraft series, this book begins with an illustrated historical narrative of the ship-type in question, followed by reviews of available kits, photographs of a range of completed models, details of appearance changes, ship drawings and a bibliography.

The ships covered are what the Imperial German Navy classified as 'large cruisers' (not adopting the term 'battlecruiser' until the 1930s for the never-built 'O'-class), beginning with the 8.2in-gunned *Blücher* of the 1906 programme and ending with the barely-begun *Ersatz Yorck* class of the War programme. *Blücher*'s historical entry repeats the old canard that her 8.2in main battery was the result of British disinformation that the *Invincible* class would have 9.2in guns. This has long since been shown to be a myth – the design history of *Blücher* shows quite clearly that this cannot have been the case – and it is sad to see yet another new book repeating it.

In this connection, one should note that Axel Grießmer's design history of German battlecruisers (which gives a full account of *Blücher*) does not feature in the bibliography, suggesting a significant hole in the authors' research. Likewise, the section on the *Mackensen* class says nothing about their convoluted design history, which included changes in both the number and calibre of the main battery. Otherwise the historical sections are sound, comprehensive and well-illustrated, and carry the story right through to the disposal of *Goeben*, by then the Turkish *Yavuz*, in the 1970s.

Miniature metal and larger resin model-kits have been produced for all the ships covered in the book, with injection-moulded plastic examples available for *Derfflinger* and *Lützow*, and even card models of *Derfflinger*, *Von der Tann* and *Goeben*. All except the last two are described and reviewed in the 'Model Products' section, as are the accessory sets that complement a number of them. The omission of the card model of *Von der Tann* is odd, as it features in the following Modelmakers' Showcase section, which includes a number of finely-made examples of a range of ships, including a striking rendition of *Hindenburg*, sunk in shallow water at Scapa Flow.

Colour drawings are provided of *Derfflinger* in 1917, illustrating the typical German paint-scheme of the era, and also of *Yavuz* (ex-*Goeben*) in 1942, showing the disruptive scheme adopted after her anti-aircraft refit in 1941. These introduce the Appearance section, which covers both colour schemes and physical modifications, and is illustrated by contemporary photographs. Line drawings of a selection of the ships covered by the book follow, but one wonders why a more comprehensive set was not provided, given that the book is aimed at model-makers. The book ends with a set of 'Selected References'. Although it has flaws, this volume will certainly be of interest to anyone wishing to model any of these fine ships.

Aidan Dodson

NAM Rodger, J Ross Dancy, Benjamin Darnell & Evan Wilson (editors), Strategy and the Sea: Essays in Honour of John B Hattendorf

The Boydell Press, Suffolk 2016; hardback, 303 pages; price £60.00.
ISBN 978-1-78327-098-9

The great, the good, and the aspiring of the naval academic world are brought together in this collection of essays embracing the very broad theme of strategy as applied to naval history in the early modern and modern periods. It is based on a series of papers delivered at All Souls College, Oxford, in 2014 to mark the retirement of the eminent US historian, Professor John B Hattendorf, from the Naval War College in Annapolis, Maryland. Some might consider a discussion of the advance planning which underpinned past naval campaigns and wars to be a rather dry subject. However, this is far from the case with these essays, which cover a broad range of topics succinctly and accessibly. Naval wars and actions do not just happen; an appreciation of the decision-making behind an offensive or a defensive strategy brings a deeper understanding of any account.

The adoption of a particular strategy can be determined by a range of factors: geographical, social and financial. Roger Knight and Jaap Bruijn show that the strategies adopted by Britain and The Netherlands in the 17th and 18th centuries were fundamentally defensive in nature, with the aim of preventing invasion. Nevertheless, the aggressively defensive actions of the British Royal Navy forced other nations either to engage in unwelcome combat or to radically alter their thinking.

The well-being and contentment of the crew would not seem to be an obvious strategic consideration in the typically harsh environment of a warship, and yet 'human resources' issues figure prominently in several of the essays. J Ross Dancy, by employing hard statistical data, shows that a deliberate policy of promoting the experience and skill of volunteers rather than reliance on impressment was chiefly responsible for the successful manpower strategy of the British Navy in its heyday in the late 18th century. Carla Rahn Phillips, examining the problems of officer recruitment in the 16th-century Spanish Empire, finds abstract factors such as honour and status to be as important as monetary gain. By the same token, with reference to modern times, Duncan Redford's paper shows that considerations such as national identity can be a key enabler in a successful recruitment drive.

It is pointed out that prior to the advent of space exploration, navies were the most technologically advanced, logistically complex systems in existence and thus made enormous demands on a country's resources. Successive writers show how the skilful harnessing of resources – or conversely, the misdirection of funds – have had a direct impact on ultimate success or failure. This assertion is as true today as it was in the 16th century.

'The Influence of Sea Power upon Three Great Global Wars, 1793–1815, 1914–1918, 1939–1945: A Comparative Analysis', is the longest and most ambitious essay in the collection. Its author, Paul Kennedy, whose work *The Rise and Fall of British Naval Mastery* is still considered a standard text on the subject, adroitly condenses lessons from the passage of these three complex and world-changing conflicts to examine the importance of sea power in the conduct and outcome of each of these wars. In particular, Kennedy makes a timely investigation into the part played by the Royal Navy in the First World War and why, ultimately, that conflict was a disappointment for the navalist who believes that command of the oceans rather than land power counts most in modern warfare.

Fittingly, the last two essays reflect on the importance of educating naval personnel about the historical context in which they work. In this respect, the writers James Goldrick and Geoffrey Till pay respect to the contribution over the past half-century of John B Hattendorf himself, whose extensive personal bibliography is published at the end of the book.

Jon Wise

Enrico Cernuschi and Alessandro Gazzi Sea Power the Italian Way

Ufficio Storico della Marina Militare, Rome 2017; softback, 192 pages, many maps and B&W illustrations; price €36.00.
ISBN 978-8899-642051

Published in English by the Italian Navy Historical Department (USMM), this ambitious book aims to provide both an overview of, and a commentary on 25 centuries of Italian sea power. The first half of the book provides a whistle-stop tour of naval warfare from Roman times to the Great War of 1914–18. These brief and easily digestible chapters are then followed by three longer chapters on the Great War, the interwar period, and the Second World War and its immediate aftermath.

The book is beautifully produced, with good quality paper and some stunning illustrations which include the paintings of the distinguished Italian naval artist Rudolf Claudus. The translation is functional and the English not always idiomatic; however, this rarely forms a barrier to comprehension.

The lack of English-language commentary on so many of the events recounted here, together with the very different perspective, make this book an invaluable contribution to the history of sea power in the Mediterranean.

John Jordan

WARSHIP GALLERY

Japan's U-boats, Sasebo, 1921

Stephen Dent introduces a series of rare photographs from **Ian Johnston**'s collection showing former U-boats being dismantled at Sasebo shortly the end of the First World War.

At the conclusion of the First World War some 176 U-boats were surrendered by Germany, and over the winter of 1918, and on into the following year, what was to happen to them became a major bone of contention amongst the victorious powers. Germany had originally objected to the hand-over, largely on the grounds that she needed to be able to defend herself in the future in the event of aggressions by any of her neighbours. But in reality she had no choice in the matter, and so in Article 188 of the Treaty of Versailles, finally signed on 28 June 1919, it was confirmed that 'Within one month of the implementation of the present Treaty all German submarines... must be delivered up to the principal Allied and associated forces'. In the end, after much debate, and no little disagreement, it was agreed that the submarines be divided up, roughly in proportion to the number of ships each of the allied countries had lost to enemy action, between Britain (105 boats), France (43), Italy (10), the United States (6), Japan (7) and Belgium (2). The many and varied fates of the majority of these vessels (a number of which were earmarked to be used for 'propaganda purposes'), and the arguments that accompanied them, falls outside the scope of this piece (though at least one is likely to feature in a future edition of *Warship*); it is the seven that Japan received that are of interest here.

As a relatively young naval nation Japan had invested heavily in the surface warships that constituted her battle fleet, only to find the conflict in the seas around Europe demonstrating that the submarine was a major new factor in naval power, but one in the development of which she was in danger of being left behind. Although a number of submarines, mostly small, had been completed to foreign designs, the acquisition of some of the latest German boats would enable Japan to start to catch up technologically with other countries. A complication was that even though all the powers in question were interested in benefiting as much as they could from the U-boats allocated to them, the terms of the Treaty meant that while the submarines themselves were handed over, this did not extend to builders plans and accompanying documents, meaning that direct copying of the boats, or major parts thereof, would not be straightforward.

The seven Japanese boats, of several different types and classes, were officially handed over at Harwich. Subsequently they were towed from Britain, via the Mediterranean and Suez Canal, by units of the Japanese destroyer force that had served in Europe during the war, arriving in Japan in June 1919. (An additional boat, *U-43*, though assigned to Japan, was broken up in Swansea.) They were then put into service in the Imperial Navy as *O-1* to *O-7*, though under agreement with Britain they were meant to be used only for research and experimental work. Shortly after the submarines' arrival, the naval attachés of the Allied and Associated powers in Tokyo were taken to see them in Yokosuka, the American attaché subsequently reporting on a number of impressive features, such as folding radio masts, net cutters, air purification apparatus, and equipment to assist with rescue and salvage in the event of accident. The Japanese, meanwhile, keen to get full benefit from the boats, had sent a naval mission to Germany with the aim of acquiring assistance in terms of specialist personnel and drawings. The Germans, on their part, in an effort to keep submarine development work going despite the Treaty – both so as to not lose expertise and to make some money – were looking to export the former to countries of a friendly or neutral disposition. German specialists soon arrived in Japan and by 1920 there were several hundred designers, engineers and former U-boat officers, often on five-year contracts commanding considerable salaries, helping the Japanese submarine programme. Despite the Treaty a number of construction drawings made the same journey, while later plans were drawn up in Japan under German supervision. However as the decade wore on this German influence dissipated as the Japanese increasingly took over all aspects of their submarine design and construction.

As for the former U-boats, after further discussions during December 1919 and January 1920 it was agreed that they were to be rendered incapable of any military service by 1 July 1921, and so during the first part of that year they were carefully dismantled at Kure, Yokosuka and Sasebo, with all armament, machinery and superstructure removed, though the hulls of several continued to be used for some years in a variety of subsidiary roles such as targets and berthing jetties. It was during this dismantling process that the photographs reproduced here were taken, showing the two boats that were decommissioned in this fashion at Sasebo Navy Yard, *O-6* and *O-3*.

O-6 (ex-*UB-125*)

UB-125 was one of the very numerous UB III class (85 completed) of what were officially designated coastal submarines. The original UB I class was intended specifically as such, with a primarily defensive role; the boats of the succeeding UB II class were considerably larger, having two engines instead of one, which made them much less vulnerable to damage or break-down. This was taken a stage further with the UB III class (otherwise known as Project 44), which was intended to carry out operational patrols all around the British Isles as well as in the Mediterranean. This greater range (on the surface) was facilitated by increased fuel bunkerage; the class also had a higher surface speed than its predecessors, but sacrificed some submerged speed and range as a result. Despite being much larger than their predecessors, these boats retained the UB designation.

UB-125 was one of a large number of boats ordered in February 1917 as a result of the declaration of unrestricted submarine warfare, which meant that relatively short building times had become of paramount importance. Ordered from AG Weser, *UB-125* was completed in May 1918 and subsequently carried out two war patrols, sinking six ships, before Germany's surrender.

After being dismantled at Sasebo, a process which was completed by June 1921, the hull of O-6 remained there, being used as a floating jetty.

O-3 (ex-*U-55*)

U-55 was one of the six boats of the *U-51* class which, despite belonging to the programme of 'Mobilisation' boats, ordered to the two most recent existing designs just weeks after the outbreak of war, was not completed in June 1916. Initially the expectation of a short conflict meant that these boats were not prioritised, while Germany's limited shipbuilding capacity was already stretched by the construction of surface vessels. U-boats required specialist skills, further reducing the number of yards that could take on the work, even though the 1915 declaration that the waters around Britain were to constitute a 'war zone' created a greatly increased demand for new U-boats. There was also a growing shortage of skilled shipbuilding and engineering personnel due to the military call-up, and problems too with the various firms

Above and overleaf: O-6, formerly *UB-125*, alongside in Sasebo, on 15 April 1921. These photos are part of a series taken presumably either as a record of the structure of the submarine as it was dismantled or in connection with the demilitarisation process agreed with Great Britain. Note that the boat still sports her German designation on the side of the bridge, though there is also a Japanese character painted above. The submarine appears to be basically in one piece, with one obvious exception: the 8.8cm gun that was fitted on the foredeck is missing. Since personnel appear to be examining where this was mounted it is possible that it was only removed shortly before these photographs were taken. The net-cutters that were fitted above the sharply pointed bow – an innovation intended to help counter anti-submarine nets – are also absent, while it is not possible to tell from these images whether the two folding radio masts have been removed or simply lowered into their housings in the deck fore and aft. Also of note are the large number of free-flooding holes in the casing forward, and the faired front end of the fin. The aftermost of the two periscopes (another new feature in this class) is raised. In the background of the photograph overleaf can be seen four destroyers and a battlecruiser of the *Kongo* class, probably *Kirishima*.

502
呉の六潜水艦解体現況
(其一) 大正 10-4-15

Previous page: Dated 3 May 1921 (and part of a separate series), this photo shows the dismantling of *O-6* now under way, with the submarine tied up close alongside. Much of the floodable upper deck casing, some of which can be seen on the quayside to the right, has been removed. This has exposed a number of the fittings situated within; prominent are the silencers for the twin 550bhp diesel engines, and the auxiliary air intakes to the rear of the fin. Of the two open hatches, the one nearest the photographer, opening forward, was for the crew, and the farther one, opening aft and with an extemporised set of sheers rigged above it, for loading the after torpedoes. Also visible, at the very stern, is the opening for the aft torpedo tube. Two destroyers, probably *Kaba* and *Kiri*, lie ahead of the submarine; the capital ship whose stern is visible in the left background is unknown, but again probably of the *Kongo* class.

Above: Also at Sasebo, but taken earlier, on 20 March 1921, this view shows *O-3* (ex-*U-55*) with the dismantling process well advanced, though the submarine can still be identified by her greater length compared to *O-6* as well as her straight bow. Most of the superstructure has been removed, including the streamlined after part of the fin, the net-cutters above the bow, the two 8.8cm deck guns, the folding radio masts, and most of the equipment housed beneath the deck. Much of this material is now strewn about the quayside: just to the left of the top of the wooden access way can be seen part of the side of the fin with 'U-55' painted on it. Among the items still prominent above the submarine's pressure hull are the diminutive conning tower, the warping capstan and motor forward, and the forward torpedo hatch. The heavily encrusted nature of the now exposed lower part of the hull is worthy of note. (All photographs courtesy of Ian Johnston)

that supplied components and machinery for the submarines which were, by their very nature, at the high-technology end of wartime production. The result of all this was a consistent pattern of delays in delivery of completed boats, of which *U-55* was one. Nevertheless, once completed she undertook 14 war patrols, being responsible for sinking some 64 ships.

O-3 was dismantled at Sasebo between March and June 1921; in 1923 the remaining hulk was subsequently briefly re-commissioned as 'Auxiliary Vessel No 2538'.

Acknowledgements: Liam Macleod and the staffs of Bath and Bristol central libararies, Aidan Dodson, John Jordan and Hans Lengerer.